Petrogenesis of Metamorphic Rocks

SPRINGER
STUDY
EDITION
geology

Petrogenesis of Metamorphic Rocks

Fifth Edition

Helmut G. F. Winkler

Springer-Verlag New York Heidelberg Berlin

Helmut G. F. Winkler
Professor Emeritus, Institute of Mineralogy and Petrology
University of Göttingen
Federal Republic of Germany

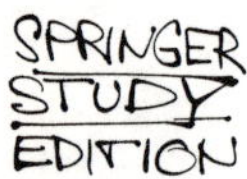

Design: Peter Klemke, Berlin

Library of Congress Cataloging in Publication Data

Winkler, Helmut G. F. 1915–
Petogenesis of metamorphic rocks.

"Springer study edition."
Includes bibliographical references and index.
1. Rocks, Metamorphic. I. Title.
QE475.A2W5613 1979 552′.4 79-14704
ISBN 0-387-90413-1

Printed in the United States of America.

9 8 7 6 5 4 3 2 1

ISBN 0-387-90413-1 Springer-Verlag New York Heidelberg Berlin
ISBN 0-540-90413-1 Springer-Verlag Berlin Heidelberg New York

Preface to the Third Edition

The first edition of this book was published in 1965 and its French translation in 1966.The revised second edition followed in 1967 and its Russian translation became available in 1969. Since then, many new petrographic observations and experimental data elucidating reactions in metamorphic rocks have made a new approach in the study of metamorphic transformation desirable and possible. It is felt that this new approach, attempted in this book, leads to a better understanding of rock metamorphism.

The concept of metamorphic facies and subfacies considers assocations of mineral assemblages from diverse bulk compositions as characteristic of a certain pressure-temperature range. As new petrographic observations accumulated, it became increasingly difficult to accommodate this information within a manageable framework of metamorphic facies and subfacies. Instead, it turned out that mineral assemblages due to reactions in common rocks of a particular composition provide suitable indicators of metamorphic conditions. Metamorphic zones, defined on the basis of mineral reactions, very effectively display the evolution of metamorphic rocks. Thus, the importance of reactions in metamorphic rocks is emphasized. Experimental calibration of mineral reactions makes it possible to distinguish reactions which are of petrogenetic significance from those which are not. This distinction provides guidance in petrographic investigations undertaken with the object of deducing the physical conditions of metamorphism.

Within a metamorphic terrain, points indicating the same reaction constitute a line or a band, here designated by the term isoreaction-grad. Points along the isoreaction-grad represent metamorphic conditions (P, T, $P_{volatiles}$) corresponding to the pertinent reaction. As the term implies, an isoreaction-grad is defined by a specific reaction and is therefore more significant than an isograd based on the appearance or disappearance of a mineral or mineral assemblage not related to a specific reaction.

Thus, this presentation of the principles of rock metamorphism differs from that of previous editions: mineral reactions in several rock groups of common composition (mafic, ultramafic, pelitic, marly, and dolomitic) are treated in separate chapters. For each compositional group, the various mineral reactions that take place during the course of metamorphism are discussed. Large divisions of metamorphic grade, also defined on the basis of mineral reactions in common rocks, provide a convenient overview of metamorphic conditions. These divisions are designated as very-low-, low-, intermediate-, and high-grade metamorphism.

The new concept is straightforward in its application. It has been used successfully in the field by the author's coworkers and others. It is hoped that this book will provide some guidance in the petrographic studies of metamorphic terrains: only rocks having certain compositions need be examined in great detail. Instead of extensive investigations, only selected petrographic observations aimed at specific targets are required to deduce physical conditions of metamorphism. In order to choose the right targets within a given metamorphic setting, Chapter 15 should be consulted because it furnishes a key for the determination of metamorphic grades and major isoreaction-grads and isograds in common rocks.

The references are predominantly to recent literature in order to save space. There is no intention to underrate the value of older publications, and the student is advised to consult the reference lists given in recent publications. For the sake of convenience, all references are grouped together at the end of each chapter.

This book stresses the chemical, mineralogical, and physicochemical aspects of metamorphism. The fabric of rocks is not treated but attention is called to the books by Spry [*Metamorphic Textures*. Pergamon Press, Oxford. 1969] and Turner and Weiss [*Structural Analysis of Metamorphic Tectonites*. McGraw-Hill, New York. 1963]. Field relations in specific metamorphic regions are not considered in detail; this has been done by Turner [*Metamorphic Petrology*. McGraw-Hill, New York. 1968]and by Miyashiro [*Metamorphism and Metamorphic Belts*. George Allen & Unwin, London. 1973]. However, this book furnishes the basis for field investigations yielding petrogenetically significant data.

I am grateful for oral and written discussions, for expert guidance on field trips, and for early communication of most recent observations to so many people in various countries that all cannot be acknowledged individually. I also express my gratitude to my young colleagues at the University of Goettingen, especially Drs. P. Metz, K.-H. Nitsch, and B. Storre, who contributed many experimental results and discussions. Most of all, I am grateful to my friend Dr. Edgar Froese of the Geological Survey of Canada. He corrected, in fact, edited, my English version of the manuscript and improved it also through critical remarks and suggestions. Mrs. Ingeborg Tradel was so kind to undertake the task of typing the English manuscript. Finally, I cordially thank my wife Ursula for her gentle understanding during the many long days extending over the several years that I spent sitting at my desk.

Göttingen, September 1973 Helmut G. F. Winkler

Preface to the Fourth Edition

I am very grateful to those who purchased the third edition and thus made it possible to issue this fourth edition after such a short time. Compared with the previous edition, no substantial changes were made; however, some of the most recent research results have been integrated into the text. Accordingly, nine figures were redrawn and five figures were added.

This edition eliminates many of the typographical errors which had been overlooked. I thank very much Drs. R. A. Binns from Australia and E. Althaus from Karlsruhe (Germany) who very kindly read the whole book of the previous edition and gave me a list of corrections that had to be made. Again, I am very grateful to my young colleagues at this Institute, especially Drs. P. Metz and B. Storre.

Göttingen, June 1975 Helmut G. F. Winkler

Preface to the Fifth Edition

This edition, like the third and fourth editions, is based on the concept of dividing metamorphic grades according to mineral reactions in common rocks. The purpose of the present edition is the inclusion of results of recent progress in metamorphic petrology. Major changes are in the chapters on the metamorphism of carbonate rocks, pelitic rocks, and ultramafic rocks, and in the chapters 15 and 18. 14 figures are new or have been corrected.

In order for this book to remain a useful guide for field geologists, the previous framework of treating the subject has been maintained. Consequently, the scope has not been expanded to include a thorough consideration of thermodynamics. A knowledge of thermodynamics is necessary to acquire a deeper understanding of experimental work and its application. For those interested in this aspect, attention is drawn to the following publications:

Fraser, D. G., ed. 1977. *Thermodynamics in Geology*. 403 pp. D. Reidel, Boston.

Froese, E. 1976. *Applications of Thermodynamics in Metamorphic Petrology*. 37 pp. Geol. Survey of Canada, Paper 75–43.

Greenwood, H. J., ed. 1977. *Application of Thermodynamics to Petrology and Ore Deposits*. 230 pp. Mineralog. Association of Canada, Short Course Handbook.

Powell, R. 1978. *Thermodynamics in Petrology—An Introduction*. 284 pp. Harper and Row, New York.

Wood, B. J. and Fraser, D. G. 1976. *Elementary Thermodynamics for Geologists*. 303 pp. Oxford University Press.

I am grateful to Drs. P. Metz, B. Storre, D. Puhan, and E. Hoffer from this Institute and specially to Dr. R. D. Schuiling, Utrecht, for having offered valuable suggestions. Again, cordial thanks are due to Dr. E. Froese, Ottawa, for comments on subject matter and improvements of the English text.

Göttingen, January 1979 Helmut G. F. Winkler

Contents

Chapter 1

Definition and Types of Metamorphism

Igneous rocks formed at relatively high temperatures of approximately 650° to 1200°C and sediments deposited at the earth's surface represent extreme ends of the temperature range realized in the processes of rock formation. In the course of later geological events such rocks may become part of a region in the earth's crust where intermediate temperatures prevail; thus they are subjected to different temperatures. Similarly, the pressure of their new environment will, in general, differ from the pressure existing at their formation. Many minerals in these rocks are no longer stable at the newly imposed conditions of temperature and pressure; they will react and form mineral assemblages in equilibrium, or tend toward equilibrium, at the new conditions. Accordingly, the chemical composition of a rock is expressed by a new mineral assemblage; it has been transformed—for example, the conversion of clay or shale to mica schist.

In a few monomineralic sedimentary rocks, such as pure limestone and sandstone, the minerals remain stable at high temperatures. Nevertheless, recrystallization of mineral grains gives a new structure to the rock: limestone is changed into marble and sandstone into metamorphic quartzite.

Such rock transformations may take place within a large temperature interval. If the process of rock weathering at the earth's surface is excluded, the temperature interval is divided into the lower temperature domain of diagenetic transformations (diagenesis) continuous with the temperature of sedimentation, and the higher temperature domain of metamorphic transformations (metamorphism).

Definition: Metamorphism is the process of mineralogical and structural changes of rocks in their solid state in response to physical and chemical conditions which differ from the conditions prevailing during the formation of the rocks; however, the changes occurring within the domains of weathering and diagenesis are commonly excluded.

Although this definition applies to all rocks, this book is not concerned with salt deposits (evaporites) because their metamorphism takes place at considerably lower temperatures and pressures than the metamorphism of silicate and carbonate rocks. Salt deposits, except occasionally anhydrite, are not present in regions of metamorphic rocks. Neither is metamorphism of sediments consisting mainly of plant matter, *i.e.*, coalifiction, treated here because it starts at considerably lower temperatures than metamorphic reactions. Nevertheless, a correlation between the rank of coalification and mineralogical changes in other sedimentary rocks has recently been recognized. As Kisch (1969) has pointed out, very low temperature rock metamorphism of the laumontite type seems to begin when coalification has already reached the rank of gasflame or gas coal; at this stage the vitrite substance contains 40 to 30% volatile matter. With increasing metamorphic intensity, the rank of coalification progresses very fast.

On the basis of geological setting, it is possible to distinguish between two types of metamorphism; one type is of local extent only, the other type is of regional dimensions. The local extent type includes *contact metamorphism,* on the one hand and, entirely different in character, *cataclastic metamorphism,* on the other hand.

Contact metamorphism takes place in heated rocks bordering larger magmatic intrusions. Contact metamorphism is static thermal metamorphism of local extent producing an aureole of metamorphic rocks around an intrusive body. Contact metamorphic rocks lack schistosity. The very fine-grained splintery varieties are called hornfelses (see chapter on nomenclature). The large temperature gradient, decreasing from the hot intrusive contact to the unaltered country rock, gives rise to zones of metamorphic rocks differing markedly in mineral constituents.

Cataclastic metamorphism is confined to the near vicinity of faults and overthrusts. Mechanical crushing and shearing cause changes in the rock fabric. The resulting cataclastic rocks are known as fault breccias, mylonites, or pseudotachylites, corresponding to diminishing grain size. A pseudotachylite is so intensely sheared that it looks like black basaltic glass (tachylite). During these changes, no heat (or not enough heat) is supplied to the rocks; therefore, prograde chemical reactions of a metamorphic nature between minerals do not occur. However, secondary retrograde alteration is common because fracture zones provide easy access for fluids. (Cataclastic metamorphism is not considered further in this book.)

A certain mineralogical change in rock that is also of local extent but is of different character from either contact or cataclastic metamorphism has been called *hydrothermal metamorphism* by Coombs (1961). Here, hot solutions of gases have percolated through fractures, causing mineralogical changes in the neighboring rock. (Such processes will not be considered any further.)

Other types of metamorphic rocks occur on a *regional scale* in areas a few hundred to many thousand square kilometers in extent; they are products of *regional metamorphism*. Two types of regional metamorphism are commonly distinguished: (1) *regional dynamothermal metamorphism* (regional metamorphism *sensu stricto*) and (2) *regional burial metamorphism*.

Regional dynamothermal metamorphism is related geographically as well as genetically to large orogenic belts. Metamorphism is brought about, as with contact metamorphism, by a supply of thermal energy, but in this case very extensive metamorphic zones are formed. Changes in mineral assemblages from zone to zone are taken to indicate a continuous increase of temperature. Temperatures of approximately 700°C, possibly even 850°C, are attained. Thermal energy is supplied to a certain part of the earth's crust, *i.e.*, at the time of metamorphism the temperature at some given depth is higher than before and after this event. The geothermal gradient, commonly expressed in degrees centigrade per kilometer, is greater than at "normal" times. In contrast to contact metamorphism, however, regional dynamothermal metamorphism takes place with penetrative movement. This is evident from the schistose structure which is so commonly developed in rocks with platy or rod-shaped minerals (chorite schist, mica schist, etc.).

For many years there was thought to be a genetic association between orogeny and dynamothermal metamorphism. However, this view should now be reexamined. All orogenic mountain belts display large volumes of metamorphic rocks, but many may have formed during one or several earlier orogenic cycles. This is particularly well demonstrated in the Tertiary orogen of the European Alps. There, metamorphism during Tertiary times affected only a restricted area; medium- and high-grade rocks having been formed in only part of the Central Swiss Alps and in a small part of the Austrian Alps. Furthermore, metamorphism succeeded the development of large-scale tectonic features. The heat was supplied to the metamorphic area after the major tectonic movements. Following these, minor penetrative deformation took place and produced the schistosity of the metamorphic rocks. This, however, is not meant to imply that all metamorphic minerals are formed strictly contemporaneously with deformation. Detailed investigations have shown that recrystallization may occur also between phases of deformation and even during postorogenic time. Although the metamorphic event commonly postdates the development of the large-scale tectonic structures, orogens are known where metamorphism took place during or, at very great depth, even before the major tectonic deformations.

Rocks produced by regional dynamothermal and local contact metamorphism differ significantly in their fabric, being schistose as opposed to nonschistose, respectively. Furthermore, an aureole of contact metamorphism is developed only if a previously low-grade metamorphic or nonmetamorphic section of the crust has been intruded and thermally metamorphosed by a magmatic body. Contact metamorphism, where easily detectable, is often due to shallow magmatic intrusions. The depth of the intrusions measured from the surface may have ranged from some hundred meters to several kilometers. Consequently, metamorphic transformations within contact aureoles took place while the load pressure was rather low, ranging from about 100 bars to 1000 or possibly even 3000 bars (the latter value corresponding to an overburden of about 11 to 12 km). Therefore, "shallow" contact metamorphism (as well as hydrothermal metamorphism) is characterized by rather low pressure. On the other hand, regional dynamothermal metamorphism takes place at higher pressures, ranging from at least 2000 to more than 10,000 bars. Intensity of pressure is the essential genetic difference between local, shallow contact metamorphism and that type of regional metamorphism termed dynamothermal; the temperatures at which metamorphic reactions take place are often the same in both cases—ranging from a comparatively low value to a maximum of about 850°C in the crust.

It is apparent that, on the basis of prevailing pressure, not only can contact metamorphism and regional dynamothermal metamorphism be distinguished, but regional dynamothermal metamorphism can also be further subdivided into several types. Whereas contact metamorphism is generally characterized by low pressures, regional dynamothermal metamorphism may occur throughout an appreciable pressure range. There are metamorphic terrains formed at intermediate, high, or very high pressures. A magmatic intrusion at rather shallow depth will impose on the adjacent country rock a very high (100°C/km or higher) geothermal gradient of local extent; the result is shallow contact metamorphism. On the other hand, the heating of larger segments of the crust may, for instance, cause a temperature of 750°C at a depth of 15 km or 25 km, corresponding to a geothermal gradient of 50°C/km or 30°C/km, respectively. Such different combinations of temperatures and pressures will be reflected in different mineral assemblages. Accordingly, there are various types of regional dynamothermal metamorphism. That is, types of comparatively low pressure, medium pressure, and high pressure metamorphism which are schematically shown by broken lines in Figure 1-1.

Regional burial metamorphism (Coombs, 1961) bears no genetic relationship to orogenesis or to magmatic intrusions. Sediments and interlayered volcanic rocks of a geosyncline may become gradually buried. The temperatures, even at great depth, are much lower than the

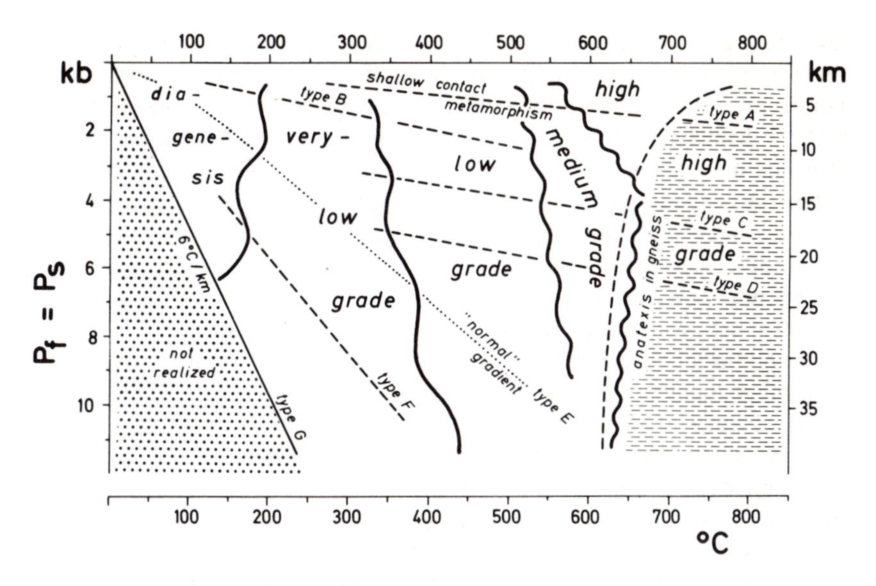

Fig. 1-1 Schematic pressure-temperature diagram for contact metamorphism and for different types of regional metamorphism. In addition to the *P-T* field of diagenesis, the fields of the different metamorphic grades explained in Chapter 7 are shown here. The *P-T* region below the lowest geothermal gradient is not realized in nature. The depths, corresponding to the pressures, are maximum values and may be somewhat less than shown (see Chapter 3).

temperatures encountered during dynamothermal metamorphism; maximum values are probably about 400° to 450°C. Penetrative movement producing a schistose structure is only active locally, if at all; generally it is absent. Therefore, the original fabric of the rocks may be largely preserved while the mineralogical composition is changed. Metamorphic changes are not distinguishable in hand specimens; only in thin sections can they be clearly recognized.

Burial metamorphism accounts for those metamorphic rocks containing the CaAl-zeolite *laumontite* (so-called zeolite facies rocks), on the one hand, and *lawsonite-glaucophane*-bearing rocks, on the other hand; a detailed description of these rocks is given later. Both these mineralogically very different rock types are formed at low and approximately equal temperatures; however, they indicate considerable differences in pressure. Rocks of the lawsonite-glaucophane type, when additionally bearing jadeitic pyroxene, originate at the highest pressures realized in the earth's crust, *i.e.,* in crustal regions with an especially low geothermal gradient. In contrast, during the formation of zeolite facies rocks, an approximately "normal" or higher geothermal gradient existed.

Despite the low temperature of metamorphism, most rocks of the lawsonite-glaucophane type are completely recrystallized in response to their metamorphic environment because of the high prevailing pressure. However, some laumontite-bearing rocks formed at low pressures as well as low temperatures may show incomplete equilibration at the imposed conditions. The reconstitution of the rocks may be complete, extensive, or, especially in more coarse-grained rocks, only incipient. At the higher temperatures of contact metamorphism and of regional dynamothermal metamorphism, rocks are completely recrystallized, even if pressures are low.

Grades of metamorphism. The distinction between very low temperature burial metamorphism and low to high temperature dynamothermal metamorphism can be made in many instances and has been widely used. However, the term burial metamorphism should be applied only in cases where the single fact of burial has been reasonably well ascertained. It should not be used synonymously with very low-grade metamorphism. The reasons for this statement are the following:

1. Burial metamorphism and very low-grade metamorphism in an orogenic belt may produce mineralogically identical rocks.
2. A continuous increase in the temperature of metamorphism from very low temperatures (corresponding to those of burial metamorphism) to high temperatures of so-called dynamothermal metamorphism has been demonstrated within a single orogenic belt.
3. The terms "burial" and "dynamothermal" imply epeirogenic and orogenic movements, respectively. The type of movement, however, has no significance in mineral reactions. These are governed for a given rock composition solely by pressure and temperature and, as shall be explained later, by the composition or even by the amount of the gas phase.

For these reasons it is sometimes impossible or even senseless to distinguish very low temperature dynamothermal metamorphism from burial metamorphism. The term "burial metamorphism" originally referred to a kind of regional metamorphism taking place in a sediment at somewhat elevated but still very low temperatures, caused by subsidence in nonorogenic terrains. But the same very low temperature range can also prevail in certain portions of orogenic belts. The significant point is that relatively low temperatures cause burial metamorphism. As in all kinds of *regional* metamorphism, the pressures may have been low, medium, or high.

Without specifying pressure, regional metamorphism may be broadly divided into types on the basis of temperature ranges:

Very low temperature metamorphism
Low temperature metamorphism
Medium temperature metamorphism
High temperature metamorphism

Very low temperature metamorphism includes burial metamorphism as well as the very low temperature rocks in a dynamothermal metamorphic sequence.

It is customary, especially in English-speaking countries, to speak of the *grade of metamorphism*. If "grade" is used in a specific sense, *e.g.*, when referring to the staurolite isograd, it refers to well-defined pressure-temperature values. A particular pressure-temperature combination describes the physical condition during the metamorphism of the very site under investigation (we shall come back to this point). If "grade" is used in a nonspecific manner, it covers a rather large range of metamorphic conditions. Thus, greenschist facies rocks formed at low temperatures are attributed to low-grade metamorphism. Since the term metamorphic grade is used commonly, the following broad division of the whole pressure-temperature field of metamorphism is suggested here:

Very low-grade metamorphism
Low-grade metamorphism
Medium-grade metamorphism
High-grade metamorphism

These divisions are primarily based on temperature and are defined in Chapter 7; however, information on pressure should be stated as well.

The following examples serve to illustrate usage of the terms:

Medium-grade and high pressure metamorphism: for rocks of the almandine-amphibolite facies with kyanite in metapelites. This corresponds to the relevant part of the broken line designated as type D in Figure 1-1.

Very low-grade and very high pressure metamorphism: for rocks of the lawsonite-jadeite-glaucophane facies of type F in Figure 1-1.

Very low-grade and low pressure metamorphism: for rocks having laumontite. This corresponds to the relevant part of the broken line designated as type B in Figure 1-1.

The temperature divisions of regional metamorphism provide a broad classification of metamorphism useful in geological fieldwork. It is simple to distinguish rocks of different metamorphic grades on the basis of mineral assemblages, *e.g.*, low-grade greenschist facies rocks from medium-grade amphibolite facies rocks.

The pressure dependence of the boundaries of the *metamorphic grades* are discussed later in Chapter 7, and it will be shown that certain common metamorphic reactions are well suited to define such boundaries. However, for clarity, the boundaries have already been drawn as wavy lines in Figure 1-1. Also indicated are different geothermal gradients or, more correctly, various temperature distributions with depth, corresponding to different types of metamorphism at low, medium, or high pressures. On continents, beyond a depth of several kilometers a geothermal gradient of 20°C/km is considered "normal"; accordingly, at a depth of 20 km the temperature would be approximately 400°C. In geosynclinal areas of rapid subsidence, the temperature rise with increased depth is smaller, whereas in active belts it is greater—occasionally considerably greater. In the vicinity of shallow-seated magmatic intrusions there exists, limited in vertical and horizontal extent, a very great temperature gradient (contact metamorphism). In Figure 1-1 the domain of diagenesis is marked off; its boundary at temperatures somewhat below 200°C is the beginning of metamorphism as defined in Chapter 2.

At high temperatures a wavy line marks the beginning of anatexis (Figure 1-1). At these pressure-temperature conditions quartzo-feldspathic rocks (mainly gneisses) begin to melt *if* water is present. Large amounts of granitic melts are formed through anatexis. According to the strict definition of metamorphism as rock transformations in the solid state, the formation of melts limits the field of metamorphism. Such a narrow definition is inadequate because:

1. The partial melting of quartzo-feldspatic rocks, leading to the formation of migmatites, indicates conditions generally attributed to high-grade metamorphism.
2. Rocks not exhibiting partial melting under the *same P-T* conditions, such as amphibolites or gneisses in the absence of a water-rich fluid phase, commonly occur in the same region together with anatectic rocks. Therefore, anatexis is considered to belong to high-grade metamorphism.

References

Coombs, D. S. 1961. *Australian J. Sci.* **24:** 203–215.

Kisch, H. J. 1969. In P. A. Schenk and I. Havenaar, eds. *Advances in Organic Geochemistry 1968*. pp. 407–425. Pergamon Press, Oxford.

Chapter 2

From Diagenesis to Metamorphism

The *Glossary of Geology* (1960) gives the following definition for *diagenesis:* "Those changes of various kinds occurring in sediments between the time of deposition and the time at which complete lithification takes place. The changes may be due to bacterial action, digestive processes of organisms, to solution and redisposition by permeating water, or to chemical replacement." Compaction and consequently reduction in pore space may be added. The change from loose sand or lime ooze to sandstone or limestone, respectively, is diagenetic, as is cementation or replacement by $CaCO_3$ or SiO_2, formation of sulfides in sediments and of silicates, such as glauconite, analcime $NaAlSi_2O_6 \cdot H_2O$, heulandite $CaAl_2Si_7O_{18} \cdot 6\ H_2O$, or its silica-rich variety called clinoptilolite. To these and other diagenetic changes, Correns (1968) further adds the formation of feldspars, mainly orthoclase but also albite and sometimes even oligoclase, in limestones and sandstones. Diagenetic formation of chlorite in clays, shales, and mudstones is also known, and Frey (1969, 1970) has deduced the following reaction for the formation of aluminum-rich chlorite:

$$\text{disordered mixed-layer illite/montmorillonite} \rightarrow \\ \text{Al-rich chlorite} + \text{illite} + \text{quartz} + H_2O$$

Thus, as Correns remarked, in diagenesis some minerals are formed which also occur in metamorphism. "Therefore, we do not have a sharp boundary towards metamorphism." This is certainly true, even if additional criteria are taken into account.

Most geologists will agree with the definition of diagenesis as given by Correns (1950):

> Diagenesis comprises all changes taking place in a sediment between sedimentation and the onset of metamorphism, except those caused by weathering.

In principle, there is no real break but a continuous sequence of rock transformations from sedimentation to high-grade metamorphism. On the other hand, it has always been very useful to distinguish between diagenetic and metamorphic rocks; therefore, a demarcation between the two categories must be defined in some way. What agreement can be reached on criteria of distinction between diagenetic and metamorphic rocks? Two approaches will be discussed: one based on sedimentary and the other on metamorphic petrology.

Concerning diagenesis, von Engelhardt (1967) makes the following interesting point:

> In the diagenetic zone, the sediments contain interstitial fluids as continuous phases which can be moved by normal flow. These fluids (most commonly aqueous solutions) are, therefore, always involved in the diagenetic reactions, the transport of substances being effected through flow or normal diffusion. Because of the intercommunication and mobility of the interstitial fluids, reactions of the open-system type prevail. . . . The diagenesis stops at a depth where in *all* sedimentary rocks the intercommunicating pore spaces have been closed up by physical or chemical processes. In this zone metamorphism begins. . . . On the whole in metamorphism, reactions of the closed-system type . . . prevail.

Thus, according to von Engelhardt, "metamorphism and diagenesis should be identified by the types of processes prevailing during these two stages of transformation of primary sediments. This way it will be possible, in almost every practical case, to distinguish between diagenetic and metamorphic changes."

In order to make this distinction, the porosity of a number of different sedimentary rocks must be determined. This is routine work in the investigation of sedimentary but not of metamorphic rocks. If necessary, it can, of course, be done. However, the observation of "no intercommunicating pore spaces in all sedimentary rocks" is a poor petrographic criterion on which to base the definition of a metamorphic rock. It is unknown whether this state has been reached before, during, or after pressure-temperature conditions prevailed, causing the formation of truly metamorphic mineral assemblages.

From the viewpoint of metamorphic petrology, *the first formation of a truly metamorphic mineral paragenesis* is the important criterion, *i.e.*, a paragenesis met with *only* in metamorphic rocks. On the other hand, from the viewpoint of sedimentary petrogenesis, the complete blocking of interconnecting pore spaces is diagnostic of the end of diagenesis. But the two conditions need not coincide, and probably do not. Therefore, in metamorphic petrology a mineralogical rather than physical criterion is required to define the beginning of metamorphism. And since by common consent the range of metamorphism is preceded by the range of diagenesis, the beginning of metamorphism also designates the end of diagenesis.

Minerals such as feldspar, chlorite, or quartz form under metamorphic as well as under diagenetic conditions. Chlorite, for example, is a stable constituent in sediments as well as in metamorphic rocks. The coexisting minerals will decide whether the rock is of metamorphic or diagenetic origin. For instance, chlorite coexists with illite-phengite in diagenetic and some metamorphic rocks, but with the CaAl-mineral prehnite or zoisite-clinozoisite only in metamorphic rocks.

In designating the beginning of metamorphism the following definition will be used:

> Metamorphism has begun and diagenesis has ended when a mineral assemblage is formed which cannot originate in a sedimentary environment.

In the early editions of this book the definition included the statement: "When mineral assemblages restricted only to sediments disappear." This sentence is now omitted because it is known that sedimentary minerals like kaolinite or montmorillonite *may* disappear from a pile of sediments already at an early stage of diagenesis.

The *first appearance* of metamorphic minerals such as laumonite, lawsonite, glaucophane, paragonite, or pyrophyllite indicates the beginning of metamorphism. However, reactions forming these minerals do not take place at identical physical conditins. Probably the first formation of paragonite and pyrophyllite occurs at somewhat higher temperatures than the first formation of laumontite and lawsonite. Some available experimental data are summarized further on.

There are many rocks which do not show any change in mineralogy at the onset of metamorphism, whereas others exhibit one or more of the critical changes. Thus rocks consisting of quartz + chlorite + illite (phengite) persist unchanged from the diagenetic stage through the very-low-grade and low-grade metamorphism to the beginning of medium-grade metamorphism, disregarding changes of crystallinity and structural order in illite. The assemblage calcite + quartz persists even to higher grades. These examples show that only rocks with a specific mineralogy may be used as indicators of the beginning of metamorphism or of higher grades of metamorphism. This is an important aspect of metamorphic petrology in general.

It follows from this discussion that

1. *Only certain rocks, and by no means all, are suitable for defining the beginning of metamorphism.*
2. *Temperature thresholds* beyond which metamorphism starts can be given for different rocks by the first appearance of a metamorphic mineral assemblage, *i.e.,* a mineral paragenesis of metamorphic origin and unknown in sedimentary rocks. Such a

boundary is not fixed but depends on the composition of a given rock.

3. *The lowest temperature* (at a given pressure) leading to a metamorphic mineral association is of special interest. It signifies the *minimum temperature* at which a truly metamorphic rock can be formed, and, according to our definition, the upper limit of diagenesis. This is the meaning of the boundary drawn in Figure 1-1 between metamorphism and diagenesis and referred to as the "beginning of metamorphism."

According to our present knowledge, the lowest temperature metamorphic minerals are laumontite and (at somewhat higher pressure) lawsonite. Reactions by which these minerals are formed furnish minimum temperatures at which truly metamorphic rocks are formed. The Working Group for the Cartography of the Metamorphic Belts of the World (Zwart *et al.*, 1967) also regards metamorphism "as starting with the occurrence of laumontite + quartz."

The exact temperatures at which laumontite ($CaAl_2Si_4O_{12} \cdot 4\ H_2O$), the only zeolite of metamorphic origin, is formed are not yet known. However, from petrographic observation in several localities all over the world, first reported from New Zealand by Coombs (1959, 1961), it is well established that the first formation of laumontite may coincide with the reaction

$$\text{analcime} + \text{quartz} = \text{albite} + H_2O$$

Campbell and Fyfe (1965) determined the equilibrium of this reaction at the very low pressure of 12 bars and at 190°C. On the basis of inferred thermodynamic data, they further calculated equilibrium data at higher H_2O pressures and concluded that the equilibrium temperature diminishes drastically from about 1 kb upward, so much that at 4 kb it is only 50°C. These calculated data were widely published and were believed to be reliable. That this is not so has recently been shown by Liou (1971a). He determined experimentally the temperatures at which analcime + quartz react to albite + H_2O. His data are about 200°C at 2 kb, 196 ± 5°C at 3 kb, and 183 ± 5°C at 5 kb. The albite formed in the experiments was structurally highly disordered, while in nature a well-ordered low albite occurs. If this is considered, the reaction temperatures will be somewhat lower; Liou suggested a temperature decrease of 30°C at 2 kb and of 70°C at 5 kb. This problem is not fully solved yet, but it is certain now that the beginning of metamorphism, as defined here, is generally between 200 and 150°C.

Further information is obtained from the formation of laumontite according to the following reaction:

$$\underset{\text{heulandite}}{CaAl_2Si_7O_{18} \cdot 6\,H_2O} = \underset{\text{laumontite}}{CaAl_2Si_4O_{12} \cdot 4\,H_2O} + \underset{\text{quartz}}{3\,SiO_2} + \underset{\text{water}}{2\,H_2O}$$

Unfortunately, rates of this reaction are so slow at ~ 200°C that the reaction could not be experimentally performed at low pressures; however, if H_2O pressure is raised appreciably, heulandite breaks down. But at this higher pressure lawsonite, $CaAl_2[(OH)_2/Si_2O_7]$, rather than laumontite is stable. The temperature at which heulandite decomposes to lawsonite + quartz + H_2O has been determined by Nitsch (1968) as 185 ± 25°C at 7 kb. This temperature will be very little changed at lower pressures, and below 3 ± 0.5 kb laumontite takes the place of lawsonite. It is very likely, therefore, that laumontite will form at temperatures somewhat below 200°C, probably as low as 175°C.

Temperatures somewhat below 200°C mark the beginning of metamorphism (indicated by a wavy line in Figure 1-1). At the lowest temperatures at which metamorphism produces laumontite, coalification processes, as reported by Kisch (1969, 1974), reach the gas-flame coal rank and sometimes the gas coal rank, characterized by less than 40 and 30% volatile matter in vitrite, respectively. This is largely confirmed by Frey and Niggli (1971) from the Swiss Alps. In addition, these authors correlated the so-called crystallinity of illite (see Chapter 7) with coal rank and metamorphism; they showed that at the beginning of metamorphism, as defined here, the first appearance of laumontite corresponds to a very poor crystallinity (7.5 on Kubler's scale) but improves from here onward markedly with increasing temperature of metamorphism. On the other hand, the correlation between illite crystallinity and coalification need not be a simple one (Wolf, 1975).

The coal rank attained at the beginning of metamorphism also sets a limit to the occurrence of oil fields; the "dead line," or the "oil phase-out zone," is reached and only gas fields may persist into areas of somewhat higher coal rank. Thus, the limit of the beginning of metamorphism, as defined here, is also of great economic importance.

References

Campbell, A. S. and Fyfe, W. S. 1965. *Am. J. Sci.* **263:** 807–816.

Coombs, 1961. *Australian J. Sci.* **24:** 203–215.

Correns, C. W. 1950. *Geochim. Cosmochim. Acta* **1:** 49–54.

——— 1968. Diagenese und Fossilisation. In R. Brinkmann, ed. *Lehrbuch der Allgemeinen Geologie,* Vol. 3. Stuttgart.

von Engelhardt, W. 1967. Interstitial solutions and diagenesis in sediments. In G. Larsen and J. C. Chillingat, eds. *Developments in Sedimentology.* Vol. 8, pp. 503–521. Elsevier, Amsterdam.

Frey, M. 1969. *Beitr. Geol. Karte Schweiz Neue Folge 137.*

———. 1970. *Fortschr. Mineral.* **47:** 22–23.

——— and Niggli, E. 1971. *Schweiz. Mineral. Petrog. Mitt.* **51:** 229–234.

Glossary of Geology. 1960. 2nd edit. American Geol. Inst., Washington.

Kisch, H. J. 1969. In P. A. Schenk and I. Havenaar, eds. *Advances in Organic Geochemistry 1968*. pp. 407–425. Pergamon Press, Oxford.

——— 1974. *Koninkl Nederl. Akademie van Wetenschappen Amsterdam.* Series B, **77:** 81–118.

Liou, J. G. 1971. *Lithos* **4:** 389–402.

Nitsch, K.-H. 1968. *Naturwiss.* **55:** 388.

Wolf, M. 1975. *Neues Jahrb. Geol. Paläont. Mh.* **1975:** 437–447.

Zwart, H. J., *et al.* 1967. *IUGS Geol. Newsletter No. 2,* 57–72.

Chapter 3

Factors of Metamorphism

General Considerations

Temperature and pressure are the physical factors that control the process of metamorphism. Metamorphism, in general, refers to the reactions between neighboring minerals of a rock in response to conditions of temperature and pressure prevailing at depth. A certain mineral paragenesis, formed at some given temperature and pressure, becomes unstable if subjected to different conditions; the minerals react to form a new paragenesis in equilibrium at the new conditions. If carbonates and H_2O- or OH-bearing minerals take part in the reaction, CO_2 and H_2O are liberated. The higher the temperature of metamorphism, the smaller the amount of CO_2 and H_2O combined in the stable minerals. Therefore, a *fluid phase composed of volatile constituents is always present during metamorphism of such rocks*.

At supercritical conditions, which commonly are realized in metamorphism except at very low temperatures, the fluid is a gas of high density with many properties of a liquid. Water, *e.g.*, at 500°C and 2000 bars has a density of 0.69 g/cm^3; at 400°C and the same pressure, its density is 0.97 g/cm^3. Volatile constituents already existed in the rocks even before metamorphism, occupying pores and minute cracks or being adsorbed on the grain boundaries. Even in the case of many igneous rocks, a sufficient amount of H_2O must have been present during metamorphism, either originally in the rocks or introduced along minute cracks. If this had not been so, the metamorphism of basalts to amphibolites or chlorite-epidote greenschists would have been impossible.

Many experiments have shown that the presence of water greatly increases the rate of recrystallization; without water some reactions could not proceed to completion even during geological periods of time. This is probably because of (1) the catalytic action of water, even if present only in small amounts, and (2) the fact that neighboring minerals in a metamorphic rock represent, in general, an association of coexisting minerals in thermodynamic equilibrium with each other—known as a mineral paragenesis.[1] The fluid phase, coexisting in equilibrium with a mineral paragenesis during metamorphism, must not be disregarded, even though petrographical investigations give no account of its amount and only rarely of its composition. Despite the presence of a mobile fluid phase containing dissolved amounts of the minerals with which it is in equilibrium, transport of material over large distances generally does not take place. There are many indications that rocks constitute a "closed" thermodynamic system during the short time required for metamorphic crystallization. Transport of material is generally limited to distances similar to the size of newly formed crystals. It has been observed frequently that minute chemical differences of former sediments are preserved during metamorphism.

Metamorphism is essentially an *isochemical*[2] process. Evidence for isochemical metamorphic transformations has become increasingly more convincing[3]; however, the highly volatile components, predominantly H_2O and CO_2, are exceptions. If they are constituents of the reacting minerals, they are essential components of the metamorphic reaction themselves, but they gradually leave the rock system while it is still under metamorphic conditions. Thus, the metamorphosed rock is partly or wholly depleted of the volatile components as compared to the parent rock. On the other hand, these components often may (not must) have access to rocks not having volatile components in their minerals (as in basalts) prior to or during metamorphism. It is suggested that met-

[1]So-called armored relics obviously do not belong to a mineral paragenesis. These are minerals completely surrounded by a rim of reaction products, thus preventing further reaction.

[2]By way of contrast, allochemical crystallization refers to metamorphic reconstitution accompanied by a change in bulk composition of the rock. This process is known as metasomatism; it may operate over a wide range of temperature and pressure conditions, even at the earth's surface. Previously, it was thought that metasomatism played a significant part in metamorphism, but at present, metasomatism is regarded as a phenomenon of more local importance in metamorphic terrains. For example, gaseous transfer of material from a crystallizing granite pluton into the adjacent country rock is a common process. This addition of material may well lead to metasomatic reactions and to the formation of minerals clearly requiring introduced constituents.

[3]As an example, Jäger (1970, 1973) proves that even in high-grade though still non-migmatitic gneisses the rock system was closed to K and Ca, thus it is highly probable that it was closed to the other elements as well. Furthermore, Dobretsov *et al.* (1967) show by statistical treatment of analytical data that no addition of Na has taken place during the formation of Na-amphibole-rich glaucophane schists from basalts. Pitcher and Berger (1972, p. 312 ff.) concluded in their study of large granite emplacements that "with the notable exception of (OH), the contact effects were isochemical, even over small distances". Ronov *et al.* (1977) showed that regional metamorphism of sediments is isochemical, apart from H_2O and CO_2.

amorphic transformations that are isochemical—with the sole exception of volatile components—be designated as *quasi isochemical.* In contrast, metamorphic differentiation, *i.e.,* metamorphic transformation on a small limited scale, is an allochemical process involving transfer of components. This can only be operative if very special conditions, specifically that of incompatible rock compositions, prevail.

Some of the H_2O and/or CO_2 liberated during a metamorphic reaction will leave the system. In fact, it is to be expected that the overpressure created by the formation of gases within the rock causes cracks, allowing part of the gas or the gas mixture to escape; thus, the fluid pressure is approximately equalized with the load pressure. Some of the internally created fluid overpressure may persist for some time. Before the maximum temperature of metamorphism reached at a particular place has noticeably declined, nearly all volatile constituents *must* have escaped; otherwise, the gradually sinking temperature would have caused a reversal of the reactions. *Retrograde metamorphism* (diaphthoresis), the transformation of higher-grade metamorphic rocks to lower-grade ones (*e.g.,* amphibolites into greenschists), is only observed in certain zones. In these zones H_2O-rich and often CO_2-bearing gases apparently were introduced along fractures and shear zones long after the highest temperature of metamorphism had considerably subsided. In such cases reactions could be reversed, producing mineral assemblages stable at the lower temperature, which contain, as a rule, more H_2O and CO_2. But products of retrograde metamorphism constitute only a small portion of all metamorphic rocks. On the other hand, incipient stages of retrograde alteration of certain minerals are frequently observed, presumably having taken place near the surface a very long time after metamorphism. With these exceptions, the mineral assemblages of metamorphic rocks generally indicate the highest grade of metamorphism, *i.e.,* they reflect the highest temperature reached during metamorphism of a certain region.

In regionally metamorphosed terrains, zones of rock may be distinguished which formed within a certain pressure-temperature field. The coexisting minerals are phases of a system in (at least) bivariant equilibrium, for the temperature and pressure can be arbitrarily varied within certain limits, *i.e.,* there are at least two degrees of freedom. The same is true of mineral assemblages in adjacent zones. However, at the junction of two zones, the conditions for a reaction boundary are realized, *i.e.,* at these conditions minerals react to form new assemblages. At the reaction boundary the system has one less degree of freedom (one more phase) than in the regions on either side of that boundary. The aim of metamorphic petrology is to determine such reaction boundaries for all reactions of significance in metamorphism. Assuming an univariant reaction, equilibrium between the right and the left side of the reaction equation (*i.e.*, the location of the reaction boundary) is determined at some given pressure by a definite temperature. There are reactions in which the equilibrium temperatures varies considerably with changes in pressure; they may serve as indicators of pressure during metamorphism. Other reactions, which are only slightly influenced by pressure, provide suitable temperature indicators. Experiments have shown that the equilibrium temperatures of numerous common reactions that release water are raised only 2° to 10°C for every increase of 1000 bars[4] in hydrostatic pressure (at low pressures the effect of pressure on the equilibrium temperature is greater).

The effective *pressure* during metamorphism is mainly *due to the load of the overlying rocks*. It is not an easy matter to estimate, on geological grounds, the previous thickness of rock cover; the original stratigraphic sequence may have been repeated several times by complex folding. The load pressure P_l increases with depth at a rate of about 250 to 300 bars/km, depending on the density of the rocks. It is reasonable to assume that the load pressure is hydrostatic in character, *i.e.*, the pressure is equal in all directions. The volatile constituents present in rock pores and minute cracks form at most conditions of metamorphism *one* fluid phase. This phase presumably is subjected to the same load pressure as the minerals of the rocks; accordingly, the assumption is made that the pressure of the fluid phase P_f is approximately equal to the load pressure P_l. It should be kept in mind, however, that in response to rising temperature, various metamorphic reactions release large amounts of H_2O and/or CO_2. In view of the small pore volume, this pressure, created within the rock formations due to metamorphic reactions, will exceed the load pressure by an amount equal to the strength of the rocks. Greater gas pressures would be released along fractures caused by the pressure. It could be assumed therefore, that the internally created gas pressure[5] will give rise to a certain overpressure so that P_f exceeds P_l at the time of the volatile-producing metamorphic reactions.

The value of the "*internally created gas overpressure*" depends on the strength of the rocks at depth. This strength decreases with increasing temperature and probably does not exceed a few hundred bars (Carmichael, 1978). Consequently, the effect of an internally created gas overpressure will be negligible, *i.e.*, the fluid pressure is practically equal to the load pressure.

---

[4]1 bar = 10^6dyne/cm^2; abbreviated *b*.
1 technical atmosphere (at) = 1 kp/cm^2 = 0.980665 bars.
1 physical atmosphere (atm) = 1.0333 kp/cm^2 = 1.01325 bars.

[5]A "fluid overpressure" has also been suggested by Rutland (1965).

On the other hand, in the case of minerals deposited from hydrothermal solutions along fractures and in adjacent wall rocks, it may be possible that $P_f < P_l$. In this case, the equilibrium temperature of a reaction liberating a gas will be lowered in proportion to the difference between the pressure acting on the solid minerals, P_s, and P_f. This situation has been discussed in detail by Thompson (1955) and Fyfe *et al.* (1958). It should be noted that in such cases the phase rule, generally expressed as $F = C - P + 2$, takes the form $F = C - P + 3$. Besides the usual two physical degrees of freedom of hydrostatic pressure and temperature, a further degree of freedom must be considered. Instead of one pressure, two pressures, P_f and P_s, are effective. Consequently, an equilibrium univariant if $P_f = P_s$ becomes bivariant if $P_f < P_s$. The situation of $P_f < P_s$ represents a special case realized only under particular conditions. In all other cases of metamorphism, it is assumed that, *as a rule, the pressure of the fluid phase* P_f *was equal to the pressure acting on the solid minerals, i.e.,* $P_f = P_s$. This condition also holds in all experimental investigations. All conclusions with regard to petrogenesis deduced from experimental results are based on this assumption.

The Composition of the Fluid Phase

In silicate rocks, the fluid phase present during metamorphism is essentially H_2O, so that $P_f = P_{H_2O}$. However, this may not be the case in graphite-bearing rocks (see Figs. 3-1 and 3-2). During the metamorphism of carbonate rocks, CO_2 is liberated. Therefore, the fluid phase will contain CO_2 in addition to the H_2O originally present in the pores of the sediments. In such cases, the pressure of the fluid phase consists of the partial pressures of H_2O and CO_2. The composition of the fluid phase is not constant because the molecular ratio of H_2O to CO_2 can be varied. In contrast to the stoichiometrically fixed ratio of CO_2 to metal oxides in the crystalline carbonates, the fluid phase may have a highly variable composition during the metamorphism of carbonate rocks. The freely variable composition of the fluid phase is, besides fluid pressure and temperature, an additional thermodynamic degree of freedom. This has the following consequence. The equilibrium temperature of, say, the reaction

$$\text{calcite} + \text{quartz} = \text{wollastonite} + CO_2$$

has been determined at some given pressure of CO_2 if $P_f = P_{CO_2}$; at this condition, the equilibrium is univariant. If, however, $P_f = p_{CO_2} + p_{H_2O}$, the equilibrium temperature is not uniquely determined at $P_f =$

const.; the smaller the mole fraction of CO_2 in the fluid phase, the lower the equilibrium temperature (see examples in Chapter 9). The equilibrium is no longer univariant but bivariant. In order to obtain some definite equilibrium temperature, the values of P_f and the mole fraction of CO_2 in the H_2O-CO_2 fluid phase must be fixed. Even if P_f could be estimated on the basis of the depth at which metamorphism took place, nothing is known about the probable composition of the fluid phase. It follows that reactions involving carbonates are in nature (at least) bivariant. Consequently, it is hardly possible to evaluate the dependence of the equilibrium temperature on the total pressure P_f. This discouraging conclusion is correct in principle, but a more promising aspect of the problem should also be considered, namely, to what extent the equilibrium temperature is dependent on the mole fraction of CO_2 and on P_f. Further on, examples will demonstrate that valuable petrological deductions can indeed be based on some reactions, even though they are bivariant.

If a sequence of sedimentary rocks consisting of carbonates and OH-bearing silicates is subjected to metamorphism, mineral reactions will liberate CO_2 and H_2O. In carbonate layers, the original water in the pores and newly formed CO_2 make up a fluid phase of unknown composition. In carbonate-free layers, water is liberated, which, together with the original water present in the sediments, forms a fluid phase consisting almost totally of H_2O. In the latter case, the composition of the fluid phase remains essentially unchanged and $P_f = P_{H_2O}$. If, however, it is assumed that during the period of metamorphism H_2O and CO_2 can diffuse so freely that the fluid phases in the various layers of carbonate rocks and carbonate-free silicate rocks become mixed, then the composition of the fluid phase present in silicate rocks would also be unknown. A reaction, univariant if $P_f = P_{H_2O}$, would become bivariant because now $P_f = p_{H_2O} + p_{CO_2}$. Even at constant P_f, there would be no uniquely determined equilibrium temperature; the greater the dilution of H_2O by CO_2, the lower the equilibrium temperature.

The question arises if it can be reasonably assumed that the fluid phases produced in various layers mix extensively during metamorphism. Many considerations speak against it. Experimental investigations of metamorphic reactions involving carbonates, and correlations of experimental results with observations in nature, clearly show that at the time of the metamorphic reactions, the fluid phase must have had a high mole fraction of CO_2. No noticeable dilution of the CO_2-rich fluid phase has taken place even though water is liberated in the surrounding noncarbonate rocks during metamorphism. Therefore, the view is advanced that, during the relatively short time required for metamorphic reactions to proceed, the fluid phase remains essentially unchanged, except where very thin layers of carbonate and carbonate-free rocks occur in the sedimentary sequence. *In the case of carbonate-free and graphite-free rocks, it may be assumed, in general, that $P_f \approx P_{H_2O}$.* Experimentally determined univariant reactions provide, at P_{H_2O} = const., unique equilibrium temperatures which are of importance in elucidating the process of metamorphism. In the case of carbonate-bearing rocks, however, the gas phase will be rich in CO_2 and always $P_f = p_{H_2O} + p_{CO_2}$, so that, even at P_f = const., the dependence of the equilibrium temperature on the composition of the fluid phase must be taken into account. Experimental studies are necessary to determine the extent of lowering of the equilibrium temperature in response to an increase in the amount of H_2O in the H_2O-CO_2 fluid phase. This effect may be large but in some cases is rather small, as examples in Chapter 9 will demonstrate.

A further consideration supports the assumption of $P_f \approx P_{H_2O}$ during metamorphism in the case of noncarbonate rocks. The average composition of all sediments (carbonates and noncarbonates) shows, according to Correns (1949), 3.86 weight percent CO_2 and 5.54 weight percent H_2O (total water content). Assume, for the sake of argument, that these amounts are totally liberated as gases and completely mixed during metamorphism. In the resulting fluid phase, H_2O and CO_2 would be present in the molecular ratio of 3.8:1. The H_2O-CO_2 fluid phase is characterized by a small mole fraction of CO_2, *i.e.*, X_{CO_2} is only 0.2. However, large amounts of carbonate are preserved in the form of marble, indicating that the calculated value of $X_{CO_2} = 0.2$ is too high. The actual value must have been considerably lower, except possibly in rocks adjacent to carbonate layers. It may be concluded that the fluid phase in metamorphic reactions not involving carbonate was essentially water; therefore $P_f \approx P_{H_2O}$.

Never, however, will the fluid phase be pure water. Apart from very small amounts of silicates (especially of quartz) dissolved in the fluid, some HCl derived from hydrolysis of chlorides from pore solutions in sediments is to be expected. The composition of the fluid will be even more complex in water-containing, noncarbonate rocks whenever *graphite is present* in the rock during metamorphism. In this case, which is common in pelitic metasediments, CO_2 and CH_4 are formed as the temperature increases. As Figure 3-1 shows, these gas species, being in equilibrium with H_2O and graphite, are also joined by CO and H_2 in much smaller amounts and by an extremely small amount of O_2 (French, 1966; Yui, 1968). By formation of these gas species the mole fraction of H_2O in the fluid phase is lowered as the temperature rises. At a constant temperature this effect diminishes with increasing total fluid pressure. Metz (1970, personal communication) has calculated that at 600°C and P_f = 3000 bars the mole fraction of water, X_{H_2O}, will have decreased from 1.0 to about 0.80, as opposed to 0.65 at P_f = 1000 bars, whenever graphite has been in equilibrium with water. But this amount of lowering of X_{H_2O} in the fluid phase does not yet have a great effect on the lowering of reaction temperatures. Thus, the breakdown temperature of muscovite in the presence of quartz, taking place at rather high temperature and being markedly dependent on H_2O pressure, will, at P_f = 1000 bars, be lowered by 30°C when graphite is present, and, at P_f = 3000 bars, by only 20°C. In other reactions that either are not so sensitive to H_2O pressure or take place at lower temperature, the effect of lowering the value of X_{H_2O} and thus of lowering the reaction temperature will be even smaller.

However, when hydrogen can diffuse to or away from the site of metamorphic reaction, the *influence of graphite* on the reduction of the partial pressure of water differs from that shown in Figure 3-1. As in

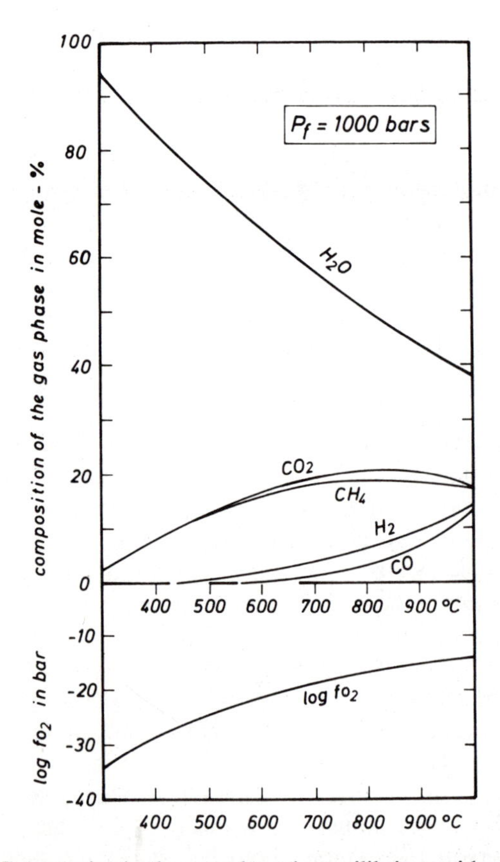

Fig. 3-1 Gas species in the gas phase in equilibrium with water and graphite. No loss of hydrogen by diffusion has been assumed, *i.e.*, the ratio of H:O remaining 2:1 (After Yui, 1968, and Metz, 1970, personal communication.)

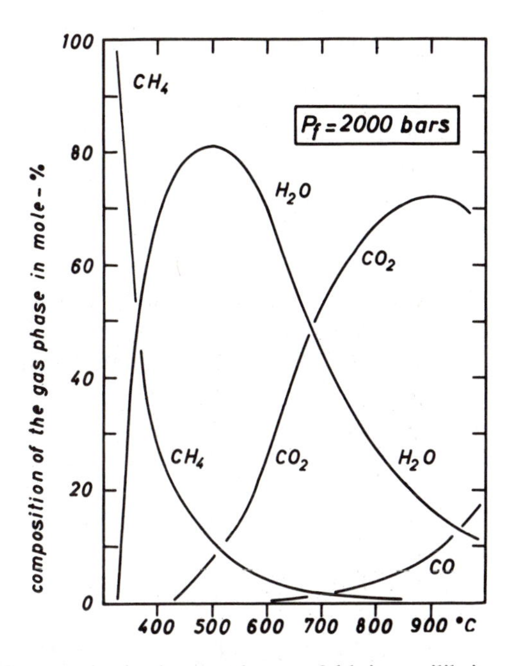

Fig. 3-2 Gas species in the gas phase at 2 kb in equilibrium with water and graphite when hydrogen fugacity is externally controlled by the quartz + fayalite + magnetite buffer. (After Eugster and Skippen, 1967.)

experimental work, this seems to be the common case in nature also. The hydrogen and oxygen fugacities are controlled externally by water and a mineral association as a buffer. Among others, a mixture of quartz + fayalite + magnetite is often used in experiments as a buffer and will probably be most commonly applicable to metamorphic reactions when temperatures are not very low.

Figure 3-2 shows graphically the calculated results of an investigation by Eugster and Skippen (1967) on the composition of the gas phase in equilibrium with water and graphite and the quartz + fayalite + magnetite buffer. Two important points can be learned from this graph:

(a) When temperatures of metamorphism are below 400°C, the mole fraction of H_2O decreases drastically from 0.67 at 400°C to zero at about 320°, at P_f = 2 kb. Methane becomes the dominant gas species. Thus, P_{H_2O} can be considerably lower than P_f in very-low grade metamorphism of graphite-bearing shales and slates.

(b) When temperatures increase above about 650°C, the mole fraction of H_2O in the gas phase decreases much more than in the case represented in Figure 3-1. At 650°C the mole fraction of

H_2O is 0.55, at 700°C it is 0.45, and at 800°C it is only about 0.25, at $P_f = 2$ kb. This clearly shows that in metapelitic rocks of high grade, and specially of the higher temperature part within high-grade metamorphism, it is important to check whether or not graphite is present. In the presence of graphite a considerable reduction of the partial pressure of water must be taken into account.

When minerals that are sensitive to reduction or oxidation, as for instance Fe-bearing minerals, are involved in metamorphism the *concentration of oxygen in the fluid gas phase* is an additional factor. This has been thoroughly studied by Eugster (1959) and Eugster and Skippen (1967). While various conditions of oxygen fugacity[6] can easily be created experimentally, natural metamorphic rocks are formed within a medium range of oxygen fugacities. The exact values are generally not known, but the common occurrence of the iron oxide magnetite instead of hematite (Fe_2O_3) or wustite (FeO) in metamorphites indicates oxygen fugacities at which magnetite is stable (shown by the stippled field in Figure 3-3). If graphite is a constituent in metamorphites, it creates in the presence of water a certain oxygen fugacity at any given temperature and total fluid pressure. The double-line curve within the stippled range of Figure 3-3, calculated for 1000 bars, shows the approximate oxygen fugacity expected during metamorphism of graphite-bearing rocks. Curves H/M, M/W, W/Fe, and M/Fe (after Eugster and Wones, 1962) are for 1 atm total pressure. For all buffers an increase in total pressure causes only a slight increase in oxygen fugacity. Curve (M + Q)/Fa (after Wones and Gilbert, 1969) is for 1000 bars total pressure. Double-lined curve shows the oxygen fugacity for the system H_2O-graphite at 1000 bars total pressure (after French, 1966).

Directed Pressure

In addition to hydrostatic pressure, *directed pressure (stress)* must be considered as a possible factor of metamorphism. Formerly it was thought that the action of shearing stress at the time of recrystallization significantly influenced mineral equilibria. This is probably not the case. Dachille and Roy (1960, 1964) have shown experimentally that the sta-

---

[6]The fugacity is the thermodynamically effective pressure of a gas species. This pressure is related to the pressure read on a gauge by the fugacity coefficient. The coefficient is dependent on temperature, pressure, and gas species, and it may be greater or smaller than 1. Only when the conditions of an ideal gas are reached does the fugacity coefficient equal 1.0.

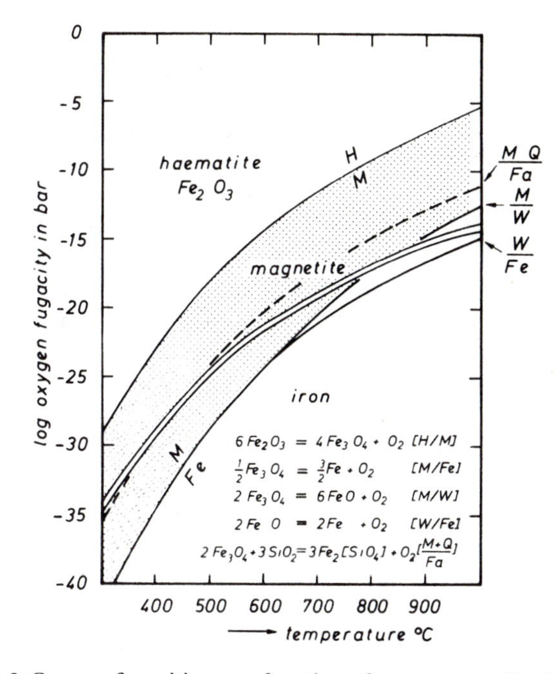

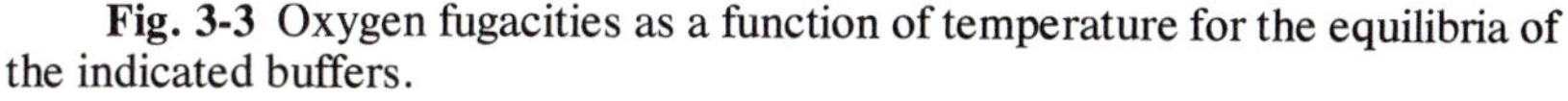
Fig. 3-3 Oxygen fugacities as a function of temperature for the equilibria of the indicated buffers.

bility field of metamorphic mineral assemblages is not influenced by shearing stress. The only effect is a considerable increase of the reaction rates leading, as a rule, toward the formation of stable parageneses. This aspect may be important in experimental investigations but is insignificant in the formation of natural mineral assemblages because of the very long time available for reactions to proceed.

Many petrographers believe that directed pressure may contribute to an appreciable increase of the effective hydrostatic pressure. Clark (1961) has coined the term "tectonic overpressure" and thinks that an overpressure of 1000 to 2000 bars could possibly be created. Flinn (1965) states that

> The magnitude of the overpressure which can exist in the rocks depends chiefly on the length of time the pressure has to persist. The higher the overpressure, the shorter the time that will elapse before it is relaxed by flow (deformation). For short periods overpressures are possible up to the rupture strength of the rock at the appropriate conditions of temperature and confining pressure.

Metamorphism takes place during long periods of time, *i.e.*, as long as the temperature is maintained and the gas phase coexisting with the newly formed mineral is present. Therefore, tectonic overpressures seem very unlikely. Furthermore, experimental evidence indicates that at geologically reasonable temperatures and strain rates the strength of rocks does not favor the creation of tectonic overpressure. In the presence of an aqueous fluid phase, rock strengths, even of quartzose rocks such as graywackes, are negligible. Thus, tectonic overpressure must be negligible (Heard, 1963; Griggs and Blasic, 1965; Brace *et al.,* 1970; Heard and Raleigh, 1972). For the Franciscan metamorphic rocks in California, where very high pressures prevailed, Ernst (1971) concluded: "Neither experimental strength-of-materials studies, detailed petrologic mapping, nor metamorphic-stratigraphic temporal relations lend support to the hypothesized production of tectonic overpressures on a regional scale."

In conclusion, it may be assumed that in most cases of metamorphism, $P_f \approx P_s$.

References

Brace, W. F., Ernst, W. G., and Kallberg, R. W. 1970. *Geol. Soc. Am. Bull.* **81:** 1325–1338.

Carmichael, D. M. 1978. *Am. J. Sci.* **278:** 769–797.

Clark, S. P. 1961. *Am. J. Sci.* **259:** 641–650.

Correns, C. W. 1949. *Einfauuhrung in die Mineralogie.* Springer-Verlag, Berlin-Göttingen-Heidelberg.

Dachille, F. and Roy, R. 1960. *In* J. H. de Boer, et al., eds. *Reactivity of Solids.* Elsevier, Amsterdam.

——— 1964. J. Geol. **72:** 243–247.

Dobretsov, N. L., *et al.* 1967. *Geochem Intern.* **4:** 772–782.

Ernst, W. G. 1971. *Am. J. Sci.* **270:** 81–108.

Eugster, H. P. 1959. Reduction and oxidation in metamorphism. *In* P. H. Abelson, ed. *Researches in Geochemistry.* John Wiley & Sons, New York.

——— and Skippen, G. B. 1967. Igneous and metamorphic reactions involving gas equilibria. *In* P. H. Abelson, ed. *Researches in Geochemistry,* Vol. 2. John Wiley & Sons, New York.

Eugster, H. P. and Wones, D. R. 1962. *J. Petrol.* **3:** 82ff.

Flinn, D. 1965. Deformation in metamorphism. *In* W. S. Pitscher and W. G. Flinn, eds. *Controls of Metamorphism.* pp. 46–72. Oliver and Boyd, Edinburgh and London.

French, B. M. 1966. *Rev. Geophys.* **4:** 223–253.

Fyfe, W. S., Turner, F. J., and Verhoogen, J. 1958. *Geol. Soc. Am. Memoir 73.*

Griggs, D. T. and Blasic, J. D. *Science* **147:** 292–295.

Heard, H. C. 1963. *J. Geol.* **71:** 162–195.

Heard, H. C. and Raleigh, C. B. 1972. *Geol. Soc. Amer. Bull.* **83:** 935–956.

Jäger, E. 1970. *Fortschr. Mineral.* **47:** 82.

——— 1973. *Eclogae geol. Helv.* **66:** 11–21.

Pitcher, W. S. and Berger, A. R. 1972. *The Geology of Donegal.* 435 pp., Wiley-Interscience, New York.

Ronov, A. B., Migdisov, A. A., and Lobach-Zhuchenko, S. B. 1977. *Geochem. Int.* **14:** 90–112.

Rutland, R. W. R. 1965. Tectonic overpressure, *In* W. S. Pitcher and G. W. Flinn, eds. *Controls of Metamorphism.* pp. 119–139. Oliver and Boyd, Edinburgh and London.

Thompson, J. B. 1955. *Am. J. Sci.* **253:** 65–103.

Wones, D. R. and Gilbert, M. C. 1969. *Am. J. Sci.* **267A:** 480ff.

Yui, S. 1968. *J. Mining Col. Akita Univ., Ser A.* **4:** 29–39.

Chapter 4

Mineral Parageneses: The Building Blocks of Metamorphic Rocks

Rocks consist of a number of different minerals. In an igneous rock various minerals crystallized from a slowly cooling magmatic melt constitute an equilibrium assemblage. Such an assemblage is called a *mineral paragenesis*. In general, all the minerals that are detectable within a single thin section belong to the mineral paragenesis that characterizes a given igneous rock. This is so because a magmatic melt is a homogenous solution of the many components that constitute such a rock. However, in metamorphic rocks derived from sediments the composition may not be the same even over a small volume. Therefore, it is well possible that all minerals observed in a single thin section do *not* belong to a single metamorphic paragenesis. Rather, two or even more mineral parageneses, *i.e.*, associations of minerals coexisting in equilibrium, may be present in the area of one thin section. This is a point of great significance. In earlier petrographic work it was believed that the determination of all the minerals of a given rock is sufficient. This is not so. *It now must be ascertained which of the minerals in a thin section are in contact. Only minerals in contact may be regarded as an assemblage of coexisting minerals, i.e., a paragenesis.*[1]

In this book a metamorphic mineral paragenesis refers to minerals in contact with each other. This point is demonstrated in Figure 4-1. The schematically drawn thin section shows four minerals; not all four, however, form a mineral paragenesis. In fact, there are two different parageneses, each consisting of three minerals in contact with each other. The two parageneses are (1) the minerals A, B, and C and (2) the minerals B, C, and D.

An example from nature is a siliceous limestone which has been metamorphosed at shallow depth and high temperature: calcite + quartz

---

[1]Alteration products are excluded, of course. In this respect, the necessity of carefully observing textural criteria is stressed.

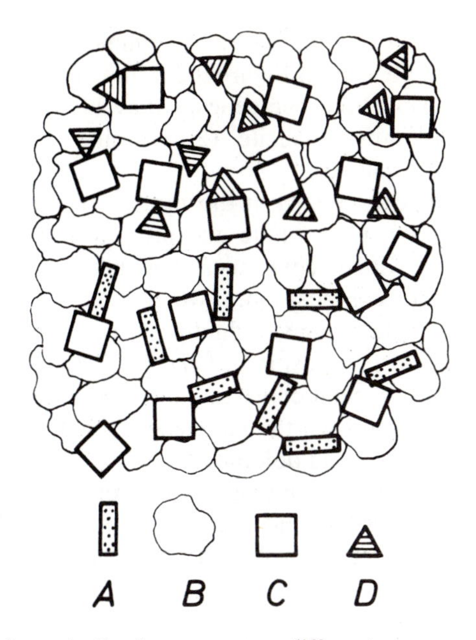

Fig. 4-1 Schematically shown are two different parageneses. They consist of (1) the minerals A, B, and C and (2) the minerals B, C, and D. Note that all four minerals together do not constitute a paragenesis because they are not in contact with each other.

have reacted to form wollastonite + CO_2. At the exact condition of the reaction equilibrium, the three solid phases—calcite + quartz + wollastonite—coexist together with the gas phase, forming a very special paragenesis. However, when conditions have been such that the equilibrium has been surpassed, either the paragenesis wollastonite + calcite or the paragenesis wollastonite + quartz is found, depending on the excess of either calcite or quartz. If a siliceous limestone consists of fine layers varying in silica content, the minerals wollastonite, calcite, and quartz may all be present in one thin section, thus apparently indicating conditions of the reaction equilibrium. But closer inspection reveals that only wollastonite + calcite and wollastonite + quartz constitute parageneses in various layers.

Another example is a marble consisting of calcite, dolomite, tremolite, talc, and quartz. If all five minerals constitute a single paragenesis this has a very special meaning: as will be shown later, at a given fluid pressure, this paragenesis determines the temperature and the composition of the CO_2-H_2O gas mixture during metamorphism. The paragenesis of the five minerals has been formed under the very special condition of an isobaric invariant point. However, detailed microscope work will most commonly reveal that not all five minerals are in mutual contact. Rather, as can be seen in Figure 9-2 in the area above curve 1, the following two or even three parageneses may be present in the specimen: (1) talc + dolomite + calcite, (2) tremolite + talc + calcite, (3) tremolite + quartz + calcite.

Paragenesis (1) crystallized in those parts of the original siliceous dolomite limestone where dolomite was present in excess, paragenesis (3) formed in areas with excess silica, and paragenesis (2) formed in areas of intermediate composition. Two or all three parageneses can exist stably side by side at a given fluid pressure within a certain *range* of temperature and of composition of the CO_2-H_2O fluid phase. Thus, the above parageneses, each consisting of three minerals, provide less petrogenetic information than the paragenesis of the five minerals talc, tremolite, quartz, calcite, and dolomite.

These examples make it obvious that careful microscopic determination of parageneses is of fundamental significance. Parageneses are the building blocks of metamorphic rocks and they may be significant petrogenetic indicators.

Chapter 5

Graphical Representation of Metamorphic Mineral Parageneses

Composition Plotting

Minerals composed of no more than three components may be represented within the plane of a triangle. As an example, wollastonite $CaSiO_3$, grossularite $Ca_3Al_2(SiO_4)_3$, anorthite $CaAl_2Si_2O_8$, sillimanite Al_2SiO_5, quartz SiO_2, and corundum Al_2O_3 consist of the three components CaO, Al_2O_3, and SiO_2 in different proportions. Because of geometric advantages, compositions are most conveniently represented in an equilateral triangle [see Ferguson and Jones (1966)]. Each corner of the triangle represents a pure component, *i.e.*, 100% CaO, 100% Al_2O_3, and 100% SiO_2. Each side allows the representation of a mixture and a mineral consisting of two components, while the interior of the triangle allows the representation of mixtures and minerals consisting of all three components.

It should be added that a line through one corner represents all compositions for which two of the components have a constant ratio. The value of the ratio is that of the point where the line from the corner cuts the opposite side of the triangle.

A few examples may illustrate the method of plotting. Sillimanite, for instance, consists of only the two components, Al_2O_3 and SiO_2, which are combined in the mole ratio 1:1; the composition of sillimanite is 50 mole% Al_2O_3 and 50 mole% SiO_2. Sillimanite plots at the middle of the side Al_2O_3-SiO_2. The mole percentages of each of the three components are marked, in steps of 20 mole%, along each of the sides of the triangle in Figure 5-1. Commercially available triangular coordinate paper is divided into 1% intervals. A line parallel to one side of the triangle represents a constant percentage of the component plotted at the opposite corner. Thus, the dotted line in Figure 5-1 represents 43 mole% CaO and the broken line represents 14 mole% Al_2O_3. At the intersection

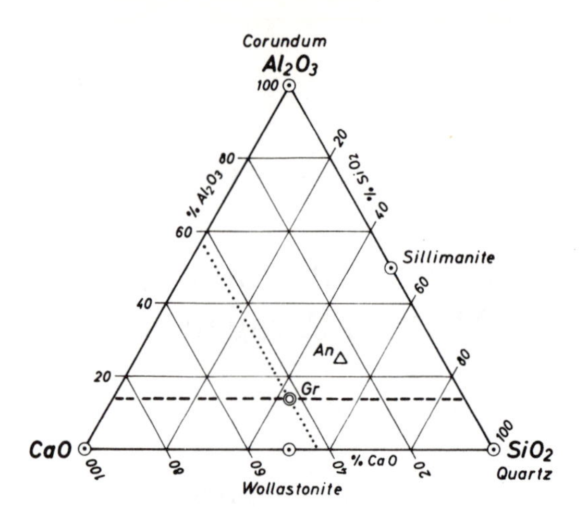

Fig. 5-1 Plotting of compositions in the equilateral triangle CaO-Al_2O_3-SiO_2.

of these two lines the composition is, of course, 43 mole% CaO and 14 mole% Al_2O_3, the remainder being 43 mole% SiO_2. This is the composition of the mineral grossularite, which has the molecular ratio 3 CaO:1 Al_2O_3:3 SiO_2, *i.e.*, 43% CaO, 14% Al_2O_3, and 43% SiO_2. It is marked by the double circle in Figure 5-1. The composition of the mineral anorthite with 1 CaO:1 Al_2O_3:2 SiO_2, *i.e.*, 25 mole% CaO, 25 mole% Al_2O_3, and 50 mole% SiO_2, has also been plotted in Figure 5-1.

From this plot of minerals in the Al_2O_3-CaO-SiO_2 concentration triangle, chemical relationships become obvious. At equilibrium, the variance of the number of degrees of freedom (F) of a chemical system equals the number of components (C) minus the number of phases (P) plus 2. This is the so-called phase rule.

$$F = C - P + 2$$

In the case of three components, three phases in equilibrium permit two degrees of freedom, *i.e.*, the temperature and the pressure may be changed independently of each other within certain limits; in other words, three phases will exist within a certain range of temperature and pressure. The coexisting minerals, when graphically represented by points, may be connected by lines known as tie lines. As shown in Figure 5-2, the connections of three minerals coexisting with each other form irregular triangles, such as sillimanite-anorthite-corundum, sillimanite-anorthite-quartz, anorthite-grossularite-corundum, anorthite-grossularite-quartz, and grossularite-wollastonite-quartz. Figure 5-2 shows that the coexistence of the minerals grossularite, wollastonite, and corundum is not possible within a range of temperature *and* pres-

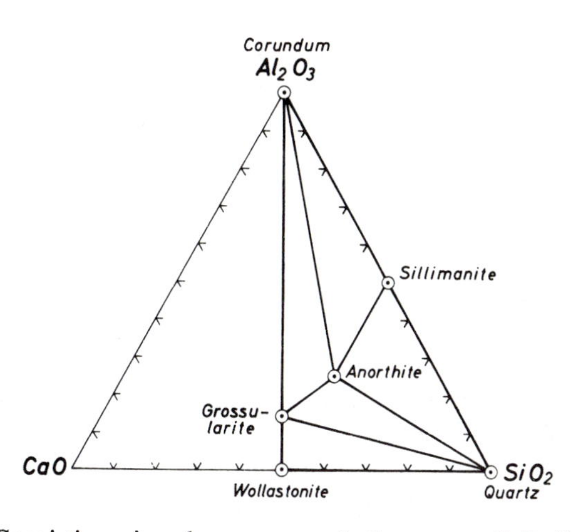

Fig. 5-2 Coexisting mineral parageneses in the system CaO-Al_2O_2-SiO_2.

sure, because any mixture of components along the line corundum-wollastonite will produce either the association grossularite + corundum or grossularite + wollastonite. For similar reasons, the minerals sillimanite, anorthite, and wollastonite, on the one hand, and corundum, sillimanite, and quartz, on the other hand, cannot coexist.

Figure 5-2 represents the various possible mineral parageneses that are stable within a certain temperature and pressure range. If, however, temperature has surpassed a certain value at a given pressure, the mineral pair grossularite + quartz becomes unstable; the connecting tie line disappears and the new tie line wollastonite + anorthite is now valid. The resulting new parageneses are shown in Figure 5-3. (Compare with

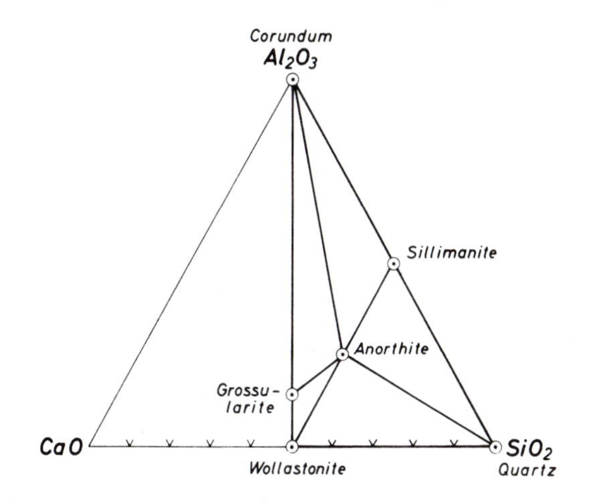

Fig. 5-3 Coexisting mineral parageneses in the system CaO-Al_2O_3-SiO_2 at higher temperatures than in Figure 5-2.

Figure 5-2.) The cross relationship between the tie lines grossularite-quartz and wollastonite-anorthite is one of the various graphical expressions that indicate a chemical reaction. In this case it is the reaction

$$\begin{array}{llll} \text{1 grossularite} & + \text{1 quartz} & = \text{2 wollastonite} & + \text{1 anorthite} \\ 1\ Ca_3Al_2[SiO_4]_3 & + 1\ SiO_2 & = 2\ CaSiO_3 & + 1\ CaAl_2Si_2O_8 \end{array}$$

Under certain conditions, systems consisting of more than three components may also be represented by an equilateral triangle. Examples are the so-called ACF and AKF diagrams which will be discussed on page 35ff. In other cases, *e.g.*, the system MgO-SiO_2-H_2O-CO_2 and the system CaO-MgO-SiO_2-H_2O-CO_2, the systems contain CO_2 and H_2O as components, which implies the presence of a CO_2-H_2O fluid phase in equilibrium with solid phases. The former system may be represented either in the triangle MgO-SiO_2-H_2O or MgO-SiO_2-CO_2. The latter system, omitting both gaseous components in the graphical representation, is shown as the CaO-MgO-SiO_2 triangle in Figure 5-4. The solid phases represented in this triangle are correctly plotted only in regard to the molecular ratios of the components CaO, MgO, and SiO_2, *i.e.*, CaO + MgO + SiO_2 taken as 100.

Figure 5-4 shows the plots of the following minerals: calcite $CaCO_3$, magnesite $MgCO_3$, brucite $Mg(OH)_2$, pericalse MgO, quartz SiO_2, dolomite $CaMg(CO_3)_2$, talc $Mg_3[(OH)_2/Si_4O_{10}]$, tremolite $Ca_2Mg_5[(OH)_2/Si_8O_{22}]$, diopside $CaMg[Si_2O_6]$, forsterite $Mg_2[SiO_4]$, and serpentine $Mg_3[(OH)_4/Si_2O_5]$. Talc contains no CaO and has a molecular ratio of 3 MgO:4 SiO_2; therefore, it is represented by a point on the MgO-SiO_2 side

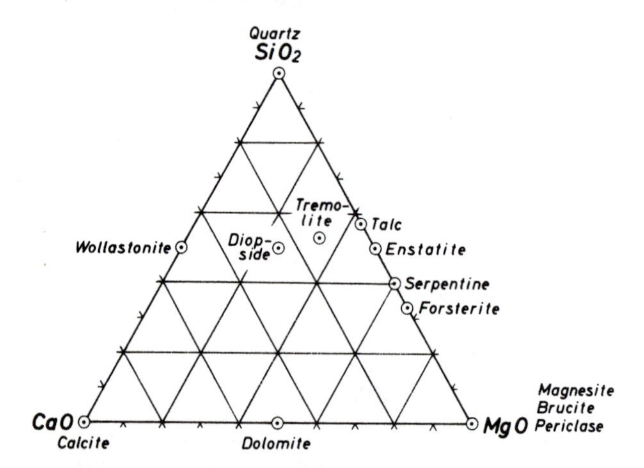

Fig. 5-4 Mineral compositions of the five-component system CaO-MgO-SiO_2-CO_2-H_2O plotted in the composition triangle CaO-MgO-SiO_2. Mg-anthophyllite, $Mg_7[(OH)_2/Si_8O_{22}]$, plots between enstatite and talc.

of the concentration triangle corresponding to 43 mole% MgO and 57 mole% SiO_2. Tremolite has a molecular ratio of 2 CaO:5 MgO:8 SiO_2; it is represented by a point within the triangle corresponding to 13 mole% CaO, 33 mole% MgO, and 53 mole% SiO_2. Figure 5-4 does not show mineral parageneses; these are fully discussed in the chapter on the metamorphism of siliceous carbonate rocks.

ACF Diagram

A correct and complete graphical representation of a mineral paragenesis is possible only if the number of minerals does not exceed four, because only four components can be represented in space at the corners of a tetrahedron. A two-dimensional representation is, of course, greatly desired and, in some special investigations, suitable projections of points within a tetrahedron onto some plane may be developed. For instance, such a method can be employed very advantageously in the study of pelitic schists, as shown by Thompson (1957) (a detailed discussion is given on pages 48 to 54). However, in order to represent mineral assemblages in rocks of diverse composition and metamorphic grade, a triangular representation developed by Eskola is used extensively. This method is necessarily a compromise, because only three components can be represented in a plane, yet the rocks contain more than three components. Nevertheless, "by means of suitable selections and restrictions" his method "allows the representation of most rocks of not too unusual composition and having an excess of silica." He continues:

> If silica is present in excess (quartz is a constituent of many metamorphic rocks) only those minerals with the highest possible SiO_2 content can be formed; consequently, the amount of SiO_2 has no influence on the mineral assemblages and need not be represented graphically. At one corner of the triangle, designated as A, that portion of Al_2O_3 (more exactly, $Al_2O_3 + Fe_2O_3$, because Fe^{3+} and Al^{3+} can substitute for each other) is plotted, which is not combined with Na and K. The second corner is defined as C = CaO and the third one as F = (Fe, Mg, Mn)O. Accessory constituents are disregarded in the graphical representation; however, before calculating the A, C, and F values, the amounts of $(Al, Fe)_2O_3$, CaO, and (Fe, Mg)O contained in the accessories are subtracted (from the chemical analysis). In this manner, the more important silicate minerals can be represented, with the exception of K and Na silicates and silica-undersaturated silicates, like olivine.

According to Eskola, the procedure for the calculation of the molecular ACF ratios of a rock is as follows: First, the chemical analysis of a rock is corrected for accessories. In order to do this, their amount must be determined (by petrographical modal analysis or x-ray phase analysis). Then an amount corresponding to 50% of the weight percent of ilmenite present is subtracted from the FeO percentage of the rock; similarly, 70 and 30% of the weight percent of magnetite are subtracted from the Fe_2O_3 and FeO percentages, respectively; 30% of the weight percent of sphene is subtracted from the CaO percentage and the weight percent of hematite is subtracted from the Fe_2O_3 percentage.

Magnetite: $FeO \cdot Fe_2O_3$ = 30 wt. % FeO and 70 wt. % Fe_2O_3
Ilmenite: $FeO \cdot TiO_2$ = 50 wt. % FeO
Sphene: $CaO \cdot TiO_2 \cdot SiO_2$ = 30 wt. % CaO (the ideal composition contains 35% CaO, but natural sphene contains less CaO)

As a next step, the weight percentages of the corrected rock analysis are converted to molecular proportions, sometimes also called molecular quotients of the oxides (Molekularzahlen), by dividing the weight percent of each oxide by its molecular weight. SiO_2, CO_2, and H_2O are disregarded. Calculation of molecular proportions into molecular percentages is not necessary.

In albite the molecular ratio $Na_2O:Al_2O_3$ is 1:1; similarly, in K feldspar the molecular ratio $K_2O:Al_2O_3$ is also 1:1. Therefore, the molecular proportions $[Na_2O]$ and $[K_2O]$ are added and an amount equal to their sum is subtracted from the molecular proportion $[Al_2O_3]$. Similarly, an amount equal to 3.3 times the molecular proportion $[P_2O_5]$ is subtracted from the molecular proportion [CaO], because in apatite $Ca_5[(OH, F, Cl)/(PO_4)_3]$ the molecular ratio $CaO:P_2O_5$ = 10:3; it contains 3.3 times as much [CaO] as $[P_2O_5]$.

As a first approximation, the scheme for calculating the ACF ratios (after making the corrections necessary for the accesories) may be summarized as follows:

$$
\begin{aligned}
[Al_2O_3] + [Fe_2O_3] - ([Na_2O] + [K_2O]) &= A \\
[CaO] - 3.3\,[P_2O_5] &= C \\
[MgO] + [MnO] + [FeO] &= F
\end{aligned}
$$

For graphical representation, these values are recalculated so that A + C + F = 100%, *i.e.,* they are expressed as molecular percentages.

Strictly speaking, the outlined scheme of calculation applies only to rocks not containing any biotite, muscovite, or paragonite. However, very many rocks do contain these minerals, and for more exact work corrections must be made. If biotite is present, a correction of the chemical analysis prior to the calculation of the molecular proportions is necessary, because biotite, like other K minerals, cannot be represented exactly in the plane of the ACF diagram.[1] Therefore, the Al_2O_3 contained in biotite must be subtracted from the Al_2O_3 content of the rock when calculating the ACF values, as was done for the Al_2O_3 contained in K feldspar and albite. But, in the case of biotite, when this correction is applied to the rock analysis, it is not sufficient; the (Fe, Mg)O contained in biotite must be subtracted as well. The scheme of calculation is based on the assumption that the total amount of K_2O is present in K feldspar; therefore, a molecular amount equal to that of $[K_2O]$ was subtracted from $[Al_2O_3]$. Because in biotite the molecular ratio $K_2O{:}Al_2O_3$ is the same as in K feldspar (1:1), it does not matter if some of the K_2O is actually combined in biotite. The scheme of calculation takes care of any biotite present as well in arriving at the A value; *i.e.*, the presence of biotite requires no change in the A value.

However, the F value, as given by the scheme of calculation, must be corrected with regard to the amount of (Fe, Mg)O contained in biotite. The following procedure is used. The amount of biotite in the rock is determined and the corresponding amounts of MgO and FeO are calculated as weight percentages, making the simplifying assumption that the weight ratio MgO:FeO in biotite is the same as that in the rock. (It is better, of course, to have the biotite chemically analyzed, and thus know the relative amounts of FeO, MgO, and MnO.) The values obtained are subtracted from the corresponding values of the chemical analysis of the rock. Then the calculation of ACF ratios according to the described scheme may be carried out if no correction with regard to muscovite or paragonite is necessary (see next section). The necessity of the biotite correction was pointed out some time ago by Eskola (1939).

Muskovite is a common mineral, *e.g.*, in phyllites and mica schists. The F value of a rock can be corrected with respect to any muscovite present only if the amounts of Fe^{2+} and Mg contained in muscovite are known through chemical analysis. In general, the simplifying assumption is made that muscovite is free of such elements and therefore the F value of the rock need not be corrected. However, the amount of muscovite must be taken into account when calculating the A value of a rock, because the molecular ratio $K_2O{:}Al_2O_3$ in muscovite is 1:3, not 1:1 as in K feldspar. First, the amount of muscovite is determined. Then the weight percent K_2O combined in muscovite is expressed as a molecular proportion (weight percent divided by molecular weight) and twice this amount is subtracted from the molecular proportion $[Al_2O_3 + Fe_2O_3]$ of the rock; note that three times the amount is not subtracted, because an

[1]Many published ACF diagrams show biotite near the F corner in order to indicate that biotite is present together with those minerals which can be represented exactly. In this book, however, a different method is employed.

amount of [Al_2O_3] equal to one molecular proportion [K_2O] has already been subtracted, according to the calculation scheme on p. 36. If paragonite is present, an analogous correction must be carried out. Only then are the molecular proportions of A, C, and F calculated and expressed as percentages.

The A, C, and F values of a rock are conveniently shown on a triangular ACF diagram. On the ACF diagram, the point representing the composition of a rock indicates the minerals to be expected if the rock belongs to a certain metamorphic zone. A rock of the same chemical composition may consist of different minerals if it belongs to some other zone. The ACF diagrams of various metamorphic zones have been compiled on the basis of petrographical observations.

It must be pointed out that the ACF diagram is only very seldom used to plot compositions of *rocks;* therefore, the described corrections for biotite, muscovite, or paragonite need only be applied in such rare cases. Commonly, the ACF diagram is used only to graphically represent various *mineral assemblages* (or a part of these) that occur in rocks of various compositions and within a limited range of metamorphic conditions. When mineral assemblages are shown, only those minerals are plotted in an ACF diagram which are composed entirely of A = Al_2O_3 + Fe_2O_3, C = CaO, F = (MgO + FeO + MnO), and SiO_2, H_2O, or CO_2; the latter three are not represented in the diagram. We stress here that combining the three oxides MgO, FeO, and MnO under a single variable is not correct, in spite of isomorphous replacement in mineral structures. This incorrect procedure creates mistakes in some cases of assemblage

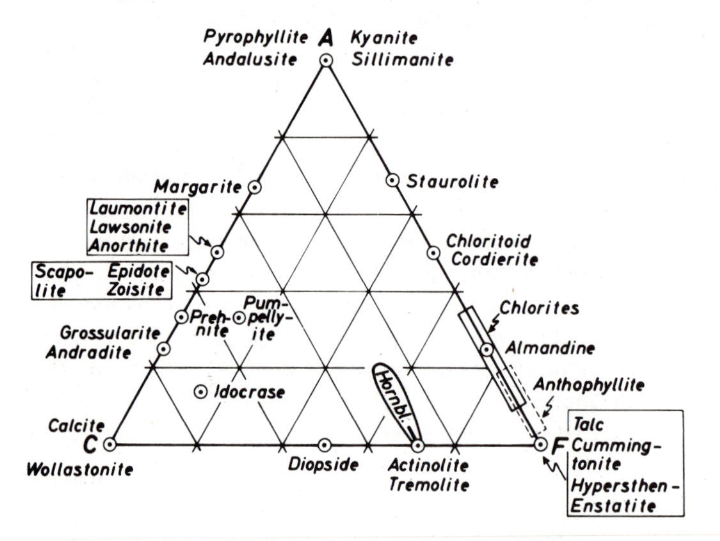

Fig. 5-5 Mineral composition plotted in the ACF diagram.

representation; this will be pointed out later when dealing with actual cases.

Table 5-1(on p.40 and 41) gives the chemical composition and ACF values of minerals that are stable in the presence of quartz; thus, forsterite, corundum, serpentine, brucite, and periclase are excluded. Figure 5-5 plots most of the minerals. It is advantageous, indeed, to know approximately the positions of the mineral plots because it helps to remember qualitatively the ratios A:F, A:C, and C:F of the minerals common in metamorphic rocks. This is important for at least a crude, quick understanding of many metamorphic reactions.

The ACF diagrams are most useful in showing mineral parageneses of Ca-rich Ca, Al, Mg, and Fe minerals, *i.e.,* of minerals occurring in metamorphic marls and mafic rocks. That part of Figure 5-5 showing minerals below the imaginary line joining laumontite, etc. and chlorite is the most useful one. The reason for this is that mineral assemblages formed by metamorphism of Al-richer rocks, *i.e.,* pelitic rocks, contain muscovite or K feldspar in appreciable amounts; therefore, they are not well represented in the upper part of the ACF diagram. Such mineral assemblages are, however, represented accurately in another kind of diagram which also distinguishes the different roles of Mg and Fe—the AFM diagram.

A′FK Diagram

It is sometimes advantageous to use an A′FK diagram in addition to the ACF. In this diagram K minerals (K feldspar, muscovite, biotite, and stilpnomelane) are represented together with minerals containing (Mg, Fe) and (Mg, Fe) + (Al, Fe^{3+}), whereas Ca minerals cannot be shown. The A′FK diagram, like the ACF diagram, is based on molecular proportions. In calculating the A′FK values of a rock, the analysis is corrected with respect to the accessories in the same manner as in calculating the ACF values. If the chemical analysis of the rock is known—but not, however, its mineralogical composition—it must be assumed that all CaO is present in anorthite, putting up with a certain error if other CaAl silicates are present. The general calculation scheme is:

$$A' = [Al_2O_3] + [Fe_2O_3] - ([Na_2O] + [K_2O] + [CaO])$$ [2]

$$K = [K_2O]$$

$$F = [FeO] + [MgO] + [MnO]$$

$$A' + K + F = 100$$

[2]Only the [CaO] combined in CaAl and $CaFe^{3+}$ silicates is subtracted from the $[Al_2O_3 + Fe_2O_3]$ value but, of course, not the [CaO] combined in carbonates and wollastonite.

Table 5-1 Minerals that can be represented in the ACF diagram.

Very rich in Al	
Pryophyllite	A = 100; $Al_2[(OH)_2/Si_4O_{10}]$
Andalusite, kyanite; sillimanite	A = 100; Al_2SiO_5
Rich in Al and Mg, Fe	
Staurolite	A = 69, F = 31; 4 $FeO \cdot 9\ Al_2O_3 \cdot 8\ SiO_2 \cdot H_2O$*
Cordierite	A = 50, F = 50; 2 $(Mg,Fe)O \cdot 2\ Al_2O_3 \cdot 5\ SiO_2$
Chloritoid	A = 50,F = 50; $FeO \cdot Al_2O_3 \cdot SiO_2 \cdot H_2O$. Fe may be replaced by Mg up to about 60 molecular percent, but commonly this replacement is only 5 to about 25%.
Rich in Al and Ca	
Margarite	A = 67, C = 33; $CaAl_2[(OH)]_2/Si_2Al_2O_{10}]$
Laumonite	A = 50, C = 50; $Ca[Al_2Si_4O_{12}]4H_2O$
Lawsonite	A = 50, C = 50; $CaAl_2[(OH)_2/Si_2O_7]H_2O$
Anorthite	A = 50, C= 50; $CaAl_2Si_2O_8$, component in plagioclase.
Scapolite	A = 43, C = 57 for the end member mejonite $Ca_8(Cl_2,SO_4,CO_3,(OH)_2)_2(Al_2Si_2O_8)_6$; the other end member is mariolite $Na_8(Cl_2,SO_4,CO_3,(OH)_2)(AlSi_3O_8)_6$
Epidote group	A = 43, C = 57; 4 $CaO \cdot 3\ (Al,Fe^{3+})_2O_3 \cdot 6\ SiO_2 \cdot H_2O$; up to one-third of the Al may be replaced by trivalent Fe. The orthorhombic *orthozoisite* contains none or very little Fe; among the monoclinic members of the epidote group, *clinozoisite* with positive optical sign and less Fe is distinguished from *epidote* proper (formerly pistacite) with negative optical sign. This latter has one-sixth to one-third of the Al replaced by Fe.
Pumpellyite	A = 34, C = 53, F = 13; similar to epidote, but contains Mg and Fe^{2+}. Approximate composition is 8 $CaO \cdot 2\ (Mg,Fe,Mn)O \cdot 5\ (Al,Fe)_2O_3 \cdot 12\ SiO_2 \cdot 7\ H_2O$
Prehnite	A = 33, C = 67; $Ca_2[(OH)_2/Al_2Si_3O_{10}]$
Rich in Ca	
Grossularite—andradite	C =75, A = 25; $Ca_3Al_2(SiO_4)_3$—$Ca_3Fe_2(SiO_4)_3$ (note that A comprises Fe^{3+} substituting for Al).
Idocrase	C = 72, A = 14, F = 14; 10 $CaO \cdot 2MgO \cdot 2\ Al_2O_3 \cdot 9\ SiO_2 \cdot 2\ H_2O$
Wollastonite	C = 100; $CaSiO_3$
Calcite	C = 100; $CaCO_3$
Rich in Mg and Fe^{2+}	
Diopside—hedenbergite	F = 50, C = 50; $CaMgSi_2O_6$—$CaFeSi_2O_6$

Dolomite	F = 50, C = 50; $CaMg(CO_3)_2$
Tremolite	F = 71.5, C = 28.5; $Ca_2(Mg,Fe)_5[(OH)_2/Si_8O_{22}]$. Only up to 20 mole% of MgO is replaced by FeO.
Actinolite	Similar to tremolite but containing more Fe, and a little Mg + Si is replaced by 2 Al.
Hornblende	Belongs to the amphibole group as tremolite and actinolite, but contains more Al in variable amounts. Compositions are shown as elongated field in Figure 5-5, extending from tremolite.
Cummingtonite-grunerite	F = 100; $(Mg,Fe)_7[(OH)_2/Si_8O_{22}]$; in grunerite more than 70 atom% of Mg is replaced by Fe^{2+}
Anthophyllite-gedrite	F = 100; similar in composition as above, but orthorhombic instead of monoclinic†
Enstatite-hypersthene	F = 100; $Mg_2Si_2O_6$—$(Mg,Fe)_2Si_2O_6$
Talc	F = 100; $Mg_3[(OH)_2/Si_4O_{10}]$
Almandine	F = 75, A = 25; $Fe_3Al_2(SiO_4)_3$
Spessartite	F = 75, A = 25; $Mn_3Al_2(SiO_4)_3$
Pyrope	F = 75, A = 25; $Mg_3Al_2(SiO_4)_3$
Chlorite	Variable composition: F = 90-65, A = 10-35; *e.g.*, $Mg_5(Mg,Al)[(OH)_8/(Al,Si)Si_3O_{10}]$ up to $(Mg,Fe)_4Al_2[(OH)_8/Al_2Si_2O_{10}]$

Minerals containing alkali that cannot be represented in an ACF diagram:

Glaukophane, crossite	Na-amphibole $Na_2(Mg,Fe)_3(Al,Fe)_2[(OH)_2/Si_8O_{22}]$
Jadeite, jadeitic pyroxene	Pyroxene containing $NaAlSi_2O_6$ as component
Stilpnomelane	$K_{<1}(Mg,Fe,Al)_{<3}[(OH)_2/Si_4O_{10}] \cdot xH_2O$; contains very little Al and K.
Muscovite	$KAl_2[(OH)_2/Si_3AlO_{10}]$
Phengite	Similar to muscovite but has more Si + Mg and less Al
Paragonite	$NaAl_2[(OH)_2/Si_3AlO_{10}]$; also as component in solid solution with muscovite.
Biotite	$K(Mg,Fe,Mn)_3[(OH)_2/Si_3AlO_{10}]$
Phlogopite	$KMg_3[(OH)_2/Si_3AlO_{10}]$

---

*The composition of staurolite, particularly the content of OH, is not exactly known; some $(OH)_4$ has been suggested to substitute for SiO_4. In staurolite FeO may be replaced by MgO up to about 20 to 30 molecular percent.

†Robinson *et al.* (1971) give the following compositions for the two end members

$$R_2^{2+}R_5^{2+}[(OH)_2/Si_8O_{22}] \text{ and } Na_{0.5}R_2^{2+}(R_{3.5}^{2+}R_{1.5}^{3+})[(OH)_2/Si_6Al_2O_{22}]$$

respectively, where R^{2+} = Mg, Fe^{2+}, Mn^{2+}, Ca, and R^{3+} = Al, Fe^{3+}, $(Ti_{0.5}^{4+} + Fe_{0.5}^{2+})$. Except at high metamorphic temperatures, members with intermediate Al and Na content exsolve to an anthophyllite-gedrite intergrowth which often can be detected by X-ray diffraction only.

This is the calculation scheme for the A′FK values generally given in textbooks, *e.g.*, by Eskola. If, however, one or more of the minerals grossularite, andradite, zoisite, epidote, hornblende, and the uncommon margarite are present besides anorthite, then, in order to be exact, the fact should be considered that the $CaO:Al_2O_3$ ratio in these minerals differs and is not equal to 1, as in anorthite. In such a case, the following procedure is necessary. The amounts of these minerals are determined as weight percentages, and thus the amounts of CaO contained in each of them is ascertained. The weight percentages of CaO in each mineral are then expressed as molecular proportions and the A′ value is calculated according to the scheme shown below. The procedure for calculating the K value does not change.

However, for more exact work, the F value must be corrected, because minerals like diopside and hornblende, not presented in the A′FK diagram, contain a certain amount of F "component." Therefore, the modified scheme for the calculation of the A′FK values is as follows:

$A' = [Al_2O_3 + Fe_2O] - [Na_2O + K_2O]$ − 1/3 of the [CaO] contained in grossularite, andradite,
− 3/4 of the [CaO] contained in the zoisite, epidote,
− the [CaO] contained in anorthite,
− twice the [CaO] contained in margarite.

$K = K_2O$

$F = [FeO] + [MgO] + [MnO]$ − (Correction with regard to hornblende and diopside)

$A' + K + F = 100$

Because biotite is represented on the A′FK diagram, the rock analysis need not be corrected with respect to biotite, as was done in calculating the ACF values.

Points representing *minerals* consisting only of A and F components occupy the same position in the A′FK diagram as in the ACF diagram. Ca minerals cannot be represented in the A′FK diagram. The diagram shows the relation of muscovite, biotite, stilpnomelane, and K feldspar to minerals consisting only of A and F "components" (pyrophyllite, andalusite, chlorite, chloritoid, cordierite, staurolite, almandine, anthophyllite, and talc). It will be shown further on that biotite and muscovite coexist with various associations of A-F minerals, whereas K feldspar coexists only in a few cases with A-F minerals.

According to the calculation scheme, K feldspar (microcline or orthoclase) is plotted at the K corner of the A′FK diagram. The point representing muscovite of ideal composition $K_2O \cdot 3\ Al_2O_3 \cdot 6\ SiO_2 \cdot 2\ H_2O$ is located so that its K:A′ ratio is 1:2 (*i.e.*, A′ = 67 and K = 33 molecular percent, respectively) because an amount of $[Al_2O_3]$ equal to one $[K_2O]$ has already been subtracted from the $[Al_2O_3 + Fe_2O_3]$ value. The actual composition of muscovite varies considerably; see Figure 5-6. The substitution scheme is (Mg, Fe) + Si for Al + Al. Muscovites which are noticeably poorer in Al and richer in SiO_2 and contain (Mg, Fe) are called phengites. Phengites are the typical so-called muscovites in very-low-grade metamorphic rocks. The composition of muscovites should be shown as a large field in the A′FK diagram, but, for the sake of simplicity, this is not done. Furthermore, muscovite may contain a small amount of Na_2O instead of K_2O, *i.e.*, some paragonite component $NaAl_2[(OH)_2/Si_3AlO_{10}]$ is dissolved in muscovite $KAl_2[(OH)_2/Si_3AlO_{10}]$. This is also disregarded in the representation.

Biotite has a variable composition as well and is represented by a field. The ideal composition of biotite is

$$K(Mg, Fe, Mn)_3[(OH)_2/Si_3AlO_{10}]$$
$$\text{or } K_2O \cdot 6\ (Mg, Fe, Mn)O \cdot Al_2O_3 \cdot 6SiO_2 \cdot 2H_2O$$

This composition corresponds to a point on the K-F side of the diagram with 14 molecular percent K and 86 molecular percent F, because an amount of $[Al_2O_3]$ equal to $[K_2O]$ has already been subtracted from the $[Al_2O_3 + Fe_2O_3]$ value. As shown in Figure 5-6, some of the (Fe, Mg) in biotite may be replaced by Al, so that biotites are represented by a small

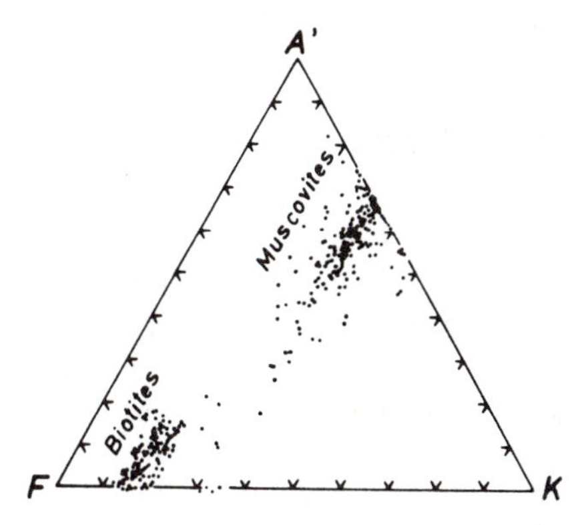

Fig. 5-6 The composition of various muscovites and biotites plotted on an A′FK diagram.

field extending parallel to the A′-F side (K is approximately constant) from the point 14% K, 86% F to about the point 14% K, 15% A′, 71% F (all values in molecular percent).

Stilpnomelane is also represented in the A′FK diagram. Its approximate composition is

$$K_{<1} (Fe^{2+}, Mg, Al)_{<3}[(OH)_2/Si_4O_{10}] \cdot xH_2O$$

The amount of Al is small and that of K very small. Many stilpnomelanes contain Fe^{3+} in appreciable amounts; however, Zen (1960) and Brown (1967, 1971) have convincingly argued that Fe^{2+} is oxidized to Fe^{3+} after the formation of stilpnomelane. Therefore, the total amount of Fe is calculated as Fe^{2+} and stilpnomelane is represented by a small field in the vicinity of the F corner of the A′FK diagram. At very low temperatures of metamorphism, stilpnomelane takes the place of biotite.

Parageneses including paragonite cannot be represented in the A′FK diagram. If paragonite is present, it may be advisable to construct an analogous AFNa diagram. It should be noted that paragonite occurs not only as a component in solid solution with muscovite or as separate crystals but also as intimate intergrowths of paragonite lamellae, 10 to 500 μ thick, in phengite or muscovite (see Albee and Chodos, 1965; Laduron and Martin, 1969).

In this book A′FK diagrams are occasionally reproduced together with ACF diagrams to provide a more complete representation. However, it must be pointed out that the A′FK diagrams have lost much of their usefulness since a better method is available for the representation of Fe, Mg, and Al-containing minerals which coexist with biotite and with muscovite or K feldspar. This method is the AFM diagram, which is discussed on pages 48 to 54.

How Are ACF and A′FK Diagrams Used?

ACF and A′FK diagrams have been constructed for a number of metamorphic zones on the basis of petrographical observations. They indicate those parageneses that may arise in rocks of different chemical composition when metamorphosed at specific conditions. The diagrams provide no information with regard to the accessories such as magnetite, hematite, apatite, ilmenite, titanite, and rutile. In addition to the parageneses given by the diagrams, other minerals are present as well. The diagrams are based on the condition that SiO_2 be present in excess in the form of quartz. Furthermore, Na minerals cannot be represented in the diagrams. This consideration is particularly important in the case of albite, either as a mineral or as a component of plagioclase. Rocks not containing any Na_2O, and therefore no albite or plagioclase, are extremely uncommon. In Na_2O-bearing rocks, albite may occur only if neither anorthite nor plagioclase is stable; otherwise, albite is dissolved as a component in plagioclase. Anorthite and plagioclase are unstable only at the lowest grades of metamorphism; the anorthite component of plagioclase forms zoisite (epidote), whereas the albite component constitutes a separate mineral. Therefore, at low grades of metamorphism, albite is present. If anorthite or anorthite-rich plagioclase is not recorded in the ACF diagram, albite is an additional mineral in Na_2O-bearing rocks. Albite may also occur in rocks metamorphosed at higher temperatures, but only if no Ca and Al are available to form plagioclase. If anorthite or plagioclase is shown in the ACF diagram and the composition of a rock plots inside a subtriangle with anorthite or plagioclase at one of its corners, albite cannot be present as a mineral.

The chemical compositions of magmatic and sedimentary rocks are of interest in the study of metamorphism because they are the primary rocks which are subjected to metamorphic conditions. Chemical analyses of these rocks are plotted on ACF and A′FK diagrams (Figure 5-7),

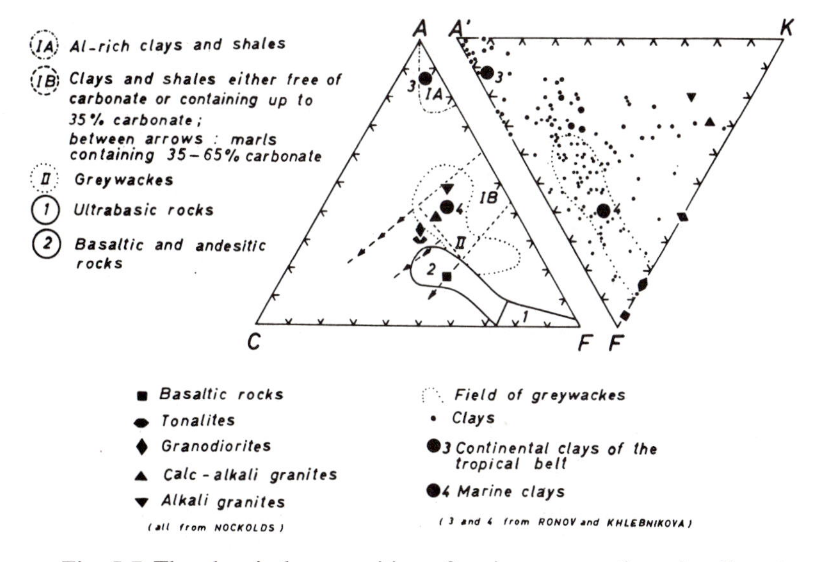

Fig. 5-7 The chemical composition of various magmatic and sedimentary rocks plotted on ACF and A′FK diagrams. Note the considerable variation of the K values of clays and shales in the A′FK diagram (Nockolds, 1954; Ronov and Khlebnikova, 1957).

using the general calculation scheme without special corrections. It is advisable to remember the fields occupied by various primary rocks. By means of ACF and A′FK diagrams for various metamorphic conditions, it is easy to know what parageneses to expect in rocks of different composition. This facilitates the understanding of metamorphic transformations.

The following instruction is important. In order to decide what minerals composed of A and F may be formed, attention must be paid to the amount of K component of a rock. Erroneous conclusions may be prevented by considering the position of a plotted rock analysis not only in the ACF diagram but in the A′FK diagram as well. The following examples illustrate that the parageneses read off the ACF diagram may be in need of some correction in the light of information supplied by the A′FK diagram:

1. Consider, as an example, an Al-rich shale in field IA of the ACF diagram (Figure 5-7). The ACF diagram of the medium-grade shallow contact metamorphism (Figure 5-8) indicates the following paragenesis for a hornfels derived from this shale:

   andalusite + some cordierite + some plagioclase + quartz

   Actually, the hornfels can have this composition only if it contains no K_2O, *i.e.*, if the point representing its composition in the A′FK diagram lies on the A′-F side. Should the point, however, have the coordinates K = 10, F = 15, the A′FK diagram indicates that muscovite has been formed as well as andalusite and cordierite. The amount of muscovite would be approximately equal to that of andalusite. Therefore, the resulting hornfels consists of

   andalusite + muscovite + some cordierite
   + some plagioclase + quartz

2. As a further example, consider the case of a sediment of granitic composition (arkose) being subjected to contact metamorphism in the inner aureole of a granitic intrusion. It is assumed that the composition of the sediment is the same as the average composition of calc-alkali granites. A point representing this composition in the ACF diagram of Figure 5-8 indicates the following paragenesis:

   cordierite + plagioclase + some anthophyllite + quartz

This does not correspond to the observed paragenesis. The plotted point in the A′FK diagram shows that it lies within the subtriangle K feldspar-biotite-muscovite—far away from the muscovite corner and closer to the K feldspar corner than to the biotite corner. This is a reflection of the high K:F ratio. Therefore, the metamorphism of an arkose of granitic composition produces the paragenesis:

K feldspar + some biotite + very little muscovite
+ plagioclase + quartz

The uncorrected position of the point representing granite in the ACF diagram (Figure 5-7) as calculated from its chemical composition is shifted, because of the biotite correction, away from the F corner until it intersects the A-C side. This rather extreme example stresses the importance of considering the plotted point of a rock composition in both the ACF and the A′FK diagrams of the particular metamorphic zone in order to understand the metamorphic parageneses.

3. Considering Fe-rich rocks, the following fact is important. Trivalent iron, particularly in sediments, may be reduced partially or totally to bivalent iron during metamorphism and thus become available for the formation of FeO-bearing silicates. In this case, the position on the ACF diagram of the point representing the composition of the original rock is shifted somewhat along a line toward F.

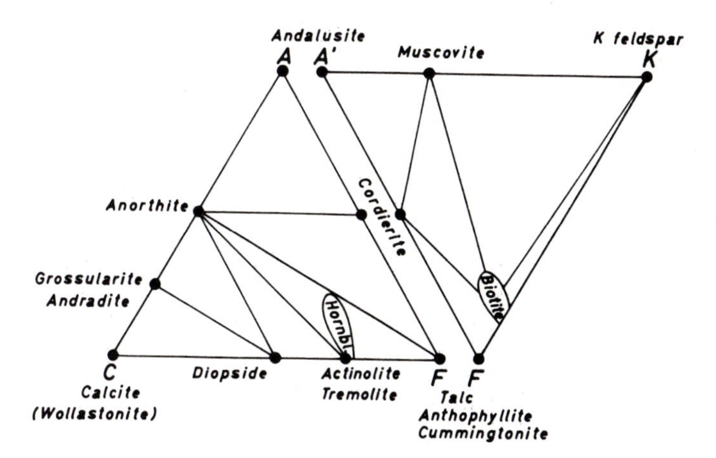

Fig. 5-8 Medium-grade shallow contact metamorphism. Parageneses shown in the ACF and AFK diagrams. (Staurolite may also have formed.)

AFM Diagrams

In ACF and A′FK diagrams, FeO and MgO (plus MnO) are grouped as one component. This simplification makes it possible to represent many observed mineral assemblages, but certainly not all of them. For example, in some rocks biotite is present together with muscovite and one of the modifications of Al_2SiO_5, whereas in other rocks biotite coexists with garnet, staurolite, quartz, and muscovite; the latter assemblages, as well as the assemblage biotite + cordierite + andalusite + muscovite + quartz, cannot be shown on A′FK diagrams, due to the simplification of treating MgO and FeO as one component. Although Mg and Fe^{2+} substitute for each other in the lattices of silicates, the extent of this isomorphous substitution is different for various coexisting silicates and, furthermore, depends on bulk composition of the rock, temperature, and pressure. This is due to the fact that FeO and MgO are actually two independent components of the system which should not be grouped together but rather should be considered as two separate components. In contrast to ACF and A′FK diagrams, FeO and MgO are treated as two components in the case of AFM diagrams derived by Thompson (1957), where A is an expression for the Al_2O_3 content, F for FeO, and M for MgO.

AFM diagrams are particularly well suited to show the dependence of the mineral assemblages of pelitic rock on the chemical composition of the rocks. With few exceptions, such mineral assemblages commonly include quartz and muscovite. Because quartz consists of only SiO_2, varying amounts of SiO_2 in the rocks are reflected merely as different amounts of quartz. The formation of other minerals in any rock is not affected by SiO_2 but entirely determined by the relative amounts of other components, as long as the presence of quartz assures the formation of SiO_2-saturated minerals. Therefore, as in the case of Eskola's ACF and A′FK diagrams, SiO_2 may be disregarded in graphical representations. Similarly, the amount of H_2O relative to other components can be neglected—either because H_2O is present as a phase, "H_2O is in excess," at the time the metamorphic assemblage formed, which makes $P_s = P_{H_2O}$, or because the amount of H_2O was externally controlled. In the latter case, P_{H_2O} is fixed not by the bulk composition of a small volume of rock but by the P_{H_2O} prevailing in the environment (environment = large volume of rock). Therefore, within any small volume of rock, P_{H_2O} is not a possible composition variable.

The chemical composition of pelitic rocks can be represented approximately in a system of six components—SiO_2-Al_2O_3-MgO-FeO-K_2O-H_2O—if the following components are neglected or taken into account by means of appropriate corrections: Fe_2O_3 and TiO_2, largely contained in biotite; Na_2O in alkali feldspar, albite, and paragonite; and CaO in plagioclase and almandine garnet. In view of the preceding discussion, it can be said that the mineral assemblage of a quartz-bearing rock depends on the relative amounts of the four components Al_2O_3, MgO, FeO, and K_2O but not on the amounts of SiO_2 and H_2O. Therefore, mineral assemblages of pelitic rocks can be represented three-dimensionally within the tetrahedron Al_2O_3-MgO-FeO-K_2O.

Minerals not containing K_2O can be plotted on one side of the tetrahedron, *i.e.*, on the plane Al_2O_3-FeO-MgO. Such minerals are chlorite, chloritoid, pyrophyllite, the modifications of Al_2SiO_5, staurolite, cordierite, and almandine garnet considering only components almandine + pyrope and subtracting components grossularite, andradite, and spessartite from the chemical analysis. The K_2O-bearing minerals biotite and stilpnomelane plot inside the tetrahedron and muscovite (of ideal composition) and K feldspar are presented along the edge Al_2O_3-K_2O of the tetrahedron.

The tetrahedron Al_2O_3-FeO-MgO-K_2O is shown in Figure 5-9. The

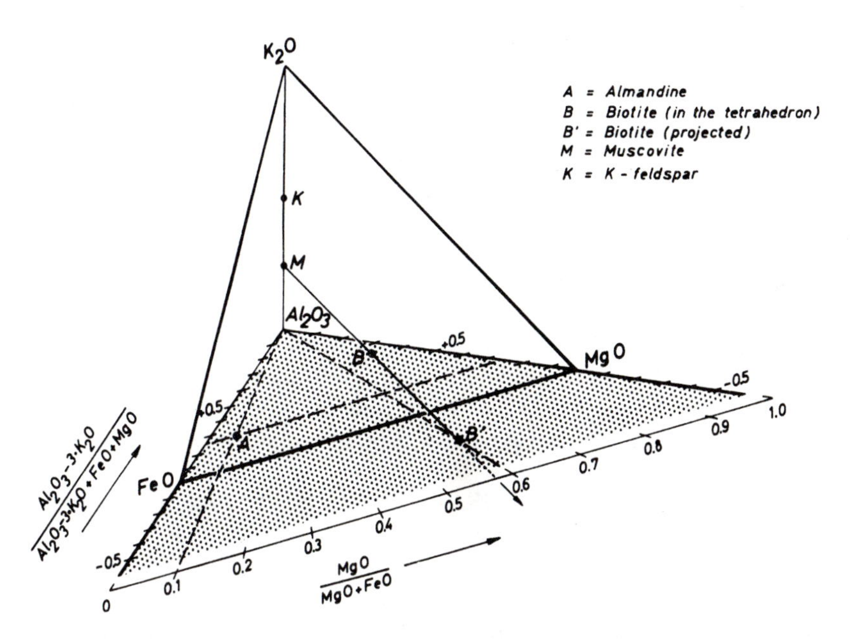

Fig. 5-9 The tetrahedron K_2O-Al_2O_3-FeO-MgO and the plane Al_2O_3-FeO-MgO extended beyond the edge FeO-MgO. All points inside the tetrahedron are projected from point M (muscovite) onto this plane. Point B lies inside the tetrahedron and is projected to B′. Point A lies in the plane Al_2O_3-FeO-MgO and therefore its location remains unchanged.

plane Al_2O_3-FeO-MgO (stippled) is extended beyond the tetrahedron. Muscovite and K-feldspar are shown by points M and K, respectively. Point A represents almandine garnet with a molecular ratio of pyrope-component-to-almandine-component of 10 to 90. Point B, inside the tetrahedron, represents a biotite with a molecular ratio MgO/(MgO + FeO) of 0.6. Stilpnomelane, also represented inside the tetrahedron, occurs in very-low-grade metamorphic rocks instead of biotite. It is obvious that the chemical composition of a *rock* consisting of both K_2O-free and K_2O-bearing minerals is also represented by a point inside the tetrahedron.

In order to achieve a planar representation, it is expedient to project all points inside the tetrahedron onto the plane Al_2O_3-FeO-MgO, on which the K_2O-free minerals chlorite, chloritoid, staurolite, cordierite, etc., are plotted in any case. Because metamorphic rocks derived from pelitic sediments commonly contain muscovite, Thompson chose point M in Figure 5-9, representing muscovite, as the projection point. Using point B as an example, it is shown in Figure 5-9 how a point lying *inside* the tetrahedron is projected along line M-B to B′; B′ lies in the same plane (shown stippled) as the corners Al_2O_3, FeO, and MgO but outside the tetrahedron, *i.e.,* B′ lies in the extended plane Al_2O_3-FeO-MgO. The plane of projection includes the corners Al_2O_3, FeO, and MgO but is not bounded by the edge FeO-MgO; it is called the AFM plane, and mineral assemblages shown in this plane are called AFM diagrams.

The AFM diagram is comparable to the A′FK diagram because the amount of K_2O relative to the other components is also taken into consideration, in the case of the AFM diagram by means of the projection through point M, onto the plane Al_2O_3-FeO-MgO. The AFM diagram, however, conveys more information than the A′FK diagram because FeO and MgO are treated as two components. It should be noted, furthermore, that every AFM diagram applies to mineral assemblages including quartz *and* muscovite in addition to those minerals shown in the diagram. In contrast to the A′FK diagram, muscovite is not shown in the AFM diagram; muscovite is present in all represented mineral assemblages.

Figure 5-10 shows the AFM plane extended beyond the line FeO-MgO. On this plane, the compositions of various minerals are plotted. As seen in Figure 5-9, point K, when projected along a straight line originating in point M, intersects the AFM plane only at infinity. This is indicated in Figure 5-10 by the arrow near the mineral "K feldspar."

The geometric projection of points inside the tetrahedron onto the AFM plane, as described in the preceding paragraphs, is carried out in actual practice by means of calculations. First of all, the weight percentages of an analysis are converted to molecular proportions of the oxides,

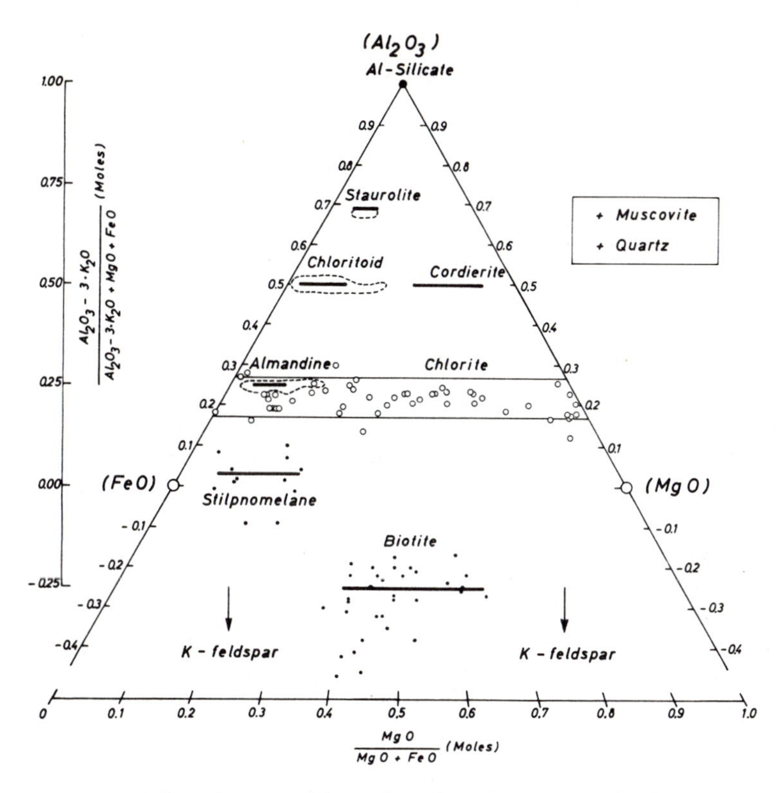

Fig. 5-10 Mineral compositions plotted on the AFM projection plane. The common range of compositions is indicated by a line. (Compiled by G. Hoschek.)

and from these the molecular proportions A and M, which serve as coordinates, are calculated as follows:

$$A = \frac{[Al_2O_3] - 3[K_2O]}{[Al_2O_3] - 3[K_2O] + [MgO] + [FeO]}$$

$$M = \frac{[MgO]}{[MgO] + [FeO]}$$

Muscovite, which serves as the projection point, contains three times as much $[Al_2O_3]$ as $[K_2O]$; this accounts for the expression $[Al_2O_3]$-$3[K_2O]$ in the formula for calculating the A value. If the A value for a biotite is calculated, it becomes apparent that this must be a negative number; this is also seen in Figures 5-9 and 5-10. In Figure 5-10 a number of biotites from metamorphic rocks are shown. The scatter of compositions is rather high, but for most biotites the A value is about −0.25, *i.e.*, they belong to the series eastonite-siderophyllite

$$K(Mg, Fe)_{2.5}Al_{0.5}[(OH)_2/Si_{2.5}Al_{1.5}O_{10}]$$

Biotites of the series phlogophite-annite

$$K(Mg, Fe)_3[(OH)_2/Si_3AlO_{10}]$$

having an A value of −0.50, are much rarer in metamorphic rocks.

The A and M values of stilpnomelane were calculated after the trivalent iron was expressed as bivalent iron and added to the analytically determined bivalent iron, because some of the Fe^{2+} in stilpnomelane, as in some chlorites, probably was oxidized to Fe^{3+} after the minerals were formed. The composition of chlorites scatters across the complete field of possible compositions, as shown in Figure 5-10; there is no indication of a region of more frequent compositions. The scatter of compositions of staurolite, chloritoid, and almandine is bounded in each case by a dashed line; the short line within each region indicates the most frequent compositions. The range of composition of cordierites is taken from Schreyer (1965a). Before calculating the A and M values of garnet, the amount of [Al_2O_3] required to form the end members spessartite $Mn_3Al_2(SiO_4)_3$ and grossularite $Ca_3Al_2(SiO_4)_3$, which are not represented in the AFM diagram, was subtracted from [Al_2O_3] of the garnet analysis. In this connection, it should be mentioned that the additional components CaO and/or MnO, not represented in the AFM diagram, may in fact cause the formation of garnet at certain physical conditions. In this case, an almandine-rich garnet containing some amount of the grossularite or spessartite end member is present as an additional mineral in the mineral assemblages, even though, on the basis of rock composition in terms of components other than CaO and MnO, this would not be expected.

The AFM diagram serves the important purpose of achieving a graphical representation of coexisting minerals. Depending on the bulk composition of the rocks and on the physical conditions of metamorphism, the compositions of coexisting biotite and cordierite or biotite and almandine are different with respect to their MgO/(MgO + FeO) ratio. Furthermore, it is possible to show in an AFM diagram mineral assemblages typical of a metamorphic zone. This can be done with greater precision than in the case of ACF and A′FK diagrams. Such applications are discussed in later chapters. The composition of a metamorphic *rock* can be shown in an AFM diagram as well, provided that certain corrections are applied to the chemical analysis before calculating the A and M values, because the chemical constituents of minerals not shown in the projection must be subtracted from the bulk rock analysis. One of the most important corrections is the subtraction of that amount of Al_2O_3 which is present in albite, in plagioclase (both albite and anorthite), and in paragonite. In principle, other corrections are the same as those discussed on p. 42 in connection with A'FK diagrams.

The AFM diagrams which have been discussed so far and which will be presented in this book apply only to metamorphic parageneses that contain muscovite and quartz as additional minerals. However, *other* AFM diagrams have been constructed for parageneses *devoid of muscovite and containing K feldspar* (Barker, 1961). In such cases, points of minerals and rocks that are situated within the K_2O-Al_2O_3-FeO-MgO tetrahedron are projected onto the Al_2O_3-FeO-MgO plane by using K feldspar as the point of projection. The value of A is accordingly calculated as

$$A = \frac{[Al_2O_3] - [K_2O]}{[Al_2O_3] - [K_2O] + [FeO] + [MgO]}$$

AFM diagrams are very useful in the representation of mineral assemblages of metamorphic rocks derived from shales and shaly sands. These are by far the commonest sediments, so that AFM diagrams may be used to advantage in the study of many metamorphic rocks. The coexistence of certain minerals is not apparent in an A'FK diagram but is readily shown in an AFM diagram, because FeO and MgO are treated as two components. This will be demonstrated in appropriate discussions.

Still another modification of an AFM projection has been devised by Reinhardt (1968, 1970), applicable to high-grade metamorphic rocks of a wider range of composition than that of metapelites. It can be used when K feldspar, plagioclase of constant composition, quartz, and magnetite (also ilmenite) are common to all rocks that are to be graphically represented. Except for those minerals, sillimanite/kyanite, cordierite, garnet, and biotite, as well as hornblende, orthopyroxene, and clinopyroxene, can be graphically shown. This diagram is particularly useful for granulites because felsic and mafic granulites may be represented in one diagram. The parameters for this kind of diagram are

$$\begin{aligned} A' &= Al_2O_3 - (K_2O + Na_2O + CaO) \\ F &= FeO - Fe_2O_3\,(- TiO_2) \\ M &= MgO \end{aligned}$$

Still another method of graphical representation of mineral assemblages in biotite-bearing, specifically mafic granulites, is given by Froese (1978).

References

Albee, A. L. 1972. *Geol Soc. Am. Bull.* **83:** 3249–3268.

Albee, A. L. and Chodos, A. A. 1965. *Geol. Soc. Am. Special Papers No. 87.*

Barker, F. 1961. *Am. Mineral.* **46:** 1166–1176.

Barth, T. F. W., Correns, C. W., and Eskola, P. 1939. Entstehung der Gesteine. Springer-Verlag, Berlin.

Brown, E. H. 1967. *Contr. Mineral Petrol.* **14:** 257–292.

—. 1971. *Contr. Mineral Petrol.* **31:** 275–299.

Eskola, P. 1939. *In* Barth, Correns, and Eskola, *Die Entstehung der Gestine.* Springer-Verlag, Berlin.

Fergusson, F. O. and Jones, T. K. 1966. *The Phase Rule.* Butterworths, London.

Froese, E. 1978. Geol. Survey Canada, Paper 78-1A, 323–325.

Laduron, D., and Martin, H. 1969. *Ann. Soc. Geol. Belgique.* **72:** 159–172.

Nockolds, S. R. 1954 *Bull. Geol. Am.* **66:** 1007–1032.

Reinhardt, E. W. 1968. *Can. J. Earth Sci.* **5:** 455–482.

— and Skippen, G. B. 1970. *Geol. Surv. Can Rept. Activities Paper 70-1.* pt. B, pp. 48–54.

Robinson, P., Ross, M., and Jaffe, H. W. 1971. *Am. Mineral.* **56:** 1005–1041.

Ronov, A. B. and Khlebnikova, Z. V. 1957. *Geochemistry.* **6:** 527–552.

Schreyer, W. 1965. *Neues Jahrb. Mineral. Abhand.* **103:** 35–79.

Thompson, J. B. 1957. *Am. Mineral.* **42:** 842–858.

Zen, E-An. 1960. *Am. Mineral.* **45:** 129–175.

Chapter 6

Classification Principles: Metamorphic Facies versus Metamorphic Grade

It was a milestone in the understanding of metamorphic petrology when Eskola (1915) published his concept of *metamorphic facies* with the intention of replacing the earlier Becke (1913) and Grubenmann (1910) concepts of metamorphic *depth zones*. Nevertheless, this concept of depth zones was elaborated and slightly modified by Grubenmann and Niggli (1924) and remained in use for decades. In the depth zone classification the physical factors of temperature, hydrostatic pressure, and directed pressure (stress) were believed to be strictly correlated in the manner demonstrated in Table 6-1.

For some time it has been known that such a correlation is not valid [for a discussion see Eskola (1939) and Winkler (1968)]. In many terrains of regional metamorphism where the temperature increased, the pressure remained nearly constant; this is well demonstrated by Turner (1968). Therefore, there was good reason to abandon the classification

Table 6-1 Correlation of physical factors in the metamorphic depth zone classification. (From Grubenmann and Niggli, 1924, p. 375.)

	Temperature	Hydrostatic pressure	Stress
Epi zone ("upper" zone)	Moderate	Mostly low	Often strong, also missing
Meso zone ("middle" zone)	Higher	Mostly higher	Often still very strong, also missing
Kata zone ("lower" zone)	High	In many cases very high	Mostly less strong, often completely missing

of metamorphism into depth zones; the epi zone, meso zone, and kata zone should no longer be used. Eskola also made a special plea to abolish the use of the expressions epi, meso, and kata. In 1939 he wrote: "As long as the expressions are kept in use it is difficult to get rid of the idea that the depth zones have something to do with depth after all." Unfortunately, this plea has not yet universally been accepted.

Eskola, having taken into consideration the earlier results on contact metamorphism by Goldschmidt (1911), strongly stressed the point that mineral assemblages rather than individual minerals are the genetically important characteristics of metamorphic rocks. The mineral assemblages, also called *mineral parageneses,* are the really significant constituents of metamorphic rocks because, as we know now for certain, only these give information on the genetic conditions during metamorphism. It is on mineral parageneses, and on relationships among mineral parageneses, that Eskola based his concept of metamorphic facies. In 1920/21 he defined the term

> metamorphic facies to designate a *group of rocks* characterized by a definite set of minerals, which under the conditions during their formation were in perfect equilibrium with each other. The quantitative and qualitative *mineral* composition in the rocks of a given facies varies gradually in correspondence with variation in the *chemical* composition of the rocks.

In 1939 he gave a somewhat modified definition: Zu einer bestimmten Fazies werden die Gesteine zusammengefügt, welche bei identischer Pauschalzusammensetzung einen identischen Mineralbestand aufweisen, aber deren Mineralbestand bei wechselnder Pauschalzusammensetzung gemäss bestimmten Regeln variiert. Very freely translated: "A certain facies comprises *all* rocks exhibiting a characteristic correlation between chemical and mineralogical composition, in such a way that *rocks of a given chemical composition have always the same mineralogical composition,* and differences in chemical composition from rock to rock are reflected in systematic differences of their mineralogical composition." Eskola continues:

> The significance of this principle is based on the observation that the mineral parageneses of metamorphic rocks in many cases conform to the laws of chemical equilibrium, but the attainment of chemical equilibrium is not essential to the definition of metamorphic facies.

This sentence expresses a significant modification of the definition of metamorphic facies as given in 1921. Taking this into consideration, the 1921 definition is clearer than that of 1939 and is therefore to be preferred. Accordingly it would read: *A metamorphic facies designates a group of rocks characterized by a definite set of minerals formed under particular metamorphic conditions.*

The quantitative and qualitative mineral composition in the rocks of a given facies varies gradually in correspondence with variation in the chemical composition of the rocks. Thus, as Tilley (1924) comments:

> a given facies may include rocks of widely different bulk composition, the variable of chemical composition is thus allowed for, whilst the *number of facies* expresses the variable physical environment under which the rocks have been formed. This conception is a distinct advance over the classification adopted by Grubenmann, and one is impressed with the elasticity of the classification, for the number of so-called facies may be indefinitely increased as progress in the study of metamorphism may require.

The practical approach to collecting the petrographic data characterizing a particular metamorphic facies is very simple in principle: An areally restricted metamorphic terrain is selected which contains rocks of widely different chemical composition, such as derived from basic lavas or tuffs, from ultrabasic rocks, from different pelitic sediments, dolomitic limestone, and marls. Because of their proximity, all these rocks were metamorphosed under practically the same physical conditions and gave rise to a set of minerals corresponding to their chemical composition. The whole *group of different rocks, i.e.,* the association of different mineral assemblages, comprises one metamorphic facies.

A classic example of a metamorphic facies is the group of chemically different rocks which Goldschmidt (1911) found in the innermost contact aureole of small high level plutons that had intruded into unmetamorphosed sediments. The various mineral parageneses found in the Oslo district have since been recognized in the inner-most contact zone of different plutons all over the world wherever very hot magma of gabbro or essexite type has intruded into shallow crustal levels. The group of mineral assemblages constituting the metamorphic contact rocks called hornfelses has been graphically represented in an ACF diagram by Eskola. Figure 6-1 (taken from Eskola, 1939) shows—indicated by the roman numerals—the 10 hornfels classes distinguished by Goldschmidt. Eskola comments that in the mineral parageneses in classes I to VII biotite is commonly present and that K feldspar and quartz may be additional constituents in all classes, except in classes VIII to X where quartz is mostly absent and calcite occurs instead.[1]

[1]This remark seems to have been forgotten, but is of significance since experiments have shown that at high temperature and low pressure metamorphism, grossularite can no longer exist with quartz; rather, the two minerals react to form anorthite + wollastonite as a co-existing pair. (Newton, 1966; Storre, pers. commun., 1970).

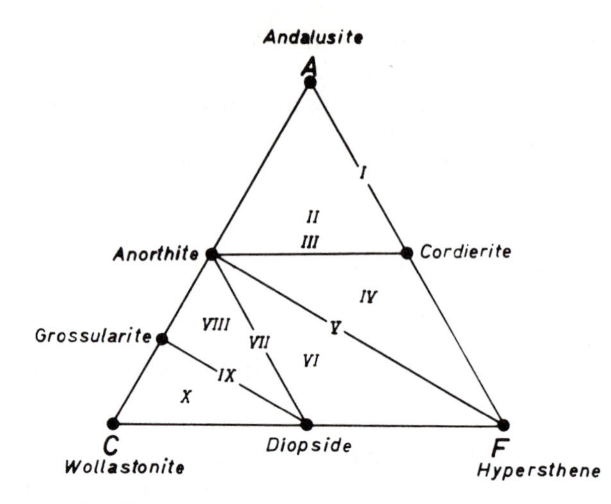

Fig. 6-1 ACF diagram of mineral assemblages grouped in Eskola's pyroxene-hornfels facies. The ten "classes of hornfelses" that Goldschmidt had distinguished are indicated by roman numerals. (After Eskola, 1939.)

This facies was given the rather unfortunate name of pyroxene-hornfels facies instead of orthopyroxene-hornfels facies or, as suggested by Winkler (1967), K feldspar-cordierite-hornfels facies. The pyroxene-hornfels facies is restricted to the innermost contact zone of hot plutons and, as Eskola (1939) had already recognized, a different facies is developed in a zone more distant from the contact; he called this zone "amphibolite facies" but Turner (1948) gave it the more appropriate name of hornblende-hornfels facies. Turner also distinguished a third hornfels facies which is developed in a zone even more distant from the contact: the albite-epidote-hornfels facies. For the relevant mineral parageneses and diagram, see Turner and Verhoogen (1960) or Winkler (1967). This shows that the facies classification easily allows an increase in the number of facies as demanded by new petrographic observations.

It is obvious that with much more detailed petrographic investigation, in regard to both variability of rocks and regional distribution, more groups of metamorphic rocks have been and will continue to be discovered which differ by a few mineral parageneses from those of the established metamorphic facies. Thus, in 1921 Eskola distinguished five facies and in 1939 their number had increased to eight. Turner and Verhoogen in 1960 recognized 10 facies and Winkler in 1967 had reason to distinguish 12 facies. However, this rather small increase in number does not correctly reflect the enormous increase of petrographic data about rocks having formed in a limited area, each under specific physical conditions, and showing incomplete agreement with that group of rocks comprising a known facies. For instance, it became evident that a group of contact metamorphic rocks exists which contains many of the mineral parageneses of the so-called pyroxene-hornfels facies (the K feldspar-cordierite-hornfels facies), but instead of the paragenesis hypersthene + plagioclase + diopside, either orthoamphibole + plagioclase + hornblende or hornblende + plagioclase + diopside was formed. The mineral parageneses for the two cases are shown by means of ACF and A′FK diagrams in Figure 6-2a and b. Since in *both* sets of mineral parageneses the same diagnostic mineral associations are present, namely, K feldspar

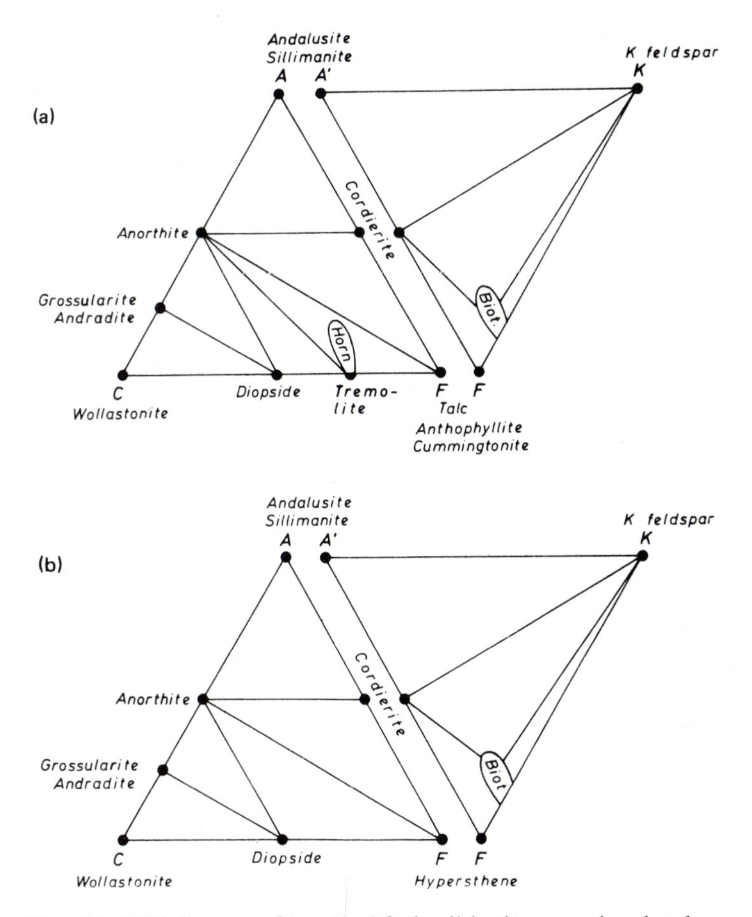

Fig. 6-2 ACF diagram of two "subfacies," both grouped under the pyroxene-hornfels facies which has been renamed K feldspar-cordierite-hornfels facies by Winkler (1967). Above: *Orthoamphibole subfacies* of the K feldspar-cordierite-hornfels facies. Below: *Orthopyroxene subfacies* of the K feldspar-cordierite-hornfels facies. In the lower temperature part, hornblende may still occur additionally if the H_2O pressure is higher than about 300 bars. The temperature interval in which hornblende along with orthopyroxene may coexist broadens considerably as the H_2O pressure is increased.

+ cordierite and K feldspar + andalusite (or sillimanite), the two groups of rocks were assigned to the same facies: the K feldspar-cordierite-hornfels facies; however, in order to point out the differences each was given the status of a so-called *subfacies*. Thus, the orthoamphibole subfacies (Figure 6-2a) and the orthopyroxene subfacies (Figure 6-2b) within the K feldspar-cordierite-hornfels facies have been distinguished by Winkler, 1967.

This procedure is correct, but the term subfacies is not and cannot be properly defined. This point is important because each group of rocks characterized by a definite set of mineral assemblages and designated as a subfacies deserves the full status of a facies. *Each subfacies is really a metamorphic facies!* In general, each group of rocks having formed under a given set of metamorphic conditions and differing from an already known group by only one mineral assemblage constitutes a new metamorphic facies. This is so because the *whole group* of mineral parageneses defines a metamorphic facies.

When petrographic data about metamorphic rocks was limited, classification into metamorphic facies was easy; it remained so for several decades and the facies classification proved its enormous capacity to incorporate a steadily increasing wealth of petrographic data. The introduction of subfacies by Turner was an elaboration of the metamorphic facies and proved very practical; it registered the accumulating knowledge and at the same time obviously aimed at preserving the general view of metamorphic rock transformations which ran great risk of getting lost in so much more detailed petrographic data.

The last attempt to organize the knowledge of the Barrovian type of metamorphism was made by Turner and Verhoogen (1960) by subdividing the greenschist facies and the almandine-amphibolite facies into three subfacies each. It later became known that regional metamorphism at much lower pressures than the Barrovian type in the Scottish Highlands showed very different mineral parageneses, and Winkler (1965, 1967) arranged the sequence of changing groups of parageneses in the form of a sequence of subfacies of the so-called Abukuma type of metamorphism. Two subfacies of the greenschist facies and three subfacies of the cordierite-amphibolite facies were established, all of which are different from those of the Barrovian type of metamorphism. Furthermore, it became quite obvious that, under conditions intermediate in pressure between the Barrovian and the Abukuma type, several more subfacies must be distinguished in the temperature range of the amphibolite facies. In addition, subfacies were established within the granulite facies and, as mentioned earlier, within the low pressure contact metamorphic hornfels facies. Although all registrations of the manifold petrographic data are correct, the greatly increased number of metamorphic facies and subfacies, which really are to be taken as facies themselves, has made it very cumbersome indeed to retain a logical and practicable classification of metamorphism.

Even if the *sequence of subfacies* observed with progressive grade in each given metamorphic terrain [the so-called facies series of Miyashiro (1961)] is taken together as a unit, the large amount of information cannot be handled without a sequence of diagrams representing the various mineral parageneses. Classification based on metamorphic facies has become impracticable. This is due to the enormous increase in petrographic information and the deduced relationships, namely, (a) that within a given temperature range of metamorphism many more mineral reactions take place than were formerly recognized, and (b) that the influence of pressure on the *sequence* of mineral reactions as well as on resulting mineral assemblages is more diverse than had been assumed.

It must be stressed that the concept of metamorphic facies is obsolete only as a means of classification. The petrographic information that is accumulated in the many subfacies is of real value and must not simply be dismissed. Mineral parageneses of different metamorphic rocks will, of course, have to be determined in any metamorphic terrain, and often at even more closely spaced intervals than has commonly been the practice. But for classifying the metamorphic rocks within a given region, a concept other than metamorphic facies is required; it should be based on petrographic criteria and have petrogenetic significance. This, of course, was also the aim of the facies classification, but in the form of a subfacies sequence it can in fact serve its purpose only in a cumbersome way.

The use of subfacies had been criticized earlier by Lambert (1965) and by Fyfe and Turner (1966). The latter authors, in a major revision of their previous ideas, proposed to abolish the subfacies, retaining, however, the facies classification. Turner (1968) is now of the opinion "that ten or a dozen facies will prove sufficient to encompass the complete gamut of metamorphic rocks; this number is small enought to permit the general petrologist to remain familiar with their essential characteristics." Turner's wish to revert in essence to Eskola's facies, to the greenschist facies, the amphibolite facies, and a few others, is understandable but this is no longer possible because the temperature-pressure field of each such "facies" is so large that in traversing it a conspicuous number of *new sets* of mineral parageneses are formed. Therefore, according to the very definition of metamorphic facies, they cannot belong to a single facies. Turner (1968) no longer calls the various subfacies by their earlier names, but nevertheless still distinguishes them—now using, in the case of the amphibolite facies, an almandine zone, staurolite-kyanite zone, sillimanite-muscovite zone, etc., each of which represents, correctly speaking, at least one facies each.

From this discussion it becomes apparent that

1. The term metamorphic facies *must* not be applied for rather large ranges of metamorphic conditions that have earlier been designated as "greenschist facies," "amphibolite facies," etc.
2. Each "subfacies" is at least one "facies" proper. However, to avoid confusion it is suggested to *abolish the term "subfacies" as well as "facies" in metamorphism.*
3. It would be expedient to have a nomenclature for those larger temperature-pressure fields that were formerly designated as "greenschist facies," "amphibolite facies," etc.
4. It is necessary to record those changes in mineral parageneses that take place when passing from one subfacies to another or from one metamorphic zone to the next.

In keeping with point 3, the entire field of metamorphic temperature-pressure conditions will be divided in only four large units of metamorphic grade:

Very-low grade
Low grade
Medium grade
High grade

In keeping with point 4 it is obvious that *only the changes in mineral assemblages are significant, not the persisting mineral parageneses.* This clearly shows the way for a detailed and yet simple metamorphic classification, because it would mean an enormous simplification if, for the purpose of classification, mineral assemblages not changing at boundaries between metamorphic zones could be disregarded. *Therefore, mineral reactions in common rocks reflecting the significant metamorphic changes will serve to classify the whole temperature-pressure field of metamorphism.* At the same time the *sequence* of certain mineral reactions in a prograde metamorphic terrain will supply additional information of significance in petrogenetic considerations.

Thus, *the terms metamorphic facies and subfacies are no longer used* in this treatise, except for comparison, and *the proposed classification of metamorphism is based on specific reactions in common rocks.* In particular:

1. Specific reactions define four large *P, T* divisions of metamorphic grade: very-low, low, medium, and high. These serve as a very useful gross grid of classification, similar to Eskola's metamorphic facies scheme.

2. Within each metamorphic grade, various metamorphic zones are established. They are also based on specific reactions in common rocks of various composition. Thus a closer spaced metamorphic grid is supplied. Therefore the entire field of metamorphic conditions is classified by mineral reactions which furnish specific mineral parageneses.
3. The sequence of specific reactions in a given metamorphic terrain may provide additional petrogenetic information.

References

Becke, F. 1913. *Akad. Wiss. Wien, Math. Naturw. Kl.* **75:** 1–53.

Eskola, P. 1915. *Bull. Comm. Geol. Finlande* **44:** 109–145.

——— 1920/21. *Norsk, Geol. Tidsskr.* **6:** 143–194.

——— 1939. *In* Barth, Correns, and Eskola, *Die Entstehung der Gesteine.* Springer-Verlag, Berlin.

Fyfe, W. S. and Turner, F. J. 1966. *Contr. Mineral. Petrol.* **12:** 354–364.

Goldschmidt, V. M. 1911. *Kristiania Vidensk. Skr. Math. Naturv. Kl.* **11.**

Grubenmann, U. 1910. *Die kristallinen Schiefer.* 2nd edit. Borntraeger, Berlin.

——— and Niggli, P. 1924. *Die Gesteinsmetamorphose* I. Borntraeger, Berlin.

Lambert, R. St. J. 1965. *Mineral Mag.* **34:** 283–291.

Miyashiro, A. 1961. *J. Petrol.* **2:** 277–311.

Newton, R. C. 1966. *Am. J. Sci.* **264:** 204–222.

Tilley, C. E. 1924. *Geol. Mag.* **61:** 167–171.

Turner, F. J. 1948. *Geol. Soc. Am. Memoir No. 30.*

——— 1968. *Metamorphic Petrology.* McGraw-Hill Book Company, New York.

——— and Verhoogen, J. 1960. *Igneous and Metamorphic Petrology.* 2nd edit., McGraw-Hill Book Company, New York.

Winkler, H. G. F. 1961. *Geol. Rundschau* **51:** 347–364.

——— 1965. *Petrogenesis of Metamorphic Rocks.* Spinger-Verlag, New York-Berlin.

——— 1967. *Petrogenesis of Metamorphic Rocks.* 2nd edit. Springer-Verlag, New York-Berlin.

——— 1968. *Geol. Rundschau* **57:** 1002–1019.

Chapter 7

The Four Divisions of Metamorphic Grade

General Considerations

Metamorphic grade has been used as an inprecise term "to signify the degree or state of metamorphism" (Tilley, 1924). For example, low-grade metamorphism implies formation at relatively low temperatures typical of the greenschist facies, and medium-grade metamorphism refers to conditions of the amphibolite facies. Although not necessary in principle for a classification of metamorphism, it is very practical in fieldwork and makes orientation much easier if the entire *P,T* range of metamorphic conditions is divided into a few large divisions of metamorphic grade:

Very-low-grade (sehr schwach)
Low-grade (schwach)
Medium-grade (mittel)
High-grade (stark)

The suggested German translations are given in order to avoid any association between metamorphic grade and depth within the earth's crust.

From a survey of metamorphism it is very evident that an increase in metamorphic grade corresponds to a progression of temperature. Earlier investigations of the *P,T* conditions represented by metamorphic facies, as summarized by Turner (1968) and Winkler (1967), have demonstrated the succession of the main metamorphic facies with increasing temperature (see Figure 7-1). Therefore, it is appropriate to arrange the sequence of metamorphic grades—very-low, low, medium, and high—according to increasing temperature as well.

The boundaries between the four metamorphic grades shall be

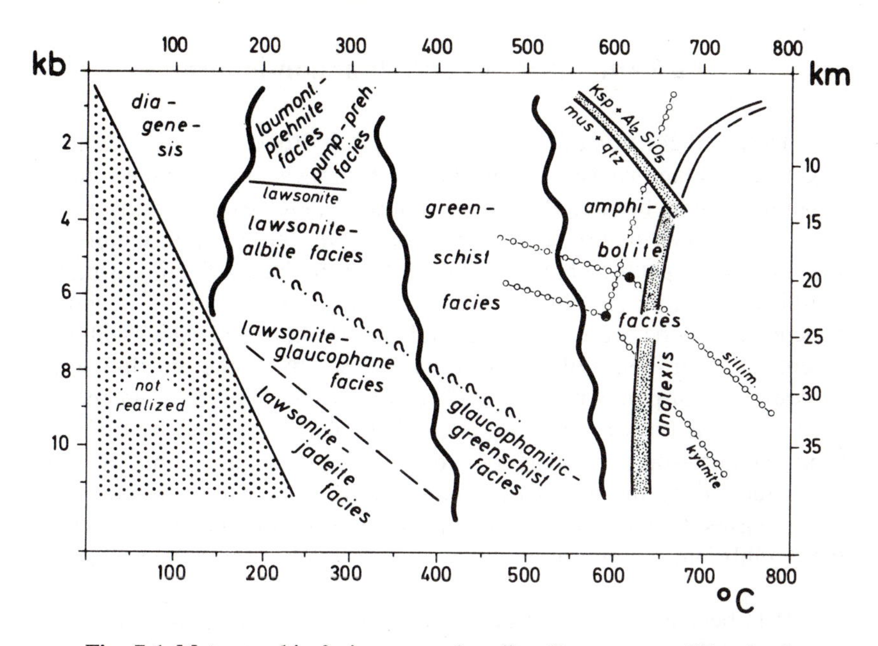

Fig. 7-1 Metamorphic facies as used earlier. Pressure condition is always $P_s = P_f$. The different case $P_s >> P_{H_2O}$ is not considered here; therefore, formation of granulites and eclogites cannot be represented in this plane.

marked by significant metamorphic changes of mineral assemblages in common rocks, *i.e.*, by specific mineral reactions. It has been the author's aim to select reactions for boundaries between metamorphic grades which reflect changes that have been recognized to take place at the beginning of the greenschist facies and at the beginning of the amphibolite facies, respectively. Thus, the boundary between "very-low grade" and "low-grade" coincides with the beginning of the greenschist facies and the boundary between "low-grade" and "medium-grade" coincides with the beginning of the amphibolite facies of Eskola. In addition, the upper part of the amphibolite facies has been assigned to "high-grade" metamorphism characterized by the coexistence of K feldspar with Al_2SiO_5 and/or almandine and cordierite. At high-grade metamorphism migmatites are formed at high water pressures and granulites at low water pressures relative to total pressure. The boundaries of the four metamorphic grades are defined by mineral reactions, and their petrographic criteria will be given in the section beginning on p. 68.

The temperature range represented by a certain grade of metamorphism may be divided into metamorphic zones on the basis of diagnostic mineral assemblages in rocks of appropriate composition. This procedure is known as the determination of isograds.

The Terms Isograd and Reaction-Isograd

Tilley (1924) coined the term "isograd" to designate a definite degree of metamorphism by the first appearance of a so-called index mineral, such as biotite, almandine, staurolite, etc. An isograd is a line on a map joining points of first appearance of a certain index mineral, *i.e.*, where a specific change in mineral assemblage reflecting a metamorphic reaction has taken place. Since a metamorphic reaction depends on temperature, pressure, and composition, an isograd will represent, in general, sets of P-T-X conditions satisfying the reaction equilibrium and not points of equal P-T-X conditions. Therefore, whenever the reaction is known, the term isograd should be replaced by the new term reaction-isograd. A reaction-isograd is a line joining points that are characterized by the equilibrium paragenesis of a specific reaction; in certain cases, two (or more) reactions taking place simultaneously define the equilibrium mineral assemblage of a reaction-isograd (see Chapter 9).

It is because of this latter fact that the term reaction-isograd will now be used instead of the term isoreaction-grad of the previous two editions. The latter term may erroneously suggest that no more than one reaction may be involved in defining such an isograd.

Petrographic work should be aimed at determining definite mineral assemblages and establishing the locations of reaction-isograds in the field. A reaction equation correctly describes the definite paragenesis which characterizes a reaction-isograd. Therefore, locating it in the field involves mapping reactants and products of the equation or—even better—the full equilibrium assemblage.

If the observed metamorphic change cannot be related to a specific reaction, the term isograd will be used in referring to this feature. A single index mineral will generally be formed by several different reactions. Only when the physical conditions of the different reactions are similar does the first appearance of an index mineral, and consequently the mapped isograd, have petrogenetic significance (*e.g.*, the staurolite isograd). If, however, the physical conditions for the formation of an index mineral are very different (*e.g.*, tremolite, diopside, and forsterite), the corresponding isograd does not have any petrogenetic meaning. Therefore, it is important to recognize significant and nonsignificant metamorphic reactions in carrying out practical work. The same distinction is required also in the case of reaction-isogrades which are characterized by the equilibrium paragenesis of a specific reaction. Only when reactions are univariant, or when multivariant, take place within a narrow temperature range at a given pressure, do they have petrogenetic significance. This point is observed throughout this book.

In the simplest case of a univariant reaction, the field locality where a specific reaction-isograd has been located represents but one of the various *P,T* combinations along the equilibrium curve drawn in the *P,T* plane. We want to know which definite *P* and *T* values from the equilibrium curve are applicable to the area under investigation. This information, however, cannot be supplied by a single reaction-isograd but requires two or more preferably intersecting reaction-isograds.

Along a reaction-isograd, both *P* and *T* will generally vary. *P,T* conditions satisfying the reaction equation will be realized along a surface within the earth's crust. As pointed out by Tilley (1924), an isograd mapped in the field is the intersection of such a surface with the present erosion surface. If, after a metamorphic event, the earth sector has been *tilted* during uplift, the line of a reaction-isograd passes at the present surface through points which represent gradually changing depths, *i.e.*, changing pressure of metamorphism. This possibility must always be kept in mind.

The mapping of isograds in the field is somewhat similar to localizing the beginning of the Barrow (1893, 1912) zones, *i.e.*, the chlorite, biotite, garnet, etc., zones. These classic isograds are due to mineral reactions in metapelites only. Later we shall discuss reactions in other rock compositions as well, because they furnish valuable additional isograds.

"For the interpretation of isograds in terms of conditions of metamorphism," Thompson and Norton (1968) wrote, "the isograds should ideally be drawn so as to minimize the effect of variation in bulk composition [see Thompson, 1957]. . . . First appearances, however, may very nearly fulfill the above requirement in areas where rocks having a considerable variation in bulk composition are intimately interstratified. In such areas the mapped isograds probably approximate fairly closely the first possible appearance of the index [the isograd] minerals . . . consistent with the local pressure, temperature, and activity of H_2O. The success of the method thus depends on both the range of bulk compositions available and the scale of the interstratification of the various rock types relative to the scale of the map." To this quoted paragraph it may be added that the chances for successful application of the isograd method are good whenever the chosen area of investigation is large enough. This often can be done because the primary purpose should not be to study and record all the different petrographic rock compositions in the area but to trace such rocks that are suitable in composition to yield information on isograds, and especially on reaction-isograds. A new approach in field and microscopic work is called for when petrogenetic information is sought concerning physical conditions of metamorphism.

The Division of Very-Low-Grade Metamorphism

Very-low-grade metamorphism can be defined to comprise the *P,T* field which is bounded on the low temperature side by the beginning of metamorphism—indicated by the first appearance of a non-sedimentary, truly metamorphic mineral (see Chapter 2)—and on the higher temperature side by a number of reactions, all of which have in common the first formation of zoisite or non-iron-rich epidote, *i.e.*, clinozoisite. Such reactions involve

1. The breakdown of the CaAl-silicate lawsonite
2. The disappearance of pumpellyite by reaction with chlorite and quartz (at pressures higher than ca. 2.5 kb).
3. The disappearance of pumpellyite by reaction with quartz and the disappearance of prehnite by reaction with chlorite and quartz (only at pressures lower than ca. 2.5 kb).

Very-low-grade metamorphism thus incorporates all rocks hitherto grouped by Winkler (1967) under

Laumontite-prehnite-quartz facies
Pumpellyite-prehnite-quartz facies
Lawsonite-albite facies
Lawsonite-jadeite-glaucophane facies

Figure 7-1 shows that the very-low grade extends from somewhat below 200°C up to about 400°C. Very-low-grade metamorphism covers a large pressure range which—concurrent with the "facies" boundaries of Figure 7-1—can well be subdivided by specific mineral reactions when a more detailed classification is required.

Pertinent facts for the boundary between very-low-grade and low-grade metamorphism have only recently been worked out. We shall first deal with lawsonite, $CaAl_2[(OH)_2/Si_2O_7] \cdot H_2O$, which is one of the diagnostic minerals of very-low-grade metamorphism. The upper stability conditions of lawsonite determine the uppermost possible boundary between very-low-grade and low-grade metamorphism. Once it was thought that this boundary would be represented by the following reaction:

$$12 \text{ lawsonite} = 6 \text{ zoisite} + 2Al_2SiO_5 + 1 \text{ pyrophyllite} + 20\ H_2O$$

However, although verified in reversed runs by Nitsch (1972), he has shown (1974) this reaction to be only metastable. The stable upper stability conditions of lawsonite are represented by the following reaction:

$$5 \text{ lawsonite} = 2 \text{ zoisite} + 1 \text{ margarite} + 2 \text{ quartz} + 8\ H_2O$$

This reaction takes place at temperatures about 30° to 20°C lower than the metastable reaction, as determined by Newton and Kennedy (1963) and by Nitsch (1974).

Margarite is a rare metamorphic mineral but it is certainly not as rare as commonly believed. Recent systematic search for margarite has revealed many occurrences (Frey and Niggli, 1971). The approximate equilibrium conditions of the reaction

$$5 \text{ lawsonite} = 2 \text{ zoisite} + 1 \text{ margarite} + 2 \text{ quartz} + 8\ H_2O$$

drawn in Figure 7-2, represents the maximum possible conditions for the boundary between very-low and low-grade metamorphism. This alone is reason enough to consider the reaction in this context. Otherwise it is most unlikely that the above reaction takes place in rocks. Certainly, margarite does not commonly form by the above reaction in metamorphic mafic rocks; margarite is typical of low-grade calcareous metapelitic rocks (see Chapter 10). Therefore, other reactions have to be considered by which lawsonite disappears and forms zoisite/clinozoisite.

In another context, Green *et al.* (1968) suggested that the following reaction takes place at very high pressures:

$$\{\text{lawsonite} + \text{jadeite}\} = \{\text{zoisite/clinozoisite} + \text{paragonite} + \text{quartz} + H_2O\}$$

The analogous reaction at lower pressures is

$$4 \text{ lawsonite} + 1 \text{ albite} = 2 \text{ zoisite/clinozoisite} + 1 \text{ paragonite} + 2 \text{ quartz} + 6H_2O$$

However, these reactions are unlikely to take place because of petro-

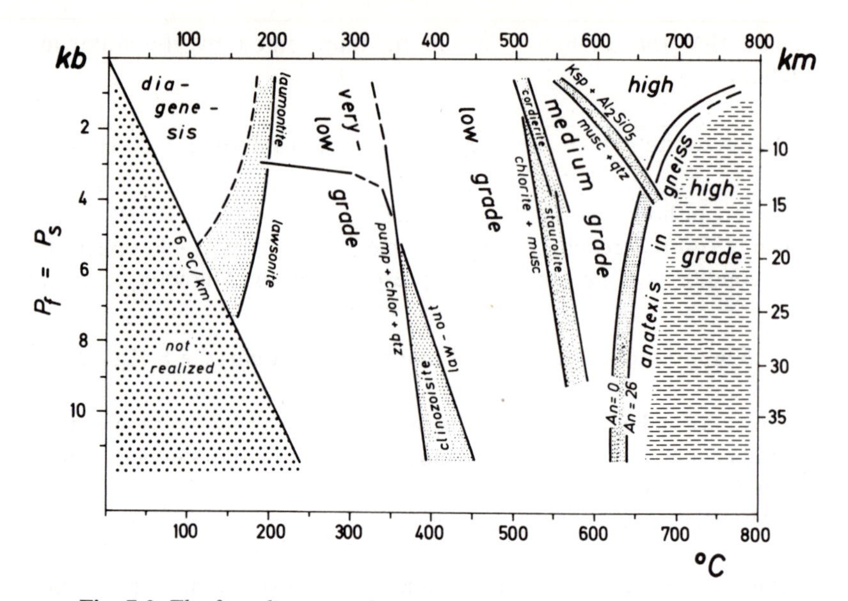

Fig. 7-2 *The four divisions of metamorphic grade;* very-low, low, medium, and high grade. Pressure condition for the graphically shown P, T data is that $P_s \approx P_{H_2O}$.

graphic reasons: lawsonite is a prominent mineral in very-low-grade mafic rocks. The greenschists derived from such rocks with progressive increase in grade have zoisite/clinozoisite instead of lawsonite, but contain white mica (among which paragonite may be present) only in accessory amounts. Therefore, other reactions must be looked for.

Lawsonite contains only slightly less CaO relative to Al_2O_3 than zoisite. This suggests that a reaction takes place between lawsonite and chlorite that forms zoisite/clinozoisite, an Al-richer chlorite variety, some quartz, and water:

$$\{\text{lawsonite} + \text{chlorite}\} =$$
$$\{\text{zoisite/clinozoisite} + \text{Al-richer chlorite} + \text{quartz} + H_2O\}$$

Further, the following reaction is probably important whenever calcite is associated with lawsonite:

$$3\ \text{lawsonite} + 1\ \text{calcite} =$$
$$2\ \text{zoisite/clinozoisite} + 1\ CO_2 + 5\ H_2O$$

This reaction [also suggested by Froese (1973, personal communication)] has been verified experimentally by Nitsch (1973, personal communication). It is sensitive to fluid composition and can take place only when the fluid is very water rich.

No equilibrium data are available for the two reactions believed to be responsible for the disappearance of lawsonite and the resultant formation of zoisite/clinozoisite at the boundary between very-low and low-grade metamorphism. However, *P,T* conditions cannot surpass the maximum stability conditions of lawsonite, shown in Figure 7-2; it is estimated that the *P-T* conditions of the reactions actually taking place in rocks will be very similar to or only slightly lower than those of the maximum lawsonite stability.

Lawsonite gives way to zoisite/clinozoisite with prograde metamorphism. Certain rocks of very-low grade may already contain a species of the epidote group, but it is invariably an Fe-rich epidote. In this case it is to be expected that the epidote will enter into solid solution with the zoisite/clinozoisite newly formed by the above reactions. Therefore, instead of an Fe-rich epidote, a mixed crystal with less Fe content will be formed when metamorphism passes from very-low grade to low grade. This change in composition can only be conveniently observed if the atomic ratio $Fe^{+3}/(Fe^{+3} + Al)$ falls below 1:7 in the newly formed mixed crystals. In this case the change of the optical character from negative for the iron-richer members to positive for the iron-poorer members is striking and diagnostically useful. According to the definition given by Deer *et al.* (1962), the positive members of the composition $Ca_2(Al, Fe^{3+})Al_2[O/(OH)/Si_2O_7/SiO_4]$ are called clinozoisite, while the optically negative members are called epidote proper.[1] Therefore, the appearance of optically positive clinozoisite or orthorhombic zoisite (which contains only small amounts of iron) indicates a change from very-low to low-grade metamorphism.

The mineralogical changes involved in the breakdown of lawsonite are identical to the change from either of the two lawsonite-bearing metamorphic facies to the greenschist facies; therefore, the boundary between very-low-grade and low-grade is the same as that for the beginning of the greenschist facies. If, instead of lawsonite, pumpellyite had been present in the rock, the disappearance of this mineral and the appearance of clinozoisite mark that boundary. Most commonly, this change will be due to the following reaction:

$$\{\text{pumpellyite} + \text{chlorite} + \text{quartz}\} = \{\text{clinozoisite} + \text{actinolite} + \text{water}\}$$

[1]It should be noted that the mineral epidote has given its name to a whole group of minerals—the epidote group. This includes, among others, orthozoisite, clinozoisite, epidote, and piemontite. What was formerly called pistacite is now designated as epidote proper.

Instead of clinozoisite, iron-richer epidote could form if enough trivalent iron is available; in such a case only the negative criterion of the disappearance of pumpellyite indicates the change from very-low grade to low grade.

Nitsch (1971) has determined the equilibrium data of the above reaction which only takes place above 2.5 ± 1.0 kb water pressure; these data are also plotted in Figure 7-2:

2.5 kb and 345° ± 20°C
4 kb and 350° ± 20°C
7 kb and 370° ± 20°C

At pressures below ca. 2.5 kb, Nitsch has further shown that the following reactions are to be expected:

{pumpellyite + quartz} =
{prehnite + clinozoisite + chlorite + water}

{prehnite + chlorite + quartz} =
{clinozoisite + actinolite + water}

These last two reactions will probably take place at quite similar temperatures, very likely near 340° ± 20°C, at 1 kb water pressure. The almost simultaneous disappearance of both prehnite and pumpellyite in *low* pressure metamorphism is well know at the border with the greenschist facies, *i.e.*, at the border between very-low and low-grade metamorphism.

It should be noted that in microscopic work lawsonite and the colorless variety of pumpellyite are not easily distinguishable from each other and from clinozoisite. This is also true for the distinction between some colored varieties of pumpellyite and epidote proper. In many cases, however, the anomalous interference colors of clinozoisite and of epidote are very helpful for quick identification. In doubtful cases, X-ray diffraction should be used.

If none of the diagnostic mineral changes can be observed because of inappropriate bulk composition of the rock, the distinction between very-low-grade and low-grade cannot be made in the field. It seems, however, that a different method may make this distinction possible whenever such commonly occurring rocks such as pelites or marly sediments are subjected to very-low-grade metamorphism. It is the so-called crystallinity of illite (or phengite), a predominantly K, Al mica from which at higher temperatures of metamorphism muscovite develops. The crystallinity is measured as the width of the 10 Å illite peak at half-peak height, as suggested by Kubler (1967, 1968). He has given convincing evidence that the width of the peak decreases, *i.e.*, crystallinity increases with the increase in temperature operative during the transformation (see also Frey, 1969). Temperature seems to be the only determining factor.

In a given locality the crystallinity must be determined by investigating the clay fraction of many samples because only the statistical mean value is significant. But this value varies continuously with temperature. Kubler has selected a certain range of crystallinity of the illite (from 7.5 to 4.0, calibrated on his standards) to define the beginning and the end of what he calls the "anchimetamorphic zone." Although the term anchimetamorphic is poorly defined, the distinction itself as made by Kubler is meaningful and seems to fit well our present system of classification.

At crystallinity 7.5, the first appearance of the nonsedimentary but metamorphic mineral pyrophyllite[2] has been observed. This indicates the "beginning of metamorphism" in shales or mudstones, *i.e.*, the beginning of very-low-grade metamorphism in the rocks investigated by Kubler. This may be incidental, and the limit may not coincide in temperature with that in other rocks, as has been mentioned earlier.

Kubler uses the crystallinity value of 4.0 (apparently the end value of the change) to designate the end of his "anchimetamorphic zone," *i.e.*, the beginning of the greenschist facies. Therefore, the change in crystallinity of illite from a statisical mean value of greater than 4.0 to the value of 4.0 is a furthur indication of the step from very-low-grade to low-grade. This new method should be widely applied in order to check its reliability.[3]

Summing up, *diagnostic minerals for very-low-grade metamorphism are* laumonite, prehnite, pumpellyite, lawsonite, and illite with crystallinity still imperfect (Kubler's value larger than 4.0). Glaucophane and jadeitic pyroxene, as such, are not restricted to very-low-grade metamorphism, but when coexisting with lawsonite, they are.

The change from very-low-grade to low-grade metamorphism is demonstrated by

1. Disappearance of lawsonite and formation of zoisite or clinozoisite.

[2]Earlier (Winkler, 1967), it was believed that pyrophyllite would not form at lower temperatures than those of the beginning of the greenschist facies, but Kubler (1967) and Frey (1969), have shown this to be incorrect. Pyrophyllite (and also chloritoid) may already occur at very-low-grade metamorphism.

[3]A useful modification of the method to determine illite crystallinity has been described by Weber (1972).

2. Disappearance of pumpellyite by reaction with chlorite and quartz, producing clinozoisite in addition to actinolite.
3. At *low pressure* metamorphism where lawsonite cannot form and reaction (2) does not take place: Disappearance of pumpellyite and then of prehnite and formation of zoisite-clinozoisite in addition to actinolite.
4. Good crystallinity of illite (phengite) equal to Kubler's value of 4.0.

In conditions 1, 2, and 3 *zoisite or clinozoisite is formed for the first time* at the cost of typical very-low-grade minerals. In 2 and 3, actinolite is also formed, but in some cases this is of little diagnostic value since actinolite may have been present already in some very-low-grade rocks.

The reactions dealt with here form a bundle or band of reaction-isograds which may be designated as the

isograd band "zoisite or clinozoisite-in,
lawsonite-out or pumpellyite-prehnite-out"

This isograd *band* defines the boundary between very-low and low-grade metamorphism; it is graphically shown in Figure 7-2.

If a relatively high amount of trivalent iron is present in the rock system, an optically negative epidote, instead of optically positive clinozoisite, will have been formed. Here, proof that one of the above reactions has taken place can only be given by the observation of the disappearance of the minerals lawsonite, pumpellyite, and prehnite in rocks of appropriate chemical compositions.

The Division of Low-Grade Metamorphism

In low-grade metamorphism, *i.e.,* from the beginning of the greenschist facies, the mineral zoisite is stable in contrast to rocks of the very-low grade. Chlorite, actinolite, white mica, iron-rich epidote, and others are not distinctive; they are already present in rocks of very-low grade. However, due to the formation of zoisite or of *non*-Fe-rich epidote, *i.e.,* clinozoisite, a very characteristic assemblage of minerals is typical of low-grade metamorphic rocks:

chlorite + zoisite/clinozoisite ± actinolite ± quartz

This assemblage, which always has been considered typical for the greenschist facies, persists over a range of increasing temperature. It is observed (together with other minerals) in metamorphic rocks that have originated from a great variety of rocks, such as basalts and tuffs, marls, certain pelites, and graywackes.

Remember that we are not interested in the many different complete sets of mineral parageneses when applying the classification of metamorphic grade; rather, we only have to make sure that in the area under investigation rocks are found that have primary chlorite + zoisite/clinozoisite as a mineral assemblage with mutual contacts (paragenesis of contacting minerals). This is all that is necessary in order to decide whether the rocks are low-grade or very-low-grade.

Within the large temperature range of the low-grade metamorphic division, a number of definite isograds can be identified, as will be shown in later chapters.

The Change from Low-Grade to Medium-Grade Metamorphism

Now we face the problem of where to mark the higher temperature boundary of the low-grade metamorphic division against the medium-grade division. In following the procedure of defining the limits of our divisions of metamorphic grade in accordance with some of the Eskola facies, we are now concerned with the changes that take place when the greenschist facies is being left and the amphibolite facies is entered. We look for mineralogical changes that characterize the beginning of the amphibolite facies; these changes should preferably occur within a large pressure range. At the same time we are looking for mineral parageneses of low-grade metamorphism (greenschist facies) which disappear at medium-grade (amphibolite facies).

Different authors have defined the beginning of the amphibolite facies in different ways. Eskola (1939) established an additional (albite-) epidote-amphibolite facies between greenschist facies and amphibolite facies by the first appearance of oligoclase-andesine instead of albite. Turner and Verhoogen, in 1960, extended the original greenschist facies to include the albite-epidote-amphibolite facies of Eskola; thus this extended greenschist facies immediately adjoins the amphibolite facies. And this boundary, in following Eskola (1920), has been defined by Turner and Verhoogen (1960) as the "sudden change in composition of plagioclase associated with epidote, from albite An_{0-7} to oligoclase or andesine An_{15-30}. This is microscopically recognizable and makes a convenient point at which to draw the high temperature boundary of the greenschist facies."

With this easily detectable change, other mineral changes were believed to occur concomitantly, namely, the first appearance of staurolite in appropriate pelitic rocks and the disappearance of chloritoid and of Mg-poor chlorite whenever quartz and muscovite are present. However, it is now well known that invariably in basic rocks the "jump" from albite to oligoclase occurs in any terrain of progressive regional metamorphism at somewhat lower temperature before staurolite appears and when chlorite + quartz + muscovite and chloritoid still exist in appropriate rocks of the same metamorphic grade. Staurolite sets in, and the assemblage Mg-poor chlorite + muscovite + quartz and chloritoid disappears when, in basic rocks of the same metamorphic grade, the plagioclase, coexisting with hornblende, already has experienced its compositional "jump" at probably a 20° to 30°C lower temperature (Wenk and Keller, 1969, Streck, 1969).

Very likely, Eskola (1920) chose the plagioclase "jump" as the boundary definition because of rather limited petrographic information; in any case, it was accepted by Turner and Verhoogen (1960) and by Wenk and Keller (1969). The latter authors use the first appearance of plagioclase An_{17} coexisting with hornblende to designate the beginning of the amphibolite facies. Much detailed work has shown that the composition An_{17} rather than An_{15} limits the compositional break between oligoclase and albite in metamorphic rocks. The two feldspar phases may coexist over a temperature range which is due to a miscibility gap analogous to that encountered in peristerites (Steck, 1976).

Wenk and Keller (1969) have shown that in the Central Alps of Switzerland the first appearance of a definite plagioclase composition, such as An_{17-20} or An_{30} or An_{50}, etc., coexisting with hornblende, is suited to characterize an isograd. In this way, a large range of metamorphic conditions can be subdivided. However, such isograds are not well defined and are, therefore, of little value.

The point relevant to our discussion is whether the beginning of the amphibolite facies, or rather the beginning of our medium-grade of metamorphism, should be defined by the break between albite and oligoclase, *i.e.*, by the first appearance of plagioclase An_{17}, or at somewhat higher temperature by the first appearance of staurolite and/or of cordierite, and by the disappearance of chlorite in the presence of muscovite and quartz and the breakdown of chloritoid. Turner (1968) now seems to favor not the plagioclase break but the appearance of an An-richer plagioclase, andesine existing with hornblende, and the formation of staurolite to designate the beginning of the amphibolite facies proper. Winkler, in 1965 and 1967, did not choose a specific composition of plagioclase to characterize the beginning of the amphibolite facies, but defined the beginning of that facies by the disappearance of chloritoid and of Mg-poor chlorite in the presence of muscovite and quartz, by the first appearance of staurolite, and—applicable only if pressures ranged from very low to about medium—the first appearance of cordierite (without almandine garnet!)[4] at the expense of chlorite, quartz, and muscovite in metapelites and metagraywackes. I still prefer these criteria, mainly for the following reasons:

Experimental research has shown that temperatures of the first formation of both cordierite and staurolite are not too dependent on H_2O pressure. The most common reactions producing these minerals are very probably the following:

$$\{\text{chlorite} + \text{muscovite} + \text{quartz}\} = \{\text{cordierite} + \text{biotite} + Al_2SiO_5 + H_2O\}$$

This reaction has been investigated from both sides. The equilibrium data obtained by Hirschberg and Winkler (1968) are as follows:

505° ± 10°C at 500 bars H_2O pressure
515° ± 10°C at 1000 bars H_2O pressure
525° ± 10°C at 2000 bars H_2O pressure
555° ± 10°C at 4000 bars H_2O pressure

Astonishingly enough, these data are valid within the ± 10°C range for a very large MgO/MgO + FeO ratio, varying in the experiments from 0.6–0.2.[5]

In order to form staurolite, chlorite containing an appreciable amount of ferrous iron is needed:

$$\{\text{chlorite} + \text{muscovite}\} = \{\text{staurolite} + \text{biotite} + \text{quartz} + H_2O\}$$

The data for this reaction, as ascertained by Hoschek (1969) with reversed runs in a system where the MgO/MgO + FeO ratio was 0.4, are

540° ± 15°C at 4000 bars H_2O pressure
565° ± 15°C at 7000 bars H_2O pressure

In the two reactions chlorite together with quartz and/or muscovite is used. Since muscovite (or rather phengite) and quartz are generally plentiful in metapelites, all chlorite is consumed at this stage of metamor-

[4]This holds only if compositions are not extremely Fe rich; if they are, almandine may form at low pressures and at these temperatures.

[5]These *P-T* data are also valid up to about 3 kb if no Fe^{2+} at all is present; at higher pressures, however, the temperature of equilibrium increases more rapidly with pressure in the Fe-free system, as Seifert (1970) has shown.

phism, provided the chlorite was relatively Fe-rich. If, however, in less common cases, a metapelite has MgO/MgO + FeO approaching 0.5 or more, the chlorite will not be used up by the reaction; rather, a relatively Mg-rich chlorite that has been formed in the reaction will coexist together with staurolite, muscovite, and quartz over a considerable temperature range. (See Chapter 14.)

The above reaction may also involve almandine garnet, as Carmichael (1970) has deduced from petrographic observations:

{chlorite + muscovite + almandine} =
{staurolite + biotite + quartz + water}

If chloritoid is also present in a low-grade metapelite it may react with andalusite or kyanite according to the following equation, also investigated by Hoschek (1967):

{chloritoid + andalusite or kyanite} =
{staurolite + quartz + H_2O}

This happens at 545° ± 20°C between 4000 and 8000 bars H_2O pressure.

Richardson (1968), who also investigated this reaction in a system completely devoid of Mg, furnished the following identical data valid for O_2 fugacity controlled by the magnetite-fayalite-quartz buffer:

520° ± 10°C at 2 kb H_2O pressure
540° ± 20°C at 5 kb H_2O pressure

If no Al_2SiO_5 mineral is present, chloritoid will produce staurolite if the pressure is high enough to allow the formation of almandine:

{chloritoid + quartz} = {staurolite + almandine + H_2O}

Calculations by Ganguly (1969) indicate that this reaction, relative to the preceding one, occurs at only slightly (perhaps 15° to 30°C) higher temperatures above possibly 4 kb.

From his investigation on chloritoid stability, Ganguly concludes that "transformations of chloritoid to staurolite can be achieved through several different reactions, depending on the bulk composition and the prevailing oxygen fugacities. It is, however, interesting to note that all these reactions are squeezed into a rather narrow temperature interval. Chloritoid-staurolite transformations should thus be regarded as potential indicators of the metamorphic grade, even in the absence of adequate data regarding the oxygen fugacities of the assemblages." If it is further taken into consideration that staurolite will be most commonly formed not from chloritoid but from Fe-rich chlorite + muscovite, and that this reaction occurs in the same narrow range of P, T conditions as the different transformations of chloritoid to staurolite, then it is obvious that not only the chloritoid-staurolite transformations but also the appearance of staurolite (without an established relationship to chloritoid) is well suited as an indicator of metamorphic grade. The various reactions can be grouped under one isograd band "staurolite-in"; this has practically almost the same significance as a reaction-isograd.

The various reactions leading to

Breakdown of Fe-rich chlorite in the presence of muscovite
Breakdown of chloritoid
Formation of cordierite (without almandine) in the presence of biotite
Formation of staurolite

all take place within a small temperature range. The P,T conditions are:

Formation of Cordierite, °C	H_2O Pressure, bars	Formation of Staurolite, °C
505 ± 10	500	
515 ± 10	1000	
525 ± 10	2000	520 ± 10
555 ± 10	4000	540 ± 15
	7000	565 ± 15

This small temperature range of about ± 20°C makes it easy to understand why in petrographic work the mineralogical changes summarized above have been commonly observed in many different metamorphic terrains to constitute a distinct change in the grade of metamorphism.

A few years ago it was still believed that those mineral changes coincide with the change from albite to oligoclase-andesine [compare the steps from the so-called quartz-albite-epidote-almandine subfacies of the greenschist facies to the staurolite-almandine subfacies of the amphibolite facies in Turner and Verhoogen (1960)]. Since it is known that the change in plagioclase composition in basic rocks takes place at somewhat lower temperatures than the entry of staurolite and cordierite and the other pertinent reactions in pelitic rocks, a decision had to be made as to which of the two alternatives should be chosen to define the limit between greenschist and amphibolite facies. Winkler (1965/1967) decided in favor of the various changes in the metapelites, not the single change in plagioclase composition. Many petrographers have done the same; others have not.

Recent experimental studies support the choice of reactions in pelitic rocks as the boundary between the low-grade and medium-grade metamorphism. For these reactions, the *P,T* conditions are known within narrow limits, whereas a similar calibration for the albite/oligoclase jump is not available. Nevertheless, this jump in the plagioclase composition of metamorphosed basalts, etc., marks an important metamorphic feature within the temperature range of low-grade metamorphism.

Summing up, the boundary between the low-grade and medium-grade division of metamorphism is defined by the band of various reaction-isograds:

1. cordierite-in (without almandine)/chlorite-out if muscovite is present
2. staurolite-in / Fe-rich chlorite-out if muscovite is present
3. staurolite-in / chloritoid-out

This implies

Formation of cordierite (without almandine)
Formation of staurolite
Disappearance of chloritoid
Disappearance of Fe-rich chlorite in the presence of muscovite.

In this connection, several comments are pertinent: A number of cases are known where staurolite and chloritoid coexist, probably within a narrow temperature range. Such a transition zone, if present, also marks the boundary between the two metamorphic grades. Similarly, cases are known where staurolite occurs together with primary chlorite and muscovite within a small temperature range. Thus, *e.g.*, Frey (1969) lists 57 different mineral parageneses of the staurolite zone of the Lukmanier area. Sixteen of these have staurolite and 4 out of the 16 still contain chloritoid, while only 2 out of the 16 rocks still contain chlorite + muscovite. It should be remarked that these seemingly aberrant samples occur on the staurolite reaction-isograd. It is well understood from the solid solution character of the involved minerals that under such circumstances chlorite + muscovite and/or chloritoid can still exist together with staurolite over a limited temperature range, while at slightly increased metamorphic degree, no chloritoid and no primary Fe-rich chlorite in contact with muscovite are observed. On the other hand, at the lower temperature side of the staurolite isograd, chloritoid and/or chlorite + muscovite (phengite) are invariably present in rocks of appropriate chemical composition.

The band of reaction-isograds "staurolite-in" and the reaction-isograd "cordierite-in" occurring in metapelites define the beginning of the medium-grade division of metamorphism for all rocks, whatever their composition, accompanying these metapelites. The experimental data relating to these reactions are similar and they seem to practically coincide in the field when pressure is low enough to permit the formation of iron-bearing cordierite. The range of T,P conditions is shown in Figure 7-2, together with data on other metamorphic grades.

If a reaction leading to the formation of a new mineral or mineral association has a univariant equilibrium, a locality on the respective reaction-isograd in the field represents the P,T conditions of one definite (although still unknown) point on the univariant equilibrium curve. However, when minerals of the solid solution type are involved—chlorite, chloritoid, staurolite, cordierite, etc.—the equilibrium of the reaction cannot be merely univariant; it must be at least divariant. This means that in the plane with P_{H_2O} and T as coordinates the univariant equilibrium line must be replaced by a divariant band. The width of such a divariant band may be quite small or rather broad. In the case of the various reactions forming staurolite or cordierite + biotite, the band of equilibrium conditions for the range of bulk compositions met with in common rock types is narrow. According to the experimental data given earlier, it is ± 10° to ± 20°C at a given H_2O pressure. Within this limit the various "staurolite-in" reaction-isograds and—in the same or different rocks—the "cordierite-in" (without almandine) reaction-isograds define the change from low-grade to medium-grade metamorphism.

Practical Determination of the Boundary

The practical determination of the boundary between low-grade and medium-grade only involves a few difficulties. In some rocks staurolite and cordierite cannot form because of an inappropriate bulk composition. Metapelites from the low-grade consisting mainly of phengite ("muscovite"), quartz, and chlorite ± chloritoid are best suited to show the change from low to medium-grade. The positive indicators, namely, the first appearance of cordierite and/or staurolite, are the most valuable. The negative indicator "chloritoid-out" can, of course, only be used if phyllites in the low-grade terrain have this mineral; it is then likely that rocks of the same bulk composition continue into the area of medium-grade metamorphism. The other negative indicator, "no chlorite touching muscovite," is potentially much more useful. This is because pelitic rocks, graywackes, and most igneous rocks of low metamorphic grade contain chlorite + muscovite. Thus, muscovite + chlorite is present in very many low-grade rocks. This assemblage disappears in medium-grade rocks if the chlorite is not too rich in Mg (see Chapter 14). Chlorite not in contact with muscovite may persist to considerably higher temperatures.

It must be remembered that the assemblage muscovite (phengite) + chlorite is not restricted to the low-grade division but is present in many rocks of the very-low-grade. However, the formation of zoisite or clinozoisite characterizes the beginning of low-grade metamorphism and the assemblage zoisite/clinozoisite + chlorite + muscovite is diagnostic for the complete temperature range of low-grade metamorphism.

Unfortunately, chlorite is easily formed as a secondary alteration product from biotite or hornblende, especially in rocks that have been subjected to postmetamorphic deformation and fracturing. As long as merely partial alteration of the host mineral into chlorite can be microscopically observed, the secondary nature of the chlorite can be established; however, there are many cases where the distinction between primary and secondary chlorite is difficult indeed. Therefore, there is the possibility that chlorite may be a postmetamorphic alteration product in medium- or high-grade metamorphic rocks. In such cases the presence of (secondary) chlorite in association with muscovite has, of course, no diagnostic meaning. Other criteria, such as the composition of plagioclase in basic rocks, the presence of diopside, forsterite, grossularite, etc., are indicators of medium- and high-grade metamorphism, although they do not determine the low-grade to medium-grade boundary. This will be discussed later.

The Change from Medium-Grade to High-Grade Metamorphism

Cordierite is stable throughout the temperature range of medium-grade metamorphism and even persists to higher temperatures. At low pressures staurolite, in the presence of quartz and muscovite, disappears before the upper limit of medium-grade is reached. On the other hand, at medium and high pressures, the stability of staurolite extends into high-grade metamorphism. Therefore, no reaction involving staurolite or cordierite can be used to characterize the upper limit of medium-grade metamorphism. Another criterion is required. It is again from worldwide petrographic observation that *we choose the breakdown of muscovite in the presence of quartz and plagioclase to define the change from medium-grade to high-grade metamorphism.*

This is a very typical change, observed at very low pressures in hornfelses of the innermost contact zone of gabbros, etc., as well as at all other pressures ranging from low to very high in schists and gneisses. It is of petrogenetic significance that the migmatic areas which all over the world are situated in the highest temperature terrain of regional metamorphism, as a rule, do not show primary muscovite in contact with quartz and plagioclase. Therefore, migmatites are properly assigned to the division of high-grade metamorphism.

Petrographically, the breakdown of muscovite in the presence of quartz and plagioclase is not only expressed by the negative criterion of the absence of muscovite; there are positive criteria as well. This is very fortunate because the postmetamorphic formation of muscovite as a secondary alteration product may otherwise cause complication. Therefore, it is of great diagnostic value that very typical new mineral parageneses are formed by the breakdown of muscovite:

K feldspar + Al_2SiO_5 mineral (andalusite, sillimanite, or kyanite)
K feldspar + cordierite
K feldspar + almandine-rich garnet

High-grade metamorphism is thus characterized by the isograd "K feldspar + Al_2SiO_5." As in the case of the "staurolite-in" isograd band, the isograd band abbreviated as "K feldspar + Al_2SiO_5" stands for a number of adjacent reaction-isograds.

At high-grade a number of reactions have to be taken into account that lead to the disappearance of muscovite in the presence of quartz. They depend on (a) pressure and (b) absence of or very low partial pressure of H_2O; this is significant in granulite terrains.

The relationships are more complicated than expected. Experimental investigation of the relevant four-component system K_2O-Al_2O_3-SiO_2-H_2O by Storre and Karotke (1972) has shown that at medium and high H_2O pressure, the assemblage muscovite + quartz is stable up to a much higher temperature than previously thought. However, this is valid only *if plagioclase is absent* (as in the above system).

Figure 7-3 shows the complete set of univariant equilibrium curves of reactions (1) to (6), which all meet in an invariant point I_1:

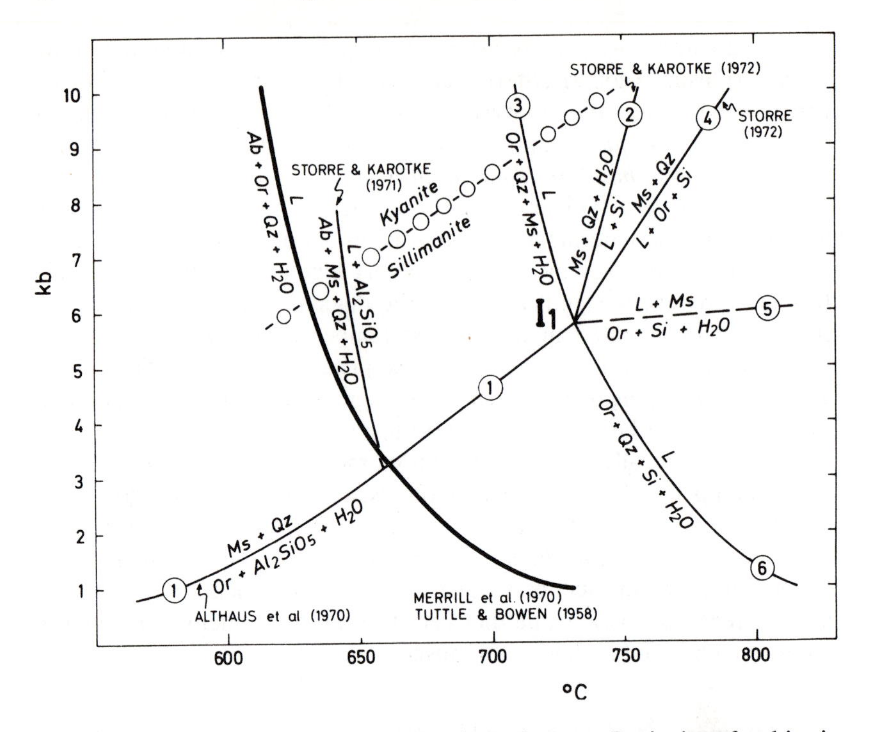

Fig. 7-3 Reactions involving muscovite and quartz. Beginning of melting in the system albite-K feldspar-quartz-water also shown by heavy line. *Or* stands for K feldspar. [After Storre (1972) and Storre and Karotke (1971, 1972)]. Extension of some of these reactions to 30 kb has been studied by Huang and Wyllie (1974).

muscovite + quartz = K feldspar + Al_2SiO_5 + H_2O (1)

muscovite + quartz + H_2O = liquid + sillimanite (2)

muscovite + quartz + K feldspar + H_2O = liquid[6] (3)

muscovite + quartz = liquid + K feldspar + sillimanite (4)

(Note: no gaseous H_2O is present here.)

K feldspar + Al_2SiO_5 + H_2O = liquid + muscovite (5)

(Note: no quartz is present here.)

K feldspar + quartz + sillimanite + H_2O = liquid[6] (6)

(Note: no muscovite is present here.)

Figure 7-3 shows that the maximum stability of the assemblage muscovite + quartz is limited by reactions (1) and (2); if no H_2O is present,

[6]The equilibrium conditions of reactions (3) and (6) are so close to those of the eutectic conditions in the system K feldspar + quartz + water (Lambert *et al.*, 1969) that graphical distinction cannot be made.

reaction (4) takes place instead of (2). Thus, muscovite + quartz may persist to very high temperatures of metamorphism. Only rare cases of such occurrences are known; they are muscovite-bearing quartzites devoid of plagioclase.

If H_2O pressure is smaller than ca. 3.5 kb, the breakdown of muscovite in the presence of quartz takes place according to reaction (1); relevant temperatures are 580°C at 1 kb and 660°C at 3 kb. *Reaction (1) defines the beginning of high-grade metamorphism as long as H_2O pressure does not exceed 3.5 kb.* At higher pressures, however, muscovite commonly does not disappear by any of these reactions. It is because of the common additional presence of plagioclase that muscovite, accompanied by quartz, plagioclase ± biotite ± K feldspar, is completely dissolved in a melt formed by anatexis of gneiss (see Chapter 18 on anatexis); this takes place at minimal temperatures of 660° at 3.5 kb and 615°C at 10 kb, *i.e.*, at temperatures well below those of reactions (2) to (6), as well as of (1) when pressures exceed 3.5 kb. During the process of anatexis, muscovite furnishes Al_2SiO_5 (sillimanite or kyanite) and K feldspar *component,* which, together with the *components* of previously crystalline quartz and plagioclase (and K feldspar, if present), constitute the anatectic melt. Muscovite disappears in the presence of quartz and plagioclase; therefore *anatexis in gneiss defines the beginning of high-grade metamorphism if H_2O pressures are larger than about 3.5 kb.*

The process of anatexis in gneiss can be given as follows:

{muscovite + quartz + plagioclase + H_2O} =
Anatectic melt consisting of the components of K feldspar, Ab-richer plagioclase, and quartz
Plus somewhat An-richer plagioclase *or* quartz, depending on their amounts previously present in the gneiss
Plus Al_2SiO_5, plus H_2O dissolved in the melt

Biotite, present in most rocks, also participates in anatectic melting, yielding K feldspar component for the melt and additional *cordierite* and/or *almandine-garnet.*

As is generally the case in prograde metamorphic reactions, the above reaction is not reversed during subsequent crystallization of the anatectic melt. It crystallizes to an assemblage of K feldspar, quartz, and plagioclase, but muscovite is not formed again. This may be due to the local separation of the remaining solids, including Al_2SiO_5, from the melt, or to a process allowing water, which is liberated during crystallization, to escape from the small volumes of crystallizing melt.

This disappearance of muscovite takes place according to the following reactions in which albite stands for plagioclase:

muscovite + quartz + albite + H_2O = liquid + Al_2SiO_5 (a)
muscovite + quartz + albite + K feldspar + H_2O = liquid (b)

These two reactions can proceed only at H_2O pressures greater than ca. 3.5 kb. The equilibrium conditions for reaction (a), determined by Storre and Karotke (1971), are also shown in Figure 7-3; those for reaction (b) are, within the limits of experimental determination, the same as the conditions for the beginning of melting in the system albite + K feldspar + quartz + H_2O (Tuttle and Bowen 1958; Merrill *et al.* 1970). This curve, shown as a heavy line in Figure 7-3, determines at water pressures greater than 3.5 kb the upper stability of muscovite in the presence of plagioclase, quartz, K feldspar, and H_2O. When K feldspar is lacking, the curve of reaction (a) applies; it lies at only slightly higher temperatures. When the composition of the plagioclase is oligoclase instead of albite, temperatures for the beginning of melting are 10° to 15°C higher.

The resulting band of conditions for the beginning of anatectic melting in muscovite-bearing gneisses is taken to represent the beginning of high-grade metamorphism whenever H_2O pressure is greater than about 3.5 kb. However, at lower pressures reaction curve (1) in Figure 7-3 (broadened to a narrow band) defines the boundary between medium and high grade, i.e., reaction muscovite + quartz = K feldspar + Al_2SiO_5 + H_2O. Here no melt is formed, and plagioclase is not a necessary reactant. However, plagioclase as well as biotite are commonly present, and then they will take part in the reaction.

The reactions to consider in the pressure range below 3.5 kb are the following cases and combinations thereof:

1 muscovite + 1 quartz = 1 K feldspar + 1 Al_2SiO_5 + H_2O (a)

6 muscovite + 2 biotite + 15 quartz
= 8 K feldspar + 3 cordierite + 8 H_2O (b)

At higher pressures than (b), or when biotite is very rich in Fe,[7] reaction (c) takes place

2 muscovite + 2 biotite + 1 quartz
= 4 K feldspar + 1 almandine + 4 Al_2SiO_5 + 4 H_2O (c)

{muscovite + quartz + Na-rich plagioclase} = {Na-bearing alkali feldspar + Na-poorer plagioclase + Al_2SiO_1 + H_2O} (d)

[7]This is to be expected from the experiments by Hsu and Burnham (1969), who synthesized almandine garnet at 2 kb when Fe/(Mg + FE) was larger than about 0.8.

In most common gneisses, combinations of reactions (a) or (b) with (d) have occurred, as well those of (a) with (b). Reaction (a) and reactions (b) + (d) + (a) have been experimentally investigated. Reaction

$$\{\text{muscovite} + \text{quartz}\} = \{\text{K feldspar} + Al_2SiO_5 + H_2O\} \qquad \text{(a)}$$

has been studied by several authors. The data obtained by Althaus *et al.* (1970) are given in Table 7-1. In comparison with the tabulated data, those given by Evans (1965) and by Chatterjee and Johannes (1974) are lower by 20°C, while the data obtained by Day (1973) are higher by 10–20°C. Therefore, the values given in Table 7-1 for reaction (a) can be taken as mean values.

Reaction (b) combined with (d) and (a) has been studied by Haack (unpublished). In order to use compositions that really occur in nature two different illite-rich clays were experimentally metamorphosed in the range of 500 to 4000 bars to the assemblage quartz-muscovite-biotite-cordierite-plagioclase ± Al_2SiO_5. The main chemical differences between the two samples are that the ratios of Ca/Na and K/Na are somewhat higher in No. 1 than in No. 2.

Table 7-1 compiles the data for reaction (a) and reaction (b) combined with (d) and (a), No. 1 and No. 2:

Table 7-1 Breakdown of muscovite in the presence of quartz (temperatures ± 10°C).

H_2O Pressure, bars	Reaction (a) ms + qtz, °C	Reactions (b) + (d) + (a), °C No. 1	No.2
1000	580	Same	560
2000	620	as	600
3000	655	reaction	635
4000	680	(a)	670

It is somewhat unexpected that reaction (a) in the simple system takes place at practically the same conditions as No. 1 in the more complex system. On the other hand, this is very fortunate; the small range of values of 20° to 10°C between No. 1 and No. 2 will probably be representative for many natural cases. In Figure 7-2, showing the boundaries of all four metamorphic grades, the limit between medium and high grade is outlined first by the band along reaction (1) muscovite + quartz = K feldspar + Al_2SiO_5 + H_2O, and second by the stippled band for the beginning of anatexis in gneisses. The limiting band has a kink; from about 3.5 to 10 kb it runs between 680° and 620°C, *i.e.*, in the range 650°

± 30°C, and from 3.5 to 2 kb, again between 680° and 620°C, while at 1 kb the temperature limit is at 580° ± 10°C.

In high-grade metamorphic rocks no primary muscovite exists with quartz and plagioclase. However, in the absence of plagioclase, muscovite + quartz are stable at high-grade metamorphism when pressures exceed 4 kb.

Granulite—High-Grade; Regional Hypersthene Zone

The different reactions leading to the disappearance of muscovite in the presence of quartz ± feldspars at the beginning of high-grade metamorphism take place under water pressure [except reaction (4) of Figure 7-3]. The data in Figure 7-3 are valid for the condition that H_2O pressure equals total pressure. It is now well known, however, that certain rocks have been subjected to the special condition of water pressure being considerably less than total pressure, such as in the formation of eclogites and granulites. While eclogites may originate within an extremely large temperature range, granulites occur only in high-grade terrains. All granulites lack primary muscovite. This, however, is neither due to anatexis nor to a temperature exceeding the equilibrium conditions of the other reactions given in Figure 7-3. The absence of muscovite in granulites must be discussed under the special condition of water pressure being very low or absent.

The point to be stressed here is that two different cases of high-grade metamorphism must be distinguished:

1. High-grade with water pressure approximately equal to load pressure
2. High-grade with water pressure very small or nonexistent and load pressure medium to high

Granulites form under condition (2). Since the most diagnostic mineral in granulites is hypersthene, it is suggested to designate *granulite high-grade metamorphism* as such or rather as *"regional hypersthene zone"* (see Chapter 16).

Pressure Divisions of the Metamorphic Grades

The *P, T* fields of various metamorphic grades are bounded by reactions which are predominantly temperature dependent and only slightly or moderately influenced by pressure. All four metamorphic grades as defined here and shown in Figure 7-2 comprise, within a certain temperature range, the total pressure range existing within the earth's crust. This pressure range can be subdivided by means of pressure-sensitive mineral reactions.

At very-low-grade the zeolite *laumontite*[8] is stable only at H_2O pressures lower than about 3 kb; at higher pressures *lawsonite* is the equivalent Ca,Al silicate. At pressures estimated as 5 kb at 200°C and 7 kb at 350°C, *glaucophane* will form, which, together with lawsonite and/or pumpellyite, constitute a prominent assemblage in mafic metamorphic rocks of very-low grade. At pressures still higher by approximately 2 kb, *jadeitic pyroxene* is formed instead of or together with albite in metamorphic graywackes at very-low grade and part of the low-grade metamorphic conditions.

These changes are outlined in Figure 7-4 by the reaction line "laumontite-out/lawsonite + quartz-in," by the estimated first appearance of glaucophane in common mafic rocks, and by the band of probable conditions for reactions involving the transformation of albite into jadeitic pyroxene + quartz (details are given later).

[8]At somewhat higher temperatures, wairakite replaces laumontite.

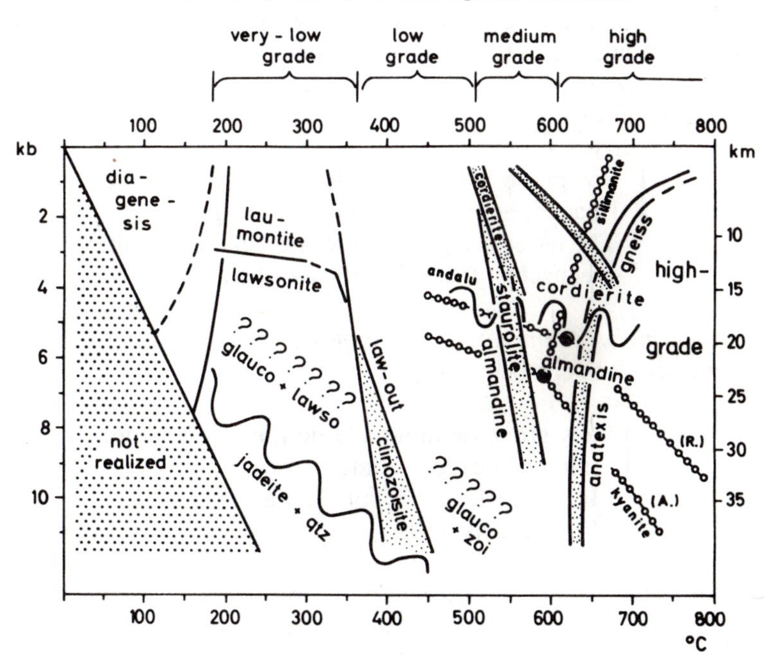

Fig. 7-4 Metamorphic grades and their pressure divisions indicated by special minerals.

This suggests the following *four pressure divisions of very-low-grade metamorphism:*

P ↓
[laumontite]-very-low-grade[8]
[lawsonite]-very-low-grade
[glaucophane + lawsonite]-very-low-grade
[jadeite + quartz]-very-low-grade

More detailed divisions can be made which later will be discussed in connection with the various metamorphic zones of very-low-grade metamorphism, but these four divisions suffice as a useful grid for fieldwork.

It is important to note that *prehnite,* with or without *pumpellyite* (which is commonly found in very-low-grade rocks formed under low pressures, *i.e.,* the prehnite-pumpellyite facies of Coombs), is by no means restricted to low pressure and therefore must not be used as pressure indicator (see Chapter 12).

In low-grade metamorphism both the low and high temperature ranges require different pressure indicators. Glaucophane remains stable in the low temperature range of low-grade metamorphism, but at these conditions is accompanied by zoisite/clinozoisite, not by lawsonite (!). This association indicates low-grade and high pressures. Still higher pressures are needed to form jadeitic pyroxene + quartz. Thus, *within the lower temperature range of low-grade metamorphism,* two divisions may be distinguished:

P ↓
low-grade (without addition, no glaucophane)
[glaucophane + clinozoisite]-low-grade
The latter is equivalent to the "glaucophanitic greenschist facies" of the early editions.

Within the *higher temperature range of low-grade* metamorphism, the following divisions can be made:

P ↓
low-grade (without addition, no almandine)
[almandine]-low-grade
If andalusite or kyanite is present, this may be stated as well.

The pressure that must be exceeded to form almandine depends very much on rock composition, especially on the ratio $Fe^{2+}/(Mg + Fe^{2+})$; the lower this value the greater the pressure required. It is estimated from experimental results (discussed later) that in common rocks the pressure must exceed about 4 kb when almandine is formed at temperatures of about 500°C.

The [almandine]-low-grade division has its continuation in medium- and high-grade metamorphism.

Within the temperature range of medium-grade, the following divisions can be distinguished with increasing pressure (see Figure 7-4):

$P \downarrow$ [cordierite]-medium-grade[9]
[almandine]-medium-grade[9]

The species of Al_2SiO_5 (andalusite, kyanite, or sillimanite) may serve as pressure indicators. Andalusite, the low pressure polymorph of Al_2SiO_5, occurs in areas of [cordierite]-medium-grade[10] and as well in some area of [almandine]-medium-grade. Only when pressures are slightly higher is the boundary andalusite/kyanite or the boundary andalusite/sillimanite surpassed, and kyanite or sillimanite, respectively, is the typical Al_2SiO_5 species. Thus, the following rather elaborate divisions within the medium-grade can be made; their approximate areas in the P,T planes are shown in Figure 7-4:

[cordierite + andalusite]-medium-grade
[almandine + andalusite]-medium-grade
[almandine + sillimanite]-medium-grade
[almandine + kyanite]-medium-grade

Judging from the results of world wide mapping, andalusite-bearing regional metamorphic terrains are much more widespread than the regions which have been metamorphosed under higher pressures [Zwart, quoted by Miyashiro (1972)].

At high-grade the same approximate divisions can be made as at the medium-grade, but in addition a third, intermediate division must be taken into account (see Chapter 14). Thus, the following three high-grade divisions are distinguished:

$P \downarrow$ [cordierite]-high-grade
[cordierite-almandine]-high-grade
[almandine]-high-grade

As in medium-grade, more detailed subdivisions can be made in high-grade with the help of the occurrence of andalusite, kyanite, or sillimanite, respectively, but precautions have to be taken. This will be pointed out briefly on the following pages.

---

[9]Rarely, an intermediate division, *i.e.,* [cordierite-almandine]-medium-grade, can be recognized; see Chapter 14.

[10]Very rare cases have been described where kyanite (instead of andalusite) occurs together with cordierite, *i.e.,* with Mg-rich cordierite (see Wenk, 1968). Special chemical composition may be the reason for this exceptional assemblage.

Summary of the coarse grid of metamorphic grades and their predominantly pressure dependent divisions.

[laumonite]-or [wairakite]-very-low-grade	low-grade	[cordierite]-medium-grade	[cordierite]-high-grade
[lawsonite]-very-low-grade	[almandine]-low-grade	[cordierite-almandine]-medium-grade	[cordierite-almandine]-high-grade
[glaucophane-lawsonite]-very-low-grade	[glaucophane-clinozoisite]-low-grade	[almandine]-medium-grade	[almandine]-high-grade
[jadeite-quartz]-very-low-grade			

Problems with the Al_2SiO_5 Species

The species of Al_2SiO_5 can be used as relative pressure indicators, but evaluation of petrographic observations requires special care because of some problems concerning their stability relations and reaction kinetics.

Let us consider first the P,T data of the stability fields of andalusite, kyanite, and sillimanite: Much work has been done to determine the three phase boundaries andalusite/kyanite, sillimanite/kyanite, and andalusite/sillimanite, and the resulting triple point where the three curves meet. Contrary to experimental investigation of other reactions, the results on Al_2SiO_5 are almost as variable as the number of laboratories working on the problem. It is not yet certain where the true phase boundaries lie.

Brown and Fyfe (1971) give the reasons for these difficulties: "Near equilibrium the ΔS of the andalusite/sillimanite reaction is only about 0.3 cal mol^{-1} deg^{-1}. This means that if we wish to know the phase boundary with a certainty of at least ± 50°C we must also use procedures which will give a free energy with an accuracy of ± 15 cal mol^{-1}. From kyanite reactions where ΔS is about 2 cal mol^{-1} deg^{-1}, again ΔG must be known within limits of at least ± 100 cal mol^{-1}. These are trivial quantities when it is noted that the lattice energies of the three polymorphs are near 5×10^5 cal mol^{-1}. It is clear when these requirements are considered that small effects, strain energy, surface energy, impurity factors, will be critical in such a system." According to these considerations, finely ground powders of the minerals should not be used in the experiments (as is conventionally done); large crystals should be used instead to determine their solubilities at different P, T conditions.

This theoretically most promising approach has been carried out by

Brown and Fyfe (1971), but their results cannot be reconciled with many geologic and petrogenetic considerations. The authors found that the andalusite-kyanite phase boundary lies at 1 kb and 400° ± 40°C, at 2 kb and 470° ± 40°C, and at 3 kb and 535° ± 40°C. If kyanite would form at such conditions, especially at such low pressures within low- and medium-grade metamorphism, it should be present in rather shallow contact metamorphic aureoles; furthermore, it should commonly accompany cordierite in medium-grade rocks; this, however, is not the case.

Two sets of results based on conventional investigations of the Al_2SiO_5 minerals are consistent with the grid of other metamorphic reactions found in nature and determined in the laboratories: They were determined by Richardson *et al.* (1968, 1969) and by Althaus (1967, 1969a,b). The two sets of the boundaries kyanite/sillimanite and kyanite/andalusite are shown in Figure 7-4, while only the andalusite/sillimanite boundary of Althaus is given.

The andalusite/sillimanite boundary as published by Althaus takes into account Weill's (1966) data at low pressures. Only this position of the andalusite/sillimanite boundary is in agreement with the common observation of the andalusite to sillimanite transition in high-grade rocks formed at low and very low pressures. On the other hand, if the boundary as given by Richardson *et al.* were valid, sillimanite could not be present in the inner contact metamorphic zone of high-level plutons. Surprisingly enough, the position of the andalusite/sillimanite boundary as given by Richardson *et al.* is most commonly cited in the literature, although its incompatibility with observations in nature was pointed out long ago by Winkler (1974, p. 91).

Instead of accepting either one of the two sets of phase boundaries, *i.e.*, the andalusite/kyanite and the kyanite/sillimanite boundary of Figure 7-4, it seems preferable to choose the band between these lines as representing conditions for the corresponding phase transformation. This is supported by Althaus (1969a,b), who found that as little as 0.5% Fe_2O_3 substituting for Al_2O_3 in sillimanite markedly raises the pressure necessary to transform it to kyanite, in comparison to Fe-free sillimanite. Further evidence bearing on this point has been obtained by Karotke and Storre (1971). The three minerals may, therefore, coexist within a *P,T* field rather than at a triple point; the center of the area is approximately at 600°C and 6 kb.

The triple point proposed by Holdaway (1971) is situated at about 500°C and 4 kb. Consequently, the stability field of andalusite is reduced in comparison to the phase diagrams given by the other authors. On the other hand, sillimanite is stable within a much larger *P,T* area. This would mean that sillimanite should form even before the beginning of medium-grade at about 4 kb and also within the lower temperature range of medium-grade metamorphism at pressures of 3–5 kb. However, observations in nature do not support these inferences. Furthermore, Holdaway's andalusite/sillimanite boundary at low pressures, like the one given by Richardson *et al.* (1969), is inconsistent with observations in high-level contact metamorphic aureoles.

Attainment of equilibrium in metamorphism has been established in very many cases. And equilibrium is generally assumed among Al_2SiO_5 minerals as well. Indeed, it is well known that with increasing temperature, andalusite-bearing rocks are followed by sillimanite-bearing rocks in the field, and the analogous sequence from kyanite- to sillimanite-bearing rocks is also well known. However, in many instances andalusite or kyanite or both are found together with sillimanite, often of the fibrolitic type (see, *e.g.*, Pitcher, 1965). In some cases, this may be indicative of the triple-point area or of a phase boundary band, but in other terrains the possibility of disequilibrium must be considered. *Persistence* of andalusite and kyanite into the stability fields of each other and of sillimanite seems to be possible. It may be assumed that andalusite, kyanite, and sillimanite each form within their own stability fields first. Suppose now that *P,T* conditions change, as in polyphase metamorphism; new mineral reactions may take place, again forming Al_2SiO_5. If such new reactions occur, that Al_2SiO_5 species is formed which is stable under the new *P,T* condition. It may well be of a different kind than that formed earlier. This earlier formed, different Al_2SiO_5 species may, after the change in *P,T* conditions, persist metastably outside its stability field, without signs of transformation, together with the other now stable Al_2SiO_5 mineral. It is suggested that this point should be seriously taken into account when making petrogenetic interpretations.

This situation may be briefly illustrated: In the eastern and northeastern part of the Central Alps in Switzerland, kyanite (at higher temperatures with sillimanite) is present, although the other metamorphic mineral assemblages indicate a pressure of less than 2 kb during the latest phase of metamorphism. In order to explain this, Winkler (1970) has suggested that in a first phase of metamorphism the formation of kyanite took place at rather high pressure, but during the same metamorphic period, quick tilting, uplift, and erosion lowered the pressure without significant loss in temperature and coexisting fluid. Therefore, all mineral parageneses that had formed in the first phase reacted again in the second lower pressure phase to adjust themselves to the new *P,T* conditions. However, kyanite did not invert to andalusite (or, in the higher temperature area, to sillimanite) because sluggish reaction rates prevented the change to the stable modification. The same may be observed in polymetamorphic areas which have undergone two or more metamorphic events. But metastable persistence seems to occur only under special circumstances which are not yet known, because complete inversion of andalusite to kyanite, and vice versa, is so well known from other areas that Chinner (1966) regards this as generally true.

Although a knowledge of the distribution of Al_2SiO_5 minerals in the field is valuable, petrogenetic conclusions must be drawn with great care, taking into account the possible metastable persistence of the Al_2SiO_5 minerals. They may appreciably limit the petrogenetic significance of the Al_2SiO_5 minerals. Interpretation of phase relations merely in terms of experimental phase boundaries, including the P,T conditions of the triple point, will be misleading. The observed Al_2SiO_5 minerals must be considered within the context of other petrogenetic data.

References

Althaus, E. 1967. *Contr. Mineral. Petrol.* **16:** 29–44.

——— 1969a. *Neues Jahrb. Mineral. Abhand.* **111:** 74–161.

——— 1969b. *Am. J. Sci.* **267:** 273–277.

———, Nitsch, K. H., Karotke, E., and Winkler, H. G. F. 1970 *Neues Jahrb. Mineral. Monatsh.* **1970:** 325–336.

Barrow, G. 1893. *Quart. J. Geol. Soc.* **49:** 330–358.

——— 1912. *Proc. Geol. Assoc.* **23:** 268–284.

Brown, E. H. and Fyfe, W. S. 1971. *Contr. Mineral. Petrol.* **33:** 227–231.

Carmichael, D. M. 1970. *J. Petrol.* **11:** 147–181.

Chatterjee, N. D. and Johannes, W. 1974. *Contr. Mineral. Petrol.* **48:** 89–114.

Chinner, G. A. 1966. *Earth Sci. Rev.* **2:** 111–126.

Day, H. W. 1973. *Am. Mineral.* **58:** 255–262.

Deer, W. A., Howie, R. A., and Zussman, J. 1962. *Rock-forming Minerals.* Vol. 1, pp. 183ff. Longmans, London.

Eskola, P. 1920/21. *Norsk. Geol. Tidsskr.* **6:** 143–194.

——— 1939. *In* Barth, Correns, and Eskola. *Die Entstehung der Gesteine.* Springer-Verlag, Berlin.

Evans, B. W. 1965. *Am. J. Sci.* **263:** 647–667.

Frey, M. 1969. *Beitr. Geol. Karte Schweiz Neue Folge 137.*

——— and Niggli, E. 1971. *Schweiz. Mineral. Petrog. Mitt.* **51:** 229–234.

Ganguly, J. 1969. *Am. J. Sci.* **267:** 910–944.

Green, D. H., Lochwood, J. P., and Kiss, E. 1968. *Am. Mineral.* **53:** 1320–1335.

Hirschberg, A., and Winkler, H. G. F. 1968. *Contr. Mineral Petrol.* **18:** 17–42.

Holdaway, M. J. 1971. *Am. J. Sci.* **271:**97–131.

Hoschek, G. 1967. *Contr. Mineral. Petrol.* **14:** 123–162.

——— 1969 *Contr. Mineral Petrol.* **22:** 208–232.

Hsu, L. C. and Burham, C. W. 1969. *Geol. Soc. Am. Bull.* **80:** 2393–2408.

Huang, W. L., and Wyllie, P. J. 1974. *Am. J. Sci.* **274:** 378–395.

Karotke, E. and Storre, B. 1971. *Fortschr. Mineral.* **49**, Beiheft 1: 24–27.

Kubler, B. 1967. *Bull. Centre Recherches Pau-SNPA* **1:** 259–278.

——— 1968. *Bull. Centre Recherches Pau-SNPA* **2:** 385–397.

Lambert, I. B., Robertson, J. K., and Wyllie, P. J. 1969. *Am. J. Sci.* **267:** 608–626.

Merrill, R. B., Robertson, J. K., and Wyllie, P. J. 1970. *J. Geol.* **78:** 558–569.

Miyashiro, A. 1972. *Tectonophysics* **13:** 141–159.

Newton, R. C. and Kennedy, G. D. 1963. *J. Geophysics. Res.* **68:** 2967–2983.

Nitsch, K.-H. 1971. *Contr. Mineral. Petrol.* **30:** 240–260.

——— 1972. *Contr. Mineral Petrol.* **34:** 116–134.

——— 1974. *Fortschr. Mineral.* **51.** (Abstract), Beiheft 1, 34.

Pitcher, W. S. 1965. *In* W. S. Pitcher and D. Flinn, eds. *Controls of Metamorphism.* pp. 327–341. Oliver and Boyd, Edinburgh and London.

Richardson, S. W. 1968. *J. Petrol.* **9:** 467–488.

———, Bell, P. M., and Gilbert, M. C. 1968. *Am. J. Sci.* **266:** 513–541.

Richardson, S. W., Gilbert, M. D., and Bell, P. M. 1969. *Am. J. Sci.* **267:** 259–272.

Seifert F. 1970. *J. Petrol.* **11:** 73–99.

Storre, B. 1972. *Contr. Mineral. Petrol.* **37:** 87–89.

——— and Karotke, E., 1971. *Fortschr. Mineral.* **49:** 56–58.

——— 1972. *Contr. Mineral. Petrol.* **37:** 343–345.

Streck, A. 1969. *Kaledonische Metamorphose.* NE-Gronland, Habilitationsschrift, Basel.

——— 1976. *Schweiz. Mineral. Petrog. Mitt.* **56:** 269–292.

Thompson, J. B. 1957. *Am. Mineral.* **42:** 842–858.

——— and Norton, S. A. 1968. Paleozoic regional metamorphism in New England and adjacent areas. In E-An Zen *et al.,* eds. *Studies of Appalachian Geology.* Interscience Publisher (John Wiley & Sons), New York.

Tilley, C. E. 1924. *Geol. Mag.* **61:** 167–171.

Turner, F. J. 1968. *Metamorphic Petrology.* McGraw-Hill Book Company, New York.

——— and Verhoogen, J. 1960. *Igneous and Metamorphic Petrology.* 2nd edit., McGraw-Hill Book Company, New York.

Tuttle, O. F., and Bowen. N. L. 1958. *Geol. Soc. Am. Memoir No. 74.*

Weber, K. 1972. *Neues Jarhb. Mineral. Monatsh.*: 267–276.

Weill, D. F. 1966. *Geochim. Cosmochim. Acta* **30:** 223–237.

Wenk, E. 1968. *Schweiz. Mineral. Petrog. Mitt.* **48:** 455–457.

——— and Keller, F. 1969. *Schweiz. Mineral. Petrol. Mitt.* **49:** 157–198.

Winkler, H. G. F. 1965. *Petrogenesis of Metamorphic Rocks.* Springer-Verlag, New York.

——— 1967. *Petrogenesis of Metamorphic Rocks.* 2nd edit. Springer-Verlag, New York.

——— 1970. *Fortschr. Mineral.* **47:** 84–105.

——— 1974. *Petrogenesis of Metamorphic Rocks.* 3rd edit. Springer-Verlag, New York.

Chapter 8

General Characteristics of Metamorphic Terrains

Metamorphic Zones in Contact Aureoles

Contact metamorphism is due to a temperature rise in rocks adjacent to magmatic intrusions of local extent which penetrate relatively shallow and cold regions of the crust. The majority of magmatic intrusions are of granitic composition. The most frequent depth (distance from the earth's surface at the time of solidification) of granitic intrusions was estimated by Schneiderhöhn (1961) to be 3 to 8 km, corresponding to a load pressure of 800 to 2100 bars. There are, of course, intrusions which solidified at greater or shallower depth; a depth of 1 km corresponds to a load pressure of 250 bars.

When a magma intrudes into colder regions, the adjacent rocks are heated. If the heat content of the intruded magma is high, *i.e.,* if the volume of the intrusion is not too small, there will be a temperature rise in the bordering country rock which lasts long enough to cause mineral reactions. The rocks adjacent to small dikes and sills are not metamorphosed (only baked), whereas larger bodies of plutonic rocks give rise to a contact aureole of metamorphic rocks. Several zones of increasing temperature are recognized in contact aureoles.

Contact metamorphic aureoles are best observed when unmetamorphosed, very-low-grade, or low-grade metamorphic rocks are overprinted by contact metamorphism. On the other hand, rocks previously subjected to medium-grade or high-grade metamorphism commonly will not show signs of contact metamorphic overprinting because the mineral assemblages are stable (or persistent) at contact metamorphic conditions.

An example of shallow contact metamorphism investigated by Melson (1966) is shown in Figure 8-1. A granite has intruded into previously unmetamorphosed country rock of sedimentary origin. Contact meta-

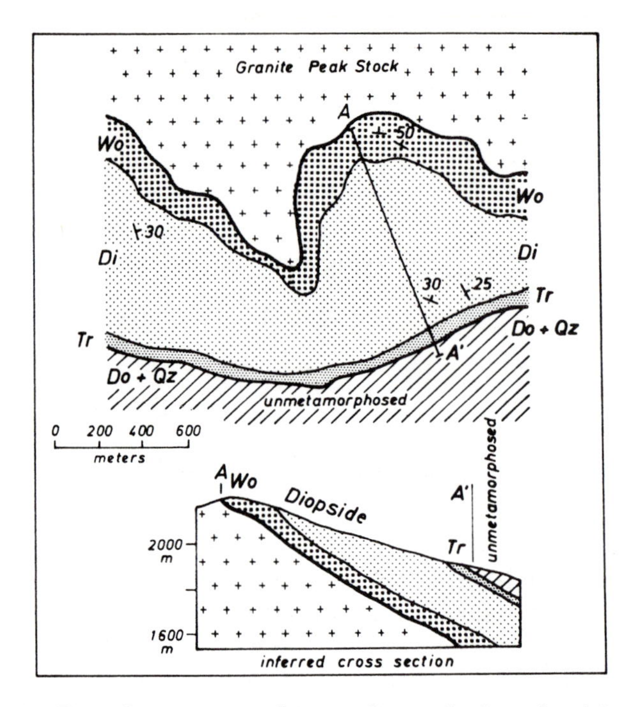

Fig. 8-1 Part of contact aureole around a granite intrusion (after Melson, 1966).

morphism has taken place and various metamorphic zones have been mapped based on transformations in dolomitic calcite-quartz beds. Progressing from the unmetamorphosed country rock toward the granite, the first metamorphic zone, here designated as the tremolite zone, is encountered where dolomite + quartz react to form tremolite + calcite. This is, in the case of the investigated terrain, only a narrow zone, obviously corresponding to only a small range of temperature. The tremolite zone is succeeded by a zone marked by the formation of diopside from the reaction of tremolite + calcite + quartz where the three minerals have been in contact. (The various reactions possible in siliceous dolomites will be fully treated in a later chapter.) The diopside zone is much broader than the tremolite zone. The map, however, gives an exaggerated impression because the intrusion dips under the investigated terrain as can be seen in the cross section of Figure 8-1. The diopside zone in which calcite + quartz also coexist is succeeded by the highest temperature inner contact zone where the latter two minerals have reacted to form wollastonite; diopside persists in this zone. Thus, the progression of the metamorphic zones, *i.e.,* the tremolite, diospide, and wollastonite zones in this case, indicates increasing temperature from the country rock to the granite contact. In rocks of different chemical composition, zones will, of course, be characterized by different mineralogical changes. Thus, shales that have been metamorphosed within the aureole of a shallow pluton will commonly exhibit a progressive sequence of zones, which are mineralogically characterized as follows:

1. White mica (mostly phengite) + chlorite + quartz
2. White mica + *biotite* + *chlorite* + quartz ± andalusite
3. *Cordierite* ± *staurolite* + *muscovite* + biotite + quartz ± andalusite
4. *Cordierite* + *K feldspar* + quartz + biotite ± andalusite or/and sillimanite

The diagnostic mineral assemblages are italic. Albite in No. 1 and 2 and anorthite-containing plagioclase in No. 3 and 4 will generally also be present.

If the country rock has previously been metamorphosed at low-grade, the sequence will not commence with No. 1 but with No. 2 or even 3. Rocks No. 1 and 2 commonly occur as spotted slates or schists with chlorite and mica constituting the spots. Rocks No. 3 and 4 are commonly nonschistose, fine-grained rocks of granoblastic fabric; they are so-called hornfelses. Rock No. 3 represents medium-grade metamorphism, while No. 4 is typical of high-grade because of the absence of primary muscovite in the stated assemblage.

The zone in direct contact with the intrusion is, of course, marked by the greatest rise in temperature. Away from the intrusion, there are zones characterized by lower temperatures. The rock volume is heated up and, therefore, the extent of the various zones of the contact aureole depends on the heat content, *i.e.*, on the size of the intrusion. The temperature rise of the country rock depends on the temperature of the intruded magma. The temperature of granitic magma intrusions is generally 700° to 800°C; of syenitic magma, about 900°C; and of gabbroic magma, about 1200°C; thus, gabbroic magma is considerably hotter than granitic magma.

On the basis of reasonable assumptions approximating natural conditions, Jaeger (1957) has calculated the temperatures in rocks adjacent to an intrusion having the shape of an infinite sheet of thickness D, *i.e.*, having the shape of large dikes or sills. As a first approximation, elongated plutonic intrusions may be treated as vertical dikes. Jaeger's calculations take into account the heat liberated on crystallization of the magma. They are based on the assumption that the heating of the country rock by the intrusion is due entirely to conductivity and that the heating by a transfer of volatile constituents can be neglected. Some of the results are summarized below.

A deep-seated gabbroic intrusion which is emplaced at a liquidus temperature of 1200°C and solidus temperature of 1050°C causes, at the contact, a rise in temperature of 727°C + T_c, where T_c is the temperature of the country rock prior to intrusion of the magma. The temperature rise is, of course, lower in rocks some distance away from the contact. If a thickness D = 1000 m be assumed, a temperature of 625°C + T_c, which almost equals the maximum possible temperature, is attained at the distance equal to 1/10 D = 100 m from the contact after several thousand years. This temperature is maintained for a period of about 10,000 years and then lowers gradually. At a distance equal to 2/10 D = 200 m, a temperature of 550°C + T_c is attained and is effective during the same length of time (though it takes longer to heat this more remote country rock). At a distance equal to 1/2 D = 500 m, the highest temperature reached is only 410°C + T_c if the temperature of the intrusion is 1200°C.

In general, it is concluded that at the immediate contact the temperature of the country rock is somewhat greater than 60% of the intrusion temperature + T_c and that at a distance equal to 1/10 of the thickness of the intrusion, the temperature of the country rock is about 50% of the intrusion temperature + T_c. At greater distances, the temperatures are lower, so that at a distance equal to 1/2 of the thickness of the intrusion the temperature of the country rock is increased by only 1/3 of the intrusion temperature. Considering magmas of different temperatures intruded at a depth of 5 to 6 km where the temperature prior to intrusion (T_c) was 150°C, the temperatures of the country rock as given in Table 8-1 may be calculated. These temperatures remain effective for an

Table 8-1

Temperature of the intrusion, °C	Temperature at the contact, °C	Temperature (°C) at various distances from the contact expressed in fractions of the thickness (D) of the intrusion		
		1/10D	2/10D	1/2D
Gabbroic magma 1200	725 +150 } 875	625 +150 } 775	550 +150 } 700	410 +150 } 560
Syenitic Magma 900	560 +150 } 710	470 +150 } 620	410 +150 } 560	300 +150 } 450
Granitic Magma 800	510 +150 } 660	420 +150 } 570	365 +150 } 515	270 +150 } 420
700	460 +150 } 610	370 +150 } 520	320 +150 } 470	235 +150 } 385

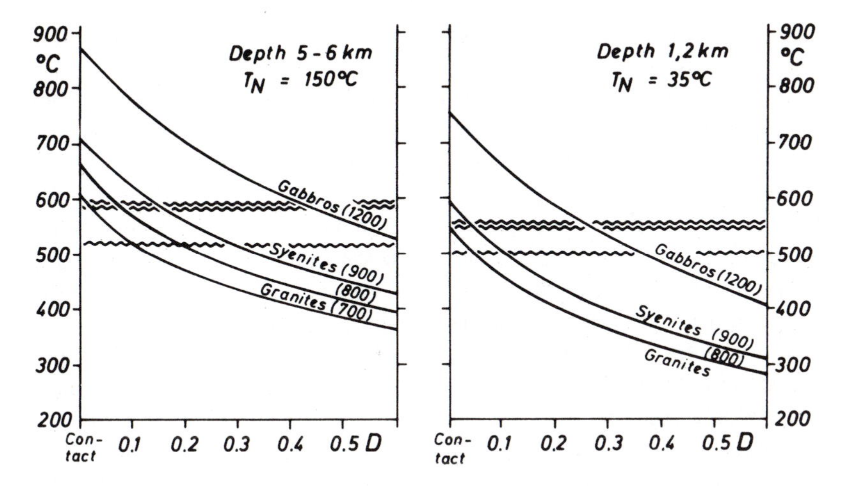

Fig. 8-2 Heating of the country rock adjacent to various intrusions at a depth of 5 to 6 and 1.2 km, respectively. Distance from the contact is given in fractions of D. D = thickness of the sheet-like intrusion. T_N same as T_C.

appreciable length of time. If the magma penetrates to a shallower depth and solidifies, *e.g.*, at a depth of 2 km, the temperatures are about 90°C lower than those stated in the table, because at this depth the temperature of the country rock prior to intrusion (T_c) was only 60°C instead of 150°C.

The maximum temperatures attained at two different depths of intrusion in rocks adjacent to igneous bodies of various temperatures and at different distances from the intrusion are shown in Figure 8-2. The distances from the contact are expressed in fractions of the thickness D of the magmatic body. At a depth of 5 to 6 km the load pressure is 1500 bars. At this pressure the beginning of medium-grade metamorphism is marked by the temperature range of 510° to 530°C and the beginning of high-grade by the range of 580° to 600°C (see Figure 7-2). At a shallower depth of 1.2 km, where the load pressure is 300 bars, the corresponding temperature ranges are about 490° to 510° and 540° to 560°C, respectively. These temperature ranges are shown in Figure 8-2. It is apparent from the figure that:

1. The width of the zone of each metamorphic grade is increased with (a) increasing temperature of the magma, (b) increasing thickness (or diameter) of the intrusion, and (c) increasing depth of intrusion. In particular, it is seen that the zone of high-grade metamorphism is considerably less extensive around granite intrusions as compared with syenite, or even more so, with gabbro intrusions;

2. High-grade metamorphism is always associated with gabbroic intrusions.
3. High-grade metamorphism is rarely developed in aureoles around granitic intrusions; generally the rocks indicate the higher temperature range of medium-grade metamorphism.

The last observation also applies to where magma, having a temperature of 700°C, solidifies at a depth of 5 to 6 km or where magma, having a temperature of 800°C, intrudes to a depth of 1.2 km; at this depth the temperature of the country rock is only 35°C. In such a case, the country rock at the contact is only heated to 545°C. In spite of the low pressure of 300 bars at a depth of 1.2 km, this temperature is not sufficiently high to develop a zone of rocks of high-grade at the contact. Only if the granitic magma has an extremely high temperature of about 900°C at such shallow depth may the temperature in the immediate vicinity of the contact reach about 600°C, which is sufficiently high to form high-grade rocks. As stated earlier, these arguments hold only if the heating of the country rock is due solely to heat conduction. The period of time during which the (nearly) maximum temperature of the country rock is sustained is proportional to the square of the thickness of the intrusion. The order of magnitude of the length of this period in years is given by the expression 0.01 D^2:

if $D =$ 1 m, the period is 3 days
if $D =$ 10 m, the period is 1 year
if $D =$ 100 m, the period is 100 years
if $D =$ 1000 m, the period is 10,000 years

In the case of intrusions several hundred to several thousand meters in thickness, the maximum temperature induced in the country rock will be maintained for very long periods of time. This means that reactions which are possible have sufficient time to proceed to completion, and equilibrium between adjacent minerals is established.

Metamorphic Zones in Regional Metamorphism

Unlike contact metamorphism, regional metamorphism is not a localized phenomenon; it is, as the name suggests, of regional extent. With the occasional exception of occurrences of very-low-grade metamorphism, the different types of regional metamorphism are confined to areas of mountain building, so that the term "regional dynamothermal metamorphism" is often used. The view is held that metamorphism as well as orogenesis ought to be regarded as due to "one and the same process."[1] The cause for both of them must be an additional supply of thermal energy at specific regions of the earth, presumably derived from great depths within the mantle. Based on his detailed investigations of plagioclases as index minerals of metamorphism in the Central Alps, Wenk (1962) pictures the presence of "thermal domes" during metamorphism in orogenic belts. "We cannot regard these thermal highs as independent phenomena, they are genetically connected to orogenesis." It is therefore "the thermal energy surging from the depths that imparts to the rock masses their special character," *i.e.*, brings about rock metamorphism. On the other hand, Niggli (1970) holds the view that in the case of the Central Alps, the increase of temperature in regions of medium- and high-grade metamorphism was caused by great subsidence caused by the load of a large pile of nappes. Generally, this may be another possibility in those cases where high temperatures (and low temperatures as well) together with high pressures have been attained; however, when high temperature metamorphism has taken place at only moderate pressures this model does not seem to be applicable.

In principle, regional metamorphism is not very different from contact metamorphism, both of them requiring a supply of thermal energy. In contact metamorphism, the original source of heat is small and now exposed as a plutonic mass, whereas in regional metamorphism, it is essentially larger, more deep-seated, and not visible.

The pressures operating during contact metamorphism generally range between 200 to 2000 bars. The effective pressure of regional metamorphism may also amount to 2000 bars but usually it is higher, often markedly higher. Whereas only confining pressure is acting during contact metamorphism, in the case of regional metamorphism it is generally supplemented by directed pressure. In contradistinction to the nearly isotropic, well-crystallized fabric of the hornfelses of contact metamorphism, regional metamorphic rocks bearing platy or prismatic minerals, like micas, chlorite, or amphibole, therefore exhibit strong schistosity.

Barrow (1893, 1912) recognized that during the metamorphism of pelitic sediments (clays and shales) certain newly formed minerals appear in a definite sequence with increasing metamorphic grade; these minerals were designated as *index minerals*. His concept has been further elaborated most of all by Tilley (1925) and Harker (1932, 1939). The following succession of index minerals with increasing metamorphic grade can be distinguished in many metamorphic terrains:

chlorite → biotite → almandine-garnet →
staurolite → kyanite → sillimanite

---

[1]Very many workers have come to this conclusion; see, *e.g.*, Bearth (1962).

Scottish Highlands

The metamorphic zones characterized by these index minerals are especially well developed on a regional scale in the Scottish Highlands. The zone boundaries have been displaced by the postmetamorphic Great Glen fault. By reversing the movement along this fault, Kennedy (1948) reconstructed the original continuous boundaries (Figure 8-3).

The development of migmatites and the occurrence of magmatic intrusions within the kyanite and sillimanite zones are very typical. In fact, Kennedy (1949) considers these to be the cause for the temperature rise, which in turn was responsible for metamorphism. However, it seems more probable that the intrusions and migmatites as well as the process of metamorphism are the results of a temperature rise in response to a geophysical process. In high-grade metamorphism conditions are established which may lead to the generation of granitic melts.

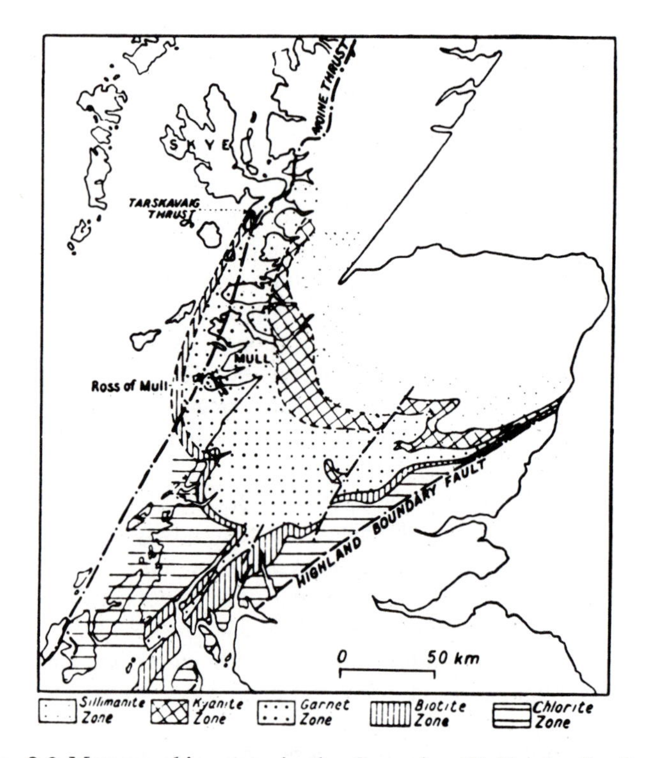

Fig. 8-3 Metamorphic zones in the Grampian Highlands, Scotland (after Kennedy, 1948). The narrow staurolite zone has been omitted; lack of appropriate rock compositions has limited the geographic extent of the staurolite zone.

Northern Michigan

A succession of zones very similar to that in the Grampian Highlands, except for a kyanite zone, has been observed in Michigan and Wisconsin by James (1955). Part of his map has been reproduced in Figure 8-4. This occurrence is extremely interesting. Four "thermal domes" reaching high-grade metamorphism (sillimanite zone) in the center have developed in a crystalline complex of generally low grade (chlorite zone). The succession of zones—sillimanite, staurolite, garnet, biotite—reflects decreasing temperature. The area of the largest "dome" is approximately 75 by 55 km. The four "thermal domes" represent four independent centers of regional metamorphism; the two closest ones are only 45 km apart and their respective biotite isograds are separated by only 4 km. In agreement with James, one can very well imagine that each of the remarkably steep thermal gradients causing the four thermal highs resulted from magmatic intrusions concealed below the surface.

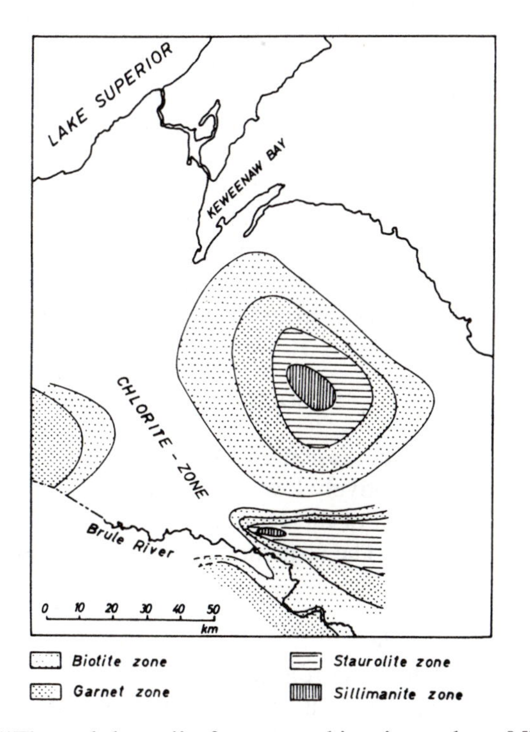

Fig. 8-4 "Thermal domes" of metamorphism in northern Michigan (after James, 1955).

New England

In New England and adjacent areas the zones distinguished by Thompson and Norton (1968) are similar. The sequence of the zones developed in pelitic rocks is

Chlorite Biotite Almandine-garnet	in low-grade
Staurolite with kyanite Sillimanite (with muscovite + quartz)	in medium-grade
Sillimanite + K feldspar (without muscovite)	in high-grade

In the east and northeast part of New England the exposed part of the crust has been metamorphosed at lower pressure; consequently, andalusite instead of kyanite has formed. Therefore, in that large area a staurolite + andalusite zone is present instead of a staurolite + kyanite zone.

While in previous and subsequent examples, the metamorphic zones have been established *after* all major tectonic deformations, this was definitely not the case in the Appalachians of New England, particularly not in its southwestern part: The main period of regional metamorphism overlapped two or more major stages of deformation, and successive nappes carried successively higher grade rocks westward over lower grade rocks. Thus, there are higher grade rocks at higher tectonic levels that are complely surrounded by lower grade rocks at lower tectonic levels. This, in fact, is a complete overturning of an isogradic surface (Thompson and Norton, 1968). It is obvious that wherever syntectonic development of metamorphic zones has taken place, interpretation of the sequence of isograds in terms of temperature and pressure may be very complicated.

Bosost, Spain

A very different succession of metamorphic zones is developed when the pressure is relatively low. An example of low pressure metamorphism in a small area of the central Pyrenees near Bosost has been studied by Zwart (1962). The sequence of the zones in pelitic rocks is as follows:

Biotite	in low grade
Staurolite-andalusite-cordierite Sillimanite-cordierite, with muscovite + quartz but not stable staurolite	in medium-grade
Sillimanite + K feldspar (without muscovite + quartz).	in high-grade

The latter zone typical of high-grade metamorphism is not present in the Bosost area itself because temperature has not risen high enough, but it is well known from other localities in the Pryenees.

A number of granite and pegmatite intrusions (up to 1000 by 300 m in size) were emplaced, particularly during the late period of metamorphism. This emplacement is essentially restricted to the area of highest temperature rise; in Bosost this is the sillimante-cordierite zone. The metamorphic zones suggest a small thermal dome (approximately 12 km in diameter) with a granite core below the present erosion surface. Zwart comes to the general conclusion that the features of the Bosost metamorphic area are best explained as a result of a heat front ascending independently of stratigraphy or structure; metamorphic zones usually cut across structural and stratigraphic boundaries. Although the period of metamorphism is contemporary with the Hercynian folding, it has been demonstrated that in detail the metamorphic reactions have little direct relation to the various folding phases.

Swiss Central Alps

The metamorphic zones distinguished in any one terrain depend, of course, on the range of temperature, prevailing pressure, and bulk composition of the rocks, *i.e.*, on the reactions taking place at the given conditions. In order to map different zones it is necessary that rock compositions allowing the formation of diagnostic mineral parageneses be sufficiently common in the field. This is not always the case. For instance, in the Swiss Central Alps an almandine-garnet zone cannot be mapped because appropriate rock compositions are lacking in widespread regions. On the other hand, siliceous dolomite beds occur distributed throughout the region which made it possible to map a tremolite + calcite isograd and a diopside + calcite isograd. Figure 8-5 shows the isograds and metamorphic zones: chloritoid, staurolite, tremolite + calcite, diopside + calcite. Furthermore, Wenk (1962) mapped zones based on ranges of plagioclase composition in carbonate-bearing pelitic rocks, the so-called Bündner Schiefer, which contain plagioclase coexisting with calcite. These zones, also shown in Figure 8-5 and taken from the publication by Wenk in Jäger *et al.* (1967), are superimposed on the isograds. For further information see Frey *et al.* (1976).

Paired Metamorphic Belts

During the last 20 to 25 years, many metamorphic terrains have been studied in detail. Miyashiro (1961) made the significant discovery that in certain regions two parallel belts of contrasting metamorphic fea-

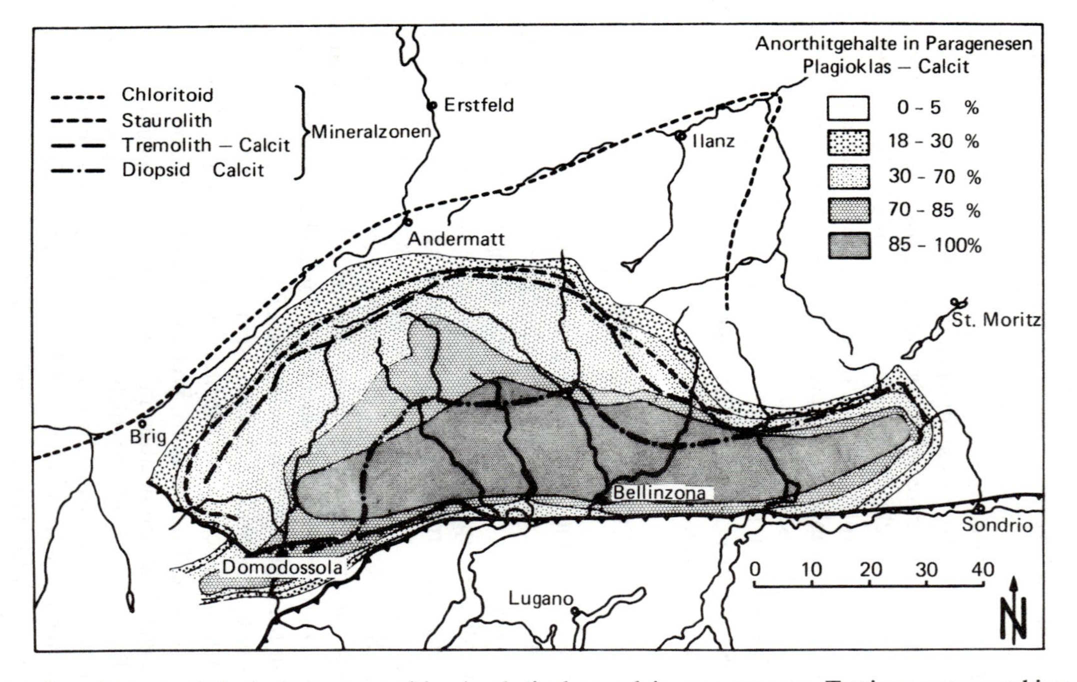

Fig. 8-5 Isograds and zones of plagioclase composition in plagioclase-calcite parageneses. Tertiary metamorphism in the Swiss Central Alps.

The E-W running fault zone (Insubrian line) cuts off the area of Tertiary metamorphism; that part of the Alps exposed to the south (Süd- or Seealpen) was metamorphosed about 300 million years earlier. In the Southern Alps *Tertiary* metamorphism probably took place below the present level of erosion while north of the Insubrian fault line greater uplift and tilting have exposed the metamorphic rocks of Tertiary age.

tures extend for several hundred kilometers; these have been called paired metamorphic belts. They are especially well developed in parts of the circum-Pacific region, such as Japan (see Figure 8-6), Celebes, New Zealand, Chile, and California. Using Miyashiro's nomenclature, a pair of metamorphic belts is composed of (a) a high pressure type and (b) a low pressure type, although the two belts may contain some areas of the medium pressure type. In our nomenclature (a) comprises [glaucophane + lawsonite] and [jadeite + quartz]-very-low-grade, which may grade into [glaucophane + clinozoisite]-low-grade, while (b), the low pressure type of Miyashiro, comprises andalusite-bearing rocks, which com-

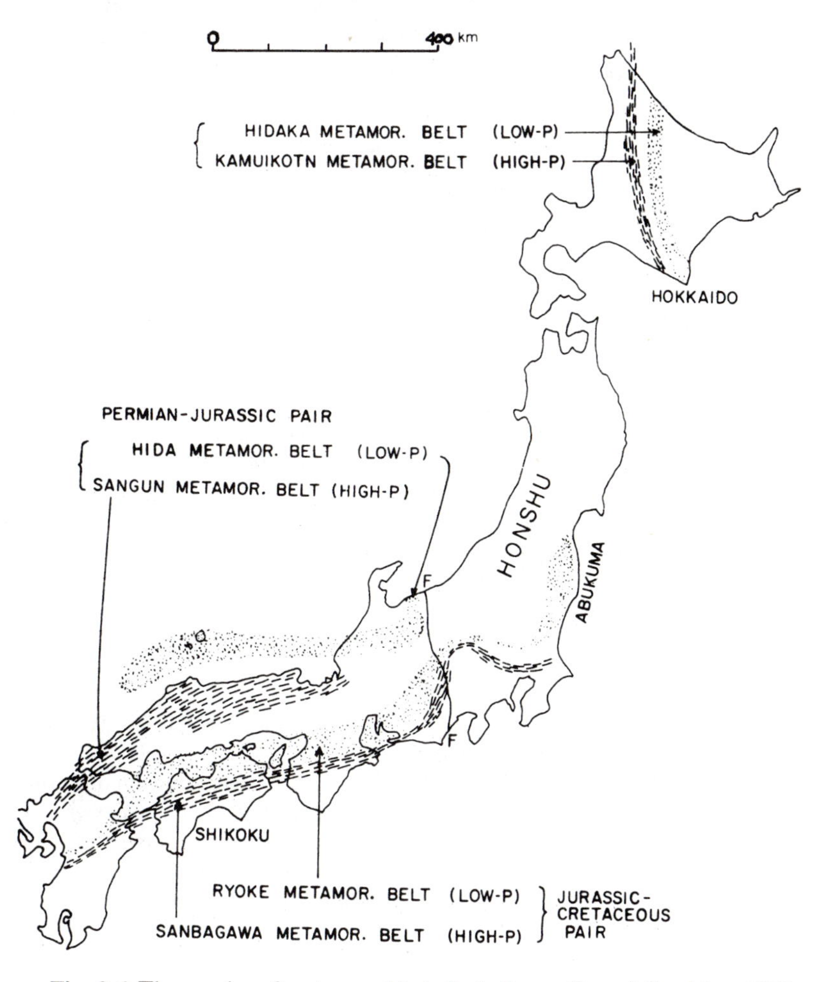

Fig. 8-6 Three pairs of metamorphic belts in Japan (from Miyashiro, 1972). [Note: Kamuikotan is the correct spelling]

monly correspond to low-grade (without almandine), [cordierite]-medium-grade, and [cordierite]-high-grade.

Quoting from Miyashiro (1972):

> The high-pressure belt is usually on the oceanic side of the low-pressure belt. The paired metamorphic belts were ascribed to the underthrusting of an ocean floor along a Benioff zone beneath island arcs and continental margins. The high-pressure (low temperature) belt was regarded as corresponding to the zone of a trench (ocean trench), which shows very low heat flow values, and the low-pressure belt to a zone of island-arc volcanism with high heat flow.

These ideas are in agreement with the hypothesis of platetectonics, which undertakes to explain worldwide relationships between metamorphism, magmatism, and orogeny (cf. Miyashiro, 1973).

References

Barrow, G. 1893. *Quart. J. Geol. Soc.* **49:** 330–358.

——— 1912. *Proc. Geol. Assoc.* **23:** 268–284.

Bearth, P. 1962. *Schweiz. Mineral. Petrog. Mitt.* **42:** 127–137.

Frey, M., Jäger, E., and Niggli, E. 1976. *Schweiz. Mineral. Petrog. Mitt.* **56:** 649–659.

Harker, A. 1932, 1939. *Metamorphism.* Methuen, London.

Jaeger, J. C. 1957, *Am. J. Sci.* **255:** 306–318.

Jäger, E., Niggli, E., and Wenk, E. 1967. *Beitr. Geol. Karte Schweiz* N.F., **134.**

James, H. L. 1955. *Geol. Soc. Am. Bull.* **66:** 1455–1488.

Kennedy, W. Q. 1948. *Geol. Mag.* **85:** 229–234.

——— 1949. *Geol. Mag.* **86:** 43–56.

Melson, W. G. 1966. *Am. Mineral.* **51:** 402–421.

Miyashiro, A. 1961. *J. Petrol.* **2:** 277–311.

——— 1972. *Am. J. Sci.* **272:** 629–656.

——— 1973. *Metamorphism and Metamorphic Belts.* George Allen and Unwin, London.

Niggli, E. 1970. *Fortschr. Mineral.* **47:** 16–26.

Schneiderhöhn, H. 1961. *Die Erzlagerstätten der Erde.* Vol. II. Die Pegmatite. Gustav Fischer Verlag, Stuttgart.

Thompson, J. B. and Norton, S. A. 1968. Paleozoic regional metamorphism in New England and adjacent areas. In E-An Zen *et al.,* eds. *Studies of Appalachian Geology.* Interscience Publisher (John Wiley & Sons), New York.

Tilley, C. E. 1952. *Quart. J. Geol. Soc.* **81:** 100–110.

Wenk, E. 1962. *Schweiz. Mineral. Petrog. Mitt.* **42:** 139–152.

Zwart, H. J. 1962. *Geol. Rundschau* **52:** 38–65.

Chapter 9

Metamorphic Reactions in Carbonate Rocks

General Considerations

The metamorphism of quartz-bearing carbonate rocks provides interesting examples of metamorphic reactions. The occurrence of siliceous dolomites and siliceous dolomitic limestones is very widespread, whereas that of magnesite-bearing sediments is much rarer.

Eskola (1922) and later Bowen (1940) made a systematic study of the sequence of reactions occurring in carbonate rocks at some given CO_2 pressure in response to rising temperature. The following minerals, well known from progressive metamorphism, are formed: tremolite, forsterite, diopside, wollastonite, periclase (brucite), monticellite, åkermanite, spurrite, mervinite, larnite, and others. To this sequence Tilley (1948) added talc as the mineral forming at even lower temperature than tremolite. In constrast to previous views, it has recently been established that the formation of talc is rather common in the metamorphism of siliceous dolomite.

Metamorphic reactions of siliceous carbonates liberate CO_2, but since water is present in the rock before metamorphism, one cannot regard CO_2 pressure and temperature as the only factors in metamorphism. Besides temperature, the total fluid pressure (being the sum of the partial pressures of CO_2 and H_2O) and the ratio of the two partial pressures (or the mole fraction of either CO_2 or H_2O) have to be taken into account. Therefore, in most reactions involving carbonates the equilibrium is (at least) bivariant. This is true even if H_2O is absent in the reaction equation, because H_2O always is a constituent of the fluid phase present in metasediments.

In the supercritical state the two components H_2O and CO_2 constitute one single fluid phase; they are miscible in all proportions. The application of bivariant equilibria to natural parageneses is more complicated than that of univariant equilibria. However, carbonate rocks are widespread and many metamorphic mineral assemblages are petrogenetically significant and, therefore, deserve a detailed treatment. This is made possible by recent progress in the experimental investigation of metamorphic reactions in rocks composed of dolomite, quartz, and either calcite or magnesite.

Any bivariant equilibrium involving the two volatile components CO_2 and H_2O is represented by a surface which is situated in a volume with the three perpendicular axes of temperature, mole fraction of CO_2 (designated as X_{CO_2}), and total fluid pressure (designated as P_f). In a bivariant equilibrium a mineral association is not restricted at a given value of P_f to a unique temperature but exists over a range of temperatures—any specific temperature depending on the composition X_{CO_2} of the fluid phase. Figure 9-1 shows schematically the equilibrium planes of three different reactions—*A, B,* and *C*. In this figure, *abcd* is an isobaric section. Such sections at constant fluid pressure have temperature as ordinate and X_{CO_2} as abscissa. This method, introduced by Greenwood (1961, 1962), is very useful in representing equilibrium conditions.

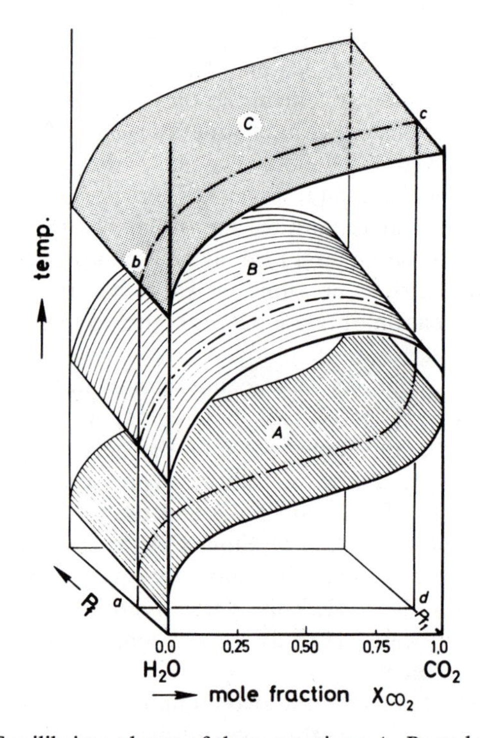

Fig. 9-1 Equilibrium planes of three reactions *A, B,* and *C* in temperature, fluid pressure, and mole fraction space. *abcd* is an isobaric section. (From Metz, 1970).

The shape of an isobaric equilibrium curve depends on the nature of the reaction. If one gas species is a reactant and the other is a product, the equilibrium curve has a point of inflection (case A). If both gas species are products, the equilibrium curve has a maximum (case B). If only one gas species, either CO_2 or H_2O, is involved in the reaction, the equilibrium temperature increases with its increasing mole fraction (case C; CO_2 is produced and H_2O is not used in the reaction. These relationships, theoretically analyzed by Greenwood (1962, 1967b), are of fundamental importance in understanding the metamorphism of carbonates. The student is, therefore, advised to study these papers.

Metamorphism of Siliceous Dolomitic Limestones

Rocks consisting of quartz, dolomite, and either calcite or magnesite remain unaffected at very-low-grade metamorphism; however, at low grade (at temperatures not well defined and therefore generally not coinciding with the beginning of that grade) some minerals begin to react, and at higher metamorphic grade numerous reactions occur in rocks of this composition. In this section we are concerned only with the formation of talc, tremolite, diopside, and forsterite and with the resulting mineral parageneses. The formation of monticellite, åkermanite, merwinite, and others taking place at higher temperatures and very low pressures is discussed in a later section. The formation of wollastonite will also be considered further on.

In dolomite, talc, tremolite, and diopside some Fe^{2+} may replace Mg in solid solution, but iron-free varieties are common as well. Extensive experimental and theoretical studies involving the iron-free minerals have been carried out. These minerals and their various assemblages may be graphically represented in the CaO-MgO-SiO_2 composition triangle (see p. 34, Figure 5-4).

Previously it was assumed that only a few reactions are necessary to describe the metamorphic transformations of siliceous dolomitic limestones. However, Turner (1968) presents a larger number, and Metz and Trommsdorff (1968) give a complete list of reactions involving the minerals quartz, dolomite, calcite, talc, tremolite, diopside, and forsterite. As established by Metz (1976), reactions (9), (10), and (12) of the fifteen reactions listed are metastable. Therefore, only twelve reactions have to be taken into account, and two of these, *i.e.*, reactions (5) and (15), occur only in rocks which contain magnesite together with quartz and dolomite prior to metamorphism.

3 dolomite + 4 quartz + 1 H_2O = 1 talc + 3 calcite + 3 CO_2 (1)

5 talc + 6 calcite + 4 quartz = 3 tremolite + 6 CO_2 + 2 H_2O (2)

2 talc + 3 calcite = 1 tremolite + 1 dolomite + 1 CO_2 + 1 H_2O (3)

5 dolomite + 8 quartz + 1 H_2O = 1 tremolite + 3 calcite + 7 CO_2 (4)

2 dolomite + 1 talc + 4 quartz = 1 tremolite + 4 CO_2 (5)

1 tremolite + 3 calcite + 2 quartz = 5 diopside + 3 CO_2 + 1 H_2O (6)

1 tremolite + 3 calcite = 1 dolomite + 4 diopside + 1 CO_2 + 1 H_2O (7)

1 dolomite + 2 quartz = 1 diopside + 2 CO_2 (8)

1 talc + 5 dolomite = 4 forsterite + 5 calcite + 5 CO_2 + 1 H_2O (9)

11 talc + 10 calcite = 5 tremolite + 4 forsterite + 10 CO_2 + 6 H_2O (10)

1 tremolite + 11 dolomite = 8 forsterite + 13 calcite + 9 CO_2 + 1 H_2O (11)

13 talc + 10 dolomite = 5 tremolite + 12 forsterite + 20 CO_2 + 8 H_2O (12)

3 tremolite + 5 calcite = 11 diopside + 2 forsterite + 5 CO_2 + 3 H_2O (13)

1 diopside + 3 dolomite = 2 forsterite + 4 calcite + 2 CO_2 (14)

4 tremolite + 5 dolomite = 13 diopside + 6 forsterite + 10 CO_2 + 4 H_2O (15)

Reactions involving dolomite and calcite are not written exactly because the solid solution between calcite and dolomite has not been taken into account (see footnote 2 on next page). In reality, the calcite is not $CaCO_3$ and the dolomite is not $CaMg(CO_3)_2$; in calcite some Ca is replaced by Mg (magnesian calcite) and dolomite contains somewhat more Ca than indicated by its ideal formula (calcian dolomite). At a given equilibrium temperature, the compositions of dolomite and calcite are determined by a solvus curve (Goldsmith and Newton, 1969). This curve has been successfully used to determine the temperature of metamorphism, *e.g.*, by Hutcheon and Moore (1973), Puhan (1976), and Bickle and Powell (1977). However, the validity of the calcite-dolomite geothermometer has been questioned by Garde (1977). Obviously, certain precautions have to be observed which are not yet fully known.

In each reaction listed above all four solid phases and the gas phase are composed of five components: CaO, MgO, SiO_2, CO_2, and H_2O.

Although H_2O is not shown as a component of the phases in reactions (8) and (14), it will commonly be a constituent of the fluid phase present during metamorphism and therefore has to be included as a component in all 15 cases. It follows from the phase rule

$$variance = components + 2 - phases$$

that a system of five phases and five components has a variance of 2; *i.e.*, all of the 15 reaction equilibria are bivariant.[1]

In P_f-X_{CO_2}-T space the equilibrium conditions are represented by a surface for each reaction. At P_f = const., each equilibrium is represented by a curve. For a constant fluid pressure, Figure 9-2 schematically shows the equilibrium curves in the T-X_{CO_2} plane. [A quantitative diagram for $P_f = 1$ kb shown in the previous editions will be revised (personal communication from Metz and Puhan, 1978) and is therefore omitted here.] Figure 9-3 gives a schematic drawing in P_f-X_{CO_2}-T space.

Figure 9-2 shows that the first reaction in a rock consisting of dolomite and quartz (with or without calcite) and containing interstitial water will lead to the formation of talc + calcite,[2] according to equation (1):

$$3\ CaMg(CO_3)_2 + 4\ SiO_2 + 1\ H_2O =$$
$$1\ Mg_3[(OH)_2/Si_4O_{10}] + 3\ CaCO_3 + 3\ CO_2$$

For this reaction H_2O is required and CO_2 is liberated; therefore, X_{CO_2} will increase with rising temperature until all dolomite or quartz has been used up. If dolomite and quartz are abundant, with rising temperature and steadily increasing X_{CO_2}, the state of the isobaric invariant point (I) will be reached.

It is evident from the experimental data that *talc* deserves special attention as the first mineral to form during the metamorphism of siliceous dolomitic limestones. This is valid as long as the X_{CO_2} value within the rock is smaller than that of the isobaric point at the given fluid pressure. If, on the other hand, a very high X_{CO_2} was previously generated by another reaction, such as

---

[1]Although in reaction (8) the number of phases is only four instead of five, the variance is still 2. This is because the number of components is less by one as well. (CaO · MgO) has to be counted as one component because dolomite and diopside have the same CaO/MgO ratio. CaO and MgO are not independent variables.

[2]The calcite experimentally formed (and occurring in nature as well) contains a few percent of $MgCO_3$ in solid solution. This has not been taken into account in the written equation. However, Gordon and Greenwood (1970) considered the effect of solid solution on the equilibrium of reaction in detail. A more general treatment of crabonate solid solution effects on equilibria in siliceous carbonate rocks is given by Skippen (1971, 1974), by Skippen and Hutcheon (1974), and by Metz (1977).

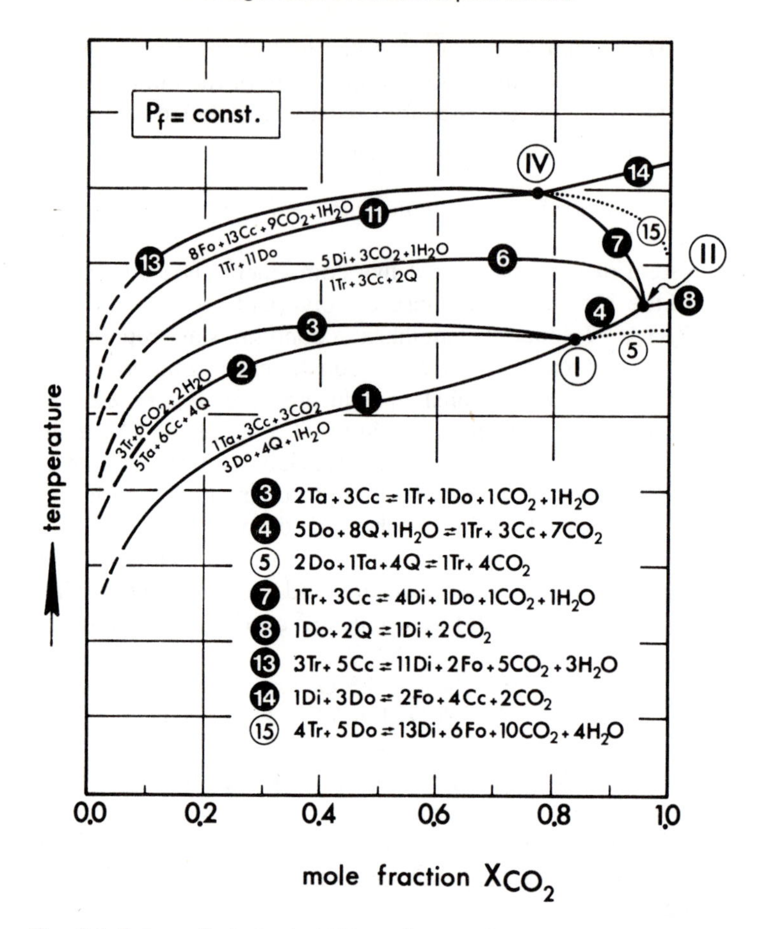

Fig. 9-2 Schematic isobaric T-X_{CO_2} diagram for reactions (1) through (15). Arabic numbers refer to reactions. Roman numbers are isobaric invariant points. The numbers are those given by Metz and Trommsdorff (1968). Reactions 9, 10, and 12, and isobaric invariant point III do not exist stably; therefore, they are not shown here. [Diagram from Käse and Metz (1979).]

$$3 \text{ dolomite} + 1 \text{ K feldspar} + 1\, H_2O = \\ 1 \text{ phlogopite} + 3 \text{ calcite} + 3\, CO_2$$

reaction (1) cannot proceed; reaction (4) takes place instead and no talc is formed.

The expected common appearance of talc is apparently not substantiated by petrographic observation. Tremolite is generally regarded as the first silicate mineral formed. However, since experimental petrology has shown that talc may form at considerably lower temperatures than tremolite, the parageneses

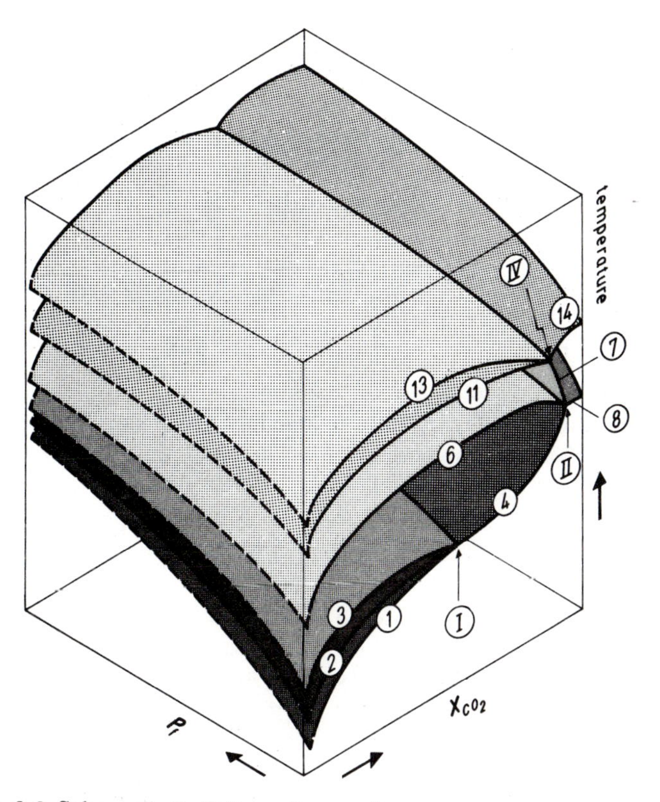

Fig. 9-3 Schematic P_f-T-X_{CO_2} diagram for reactions (1) through (14). (By kind permission of D. Puhan and P. Metz, 1978.)

$$\text{talc} + \text{calcite} + \text{dolomite} \qquad (1a)$$
$$\text{talc} + \text{calcite} + \text{quartz} \qquad (1b)$$

have been recorded more frequently. These assemblages are easily overlooked because talc may be mistaken for white mica in thin section if no special care in mineral determination is exercised.

Talc as the first regional metamorphic product formed according to reaction (1) has been observed by Guitard (1966) and Steck (1969), in a large area of the Damara Belt in Southwest Africa by Puhan and Hoffer (1973), and in the Central Alps by Trommsdorff (1972). Careful microscopic investigation even documented the equilibrium paragenesis of reaction (1):

$$\text{talc} + \text{calcite} + \text{dolomite} + \text{quartz} \qquad (1)$$

All four minerals were observed in mutual contact, thus indicating the existence of equilibrium. Of course, the gas phase present at the time of reaction has since disappeared.

It is readily seen that the equilibrium paragenesis talc + calcite + dolomite + quartz may be found over a large areal extent because it is stable at $P_f = 5$ kb over a temperature range of about 100°C and a corresponding range of X_{CO_2} (see curve 1 in Figure 9-5); the same is also true for lower pressures, as shown by the experimental work of Metz and Puhan (1970, 1971). Therefore, because of bivariance, this reaction equilibrium does not indicate unique *P-T* conditions. However, wherever it is found in the field, it may provide a guide to occurrences of the petrogenetically significant paragenesis of the isobaric invariant point (I). Starting from the assemblage of reaction (1), the invariant point (I) can only be reached in a direction of increasing T and X_{CO_2} by buffering the fluid phase (see Greenwood, 1975). The assemblage of the invariant point

(I) tremolite + talc + calcite + dolomite + quartz

has been observed by Jansen *et al.* (1978) in Naxos, by Puhan and Hoffer (1973) at several localities in the Damara Belt, and by Trommsdorff (1972) in the Central Alps. In these field studies, a knowledge of experimentally and theoretically determined phase relations was of great help in looking for relevant assemblages.

An isobaric invariant point is generated by the intersection of two isobaric univariant curves. For instance, in Figure 9-2 reactions (1) and (2) intersect at point (I). The equilibrium assemblages of reactions (1) and (2) coexist at point (I). Reaction (2) is one of several reactions leading to the formation of tremolite:

$$5\ \mathrm{Ta} + 6\ \mathrm{Cc} + 4\ \mathrm{Q} = 3\ \mathrm{Tr} + 6\ \mathrm{CO_2} + 2\ \mathrm{H_2O} \tag{2}$$

$$5\ \mathrm{Mg_3[(OH)_2/Si_4O_{10}]} + 6\ \mathrm{CaCO_3} + 4\ \mathrm{SiO_2} = 3\ \mathrm{Ca_2Mg_5[(OH)_2/Si_8O_{22}]} + 6\ \mathrm{CO_2} + 2\ \mathrm{H_2O}$$

As shown in Figure 9-2, reaction (2) may proceed within a very large range of X_{CO_2} values and it covers a temperature range similar to that of reaction (1). The following parageneses are produced by reaction (2), which is rare because of compositional restrictions:

$$\mathrm{Tr + Ta + Cc} \tag{2a}$$

$$\mathrm{Tr + Q + Cc} \tag{2b}$$

Paragenesis Tr + Ta + Q is not possible in metamorphosed siliceous dolomitic limestones and siliceous dolomites. Magnesite is required as a constituent in the original rocks to form this paragenesis.

The stability fields of these parageneses, as well as most others to be discussed later, depend on X_{CO_2} and very markedly on P_f as well. Therefore, it is advisable to document whenever possible not only the reactants or the products but also the paragenesis corresponding to a specific reaction equilibrium. In the case of reaction (2) this paragenesis is

$$\text{Tr} + \text{Ta} + \text{Cc} + \text{Q} \tag{2}$$

A really valuable indicator of metamorphic conditions is the paragenesis of the isobaric invariant point (I), where reaction curves (1) and (2) and/or (4) intersect, giving rise to the paragenesis

$$\text{(I) Tr} + \text{Ta} + \text{Cc} + \text{Do} + \text{Q}$$

As mentioned previously, this assemblage has been found in nature.

Puhan and Metz (1973) have determined experimentally the X_{CO_2} and T conditions of point (I) at various P_f; however, the data may require further refinement because the authors used natural tremolite which contained some fluorine (personal communication, 1978). For petrogenetic considerations, the relationship between P_f and T is of particular interest because it may be possible to estimate either one from other observations and thus determine the other variable; *e.g.*, it may be possible, by means of the calcite-dolomite geothermometer, to determine the metamorphic temperature independently.

Discussion of Reaction-Isograds

It is appropriate at this stage to apply the concept of petrogenetically significant and nonsignificant reaction-isograds to the preceding phase relations. In P_f-T-X_{CO_2} space, the assemblage (I) is represented by a univariant curve and the assemblages (1) and (2) are represented by bivariant surfaces. In the field, occurrences of assemblage (I) will constitute a reaction-isograd *line*. This is true, in general, of univariant equilibria. However, assemblages constituting bivariant equilibria are stable over a T range at any given P_f [*e.g.*, reaction (1)]. Consequently, occurrences of a bivariant assemblage do not lie along a reaction-isograd line but are stable within a reaction-isograd *area*. In this case the appearance of assemblage (1) is of no petrogenetic significance because the reaction temperature, at any given P_f, depends strongly on the composition of the fluid phase. This composition may vary from one locality to another.

The equilibrium conditions of reaction (2) are quite similar to those of reaction (1), as can be seen from Figure 9-5. Therefore, even the equilibrium paragenesis Ta + Cc + Q + Tr of reaction (2) characterizes a reaction-isograd area or band rather than a line. Neither the first appearance nor the areal distribution of paragenesis (2) in the field is of much petrogenetic significance. This is true for the bivariant equilibrium parageneses of most reactions to be discussed later. However, a few parageneses of bivariant reaction equilibria are known which do provide significant petrogenetic information. An example is reaction (8) which takes place only within a restricted range of X_{CO_2} and within a very small temperature range. Equilibrium data for reaction (8) are available only at fluid pressures of 1 and 2 kb (Slaughter *et al.,* 1975). Another favorable case is reaction (14) which at 5 kb fluid pressure takes place within a temperature range of only 35°C in response to changes in the mole fraction of CO_2. As shown in Figure 9-4, a decrease of fluid pressure makes the temperature range considerably smaller, according to data by Käse and Metz (1979). Therefore, the equilibrium paragenesis of reaction (14), *i.e.,* Fo + Cc + Di + Do, could well serve as a reaction-isograd in the field.

Clearly, it is always petrogenetically significant to search for the parageneses of the isobaric invariant points shown in Figures 9-2 and-9-5:

(I) Ta + Cc + Tr + Do + Q
(II) Tr + Cc + Di + Do + Q
(IV) Fo + Cc + Tr + Di + Do

Up to 5 kb pressure, the positions of (IV) are known (Käse and Metz, 1979). Its P_f-T values are shown graphically in Figure 9-4, together with the stability field of the paragenesis of reaction (14). Compared with the boundaries of metamorphic grades, it is interesting to note that the parageneses of (IV) and (14) will be found in the field in two different settings:

a. At fluid pressures below about 3.5 to 4 kb, parageneses (IV) and (14) will be found somewhat *before* the boundary between medium- and high-grade metamorphism, as detected in metapelites and graywackes,
b. At fluid pressures above about 4 kb, the parageneses (IV) and (14) occur *after* high-grade metamorphism has commenced.

Other Reactions

Compared with the schematic diagram of Figure 9-2, Figure 9-5 shows in a more quantitative way the various reactions taking place in rocks derived from siliceous dolomitic limestones and siliceous dolo-

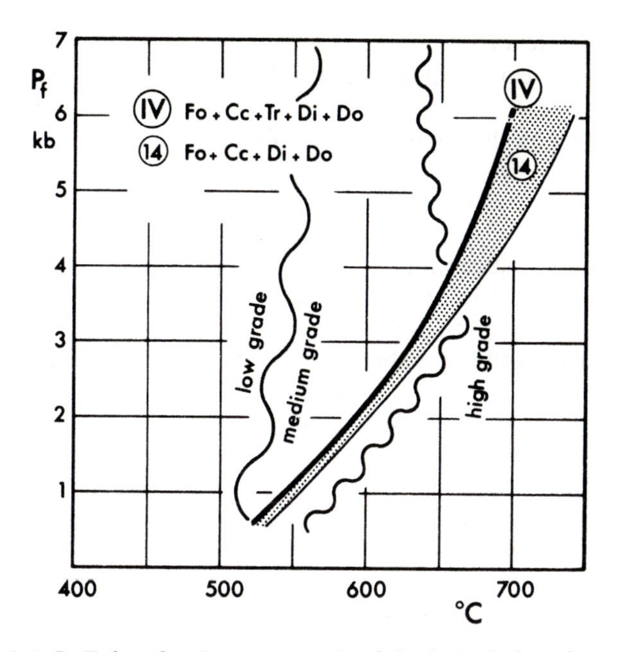

Fig. 9-4 P_f-T data for the paragenesis of the isobaric invariant point (IV) and for the stability range of the bivariant equilibrium paragenesis (14).

mites. Although much work has been done by various authors, differences in the results call for caution until these differences have been resolved.[1] Therefore, the positions of the isobaric curves in Figure 9-5 are shown mainly by broken lines, indicating uncertainty. Only the curves of reaction (11) by Metz (1976) and reaction (14) by Käse and Metz (1979) are believed to be reliable for a fluid pressure of 5 kb.

We shall refer to Figure 9-5 in the following short discussion of reactions (3) to (14).

Reaction

$$2\ \mathrm{Ta} + 3\ \mathrm{Cc} = 1\ \mathrm{Tr} + 1\ \mathrm{Do} + 1\ \mathrm{CO_2} + 1\ \mathrm{H_2O} \qquad (3)$$

takes place only when quartz is absent due to the fact that all quartz has been consumed in reaction (2) or (1). The equilibrium data of (3) seem to be very close to reaction (2) (personal communication of Puhan and Metz, 1978).

[1]The following studies, in addition to papers cited in this chapter, are relevant: Gordon and Greenwood (1970), Skippen (1971, 1974), Slaughter *et al.* 1975).

Reaction

$$5\ Do + 8\ Q + 1\ H_2O = 1\ Tr + 3\ Cc + 7\ CO_2 \quad (4)$$

is represented by the important assemblage

$$Do + Q + Tr + Cc \quad (4)$$

Reaction

$$2\ Do + 1\ Ta + 4\ Q = 1\ Tr + 4\ CO_2 \quad (5)$$

is represented by the assemblage

$$Do + Q + Tr + Ta \quad (5)$$

However, this assemblage is possible only if the original rock consisted of magnesite + dolomite + quartz, *i.e.*, did not contain calcite.

Reaction

$$1\ Do + 2\ Q = 1\ Di + 2\ CO_2 \quad (8)$$

is represented by the diopside-bearing assemblage

$$Do + Q + Di \quad (8)$$

As mentioned before, this paragenesis is petrogenetically significant.

Finally, the assemblage of the isobaric invariant point (II) is characterized by the following assemblage of five minerals:

$$(II)\ Do + Q + Di + Cc + Tr$$

This paragenesis (II) will be very significant as soon as equilibrium data become available at various fluid pressures. The isobaric univariant lines of reactions (4), (8), (6), and (7) radiate from the isobaric invariant point (II). Reaction (6), in addition to reaction (8), is believed to be of great importance in the first formation of diopside.

$$1\ Tr + 3\ Cc + 2\ Q = 5\ Di + 3\ CO_2 + 1\ H_2O \quad (6)$$

$$1\ Ca_2Mg_5[(OH)_2/Si_8O_{22}] + 3\ CaCO_3 + 2\ SiO_2 = 5\ CaMgSi_2O_6 + 3\ CO_2 + 1\ H_2O$$

The equilibrium paragenesis of this reaction is

$$Di + Tr + Cc + Q \quad (6)$$

In isobaric T-X_{CO_2} sections, the equilibrium curve of reaction (6) passes through a very flat maximum at $X_{CO_2} = 0.75$. Therefore, at values of $X_{CO_2} > 0.4$, the isobaric equilibrium temperature changes very little. Paragenesis (6) may therefore be of petrogenetic significance.

Reaction (7) is

$$1\ \text{Tr} + 3\ \text{Cc} = 4\ \text{Di} + 1\ \text{Do} + 1\ CO_2 + 1\ H_2O \qquad (7)$$

Its equilibrium paragenesis

$$\text{Di} + \text{Tr} + \text{Cc} + \text{Do} \qquad (7)$$

is distinct from paragenesis (6) by the absence of quartz and the presence of dolomite instead. The isobaric equilibrium line of reaction (7) runs between the two isobaric invariant points (II) and (IV) as seen in Figure 9-5. Depending on fluid composition, isobaric equilibrium temperatures vary considerably.

The first formation of forsterite in metamorphosed siliceous dolomites takes place at relatively high temperatures according to reactions (11) and (14). Reaction (11) is the following:

$$1\ \text{Tr} + 11\ \text{Do} = 8\ \text{Fo} + 13\ \text{Cc} + 9\ CO_2 + 1\ H_2O \qquad (11)$$

$$1\ Ca_2Mg_5[(OH)_2/Si_8O_{22}] + 11\ CaMg(CO_2)_2 = 8\ Mg_2SiO_4 + 13\ CaCO_3 + 9\ CO_2 + 1\ H_2O$$

The equilibrium paragenesis is

$$\text{Fo} + \text{Cc} + \text{Tr} + \text{Do} \qquad (11)$$

Reaction (11) has been investigated by Metz (1967, 1976) at 1, 3, and 5 kb fluid pressure. Disregarding fluid compositions of less than $X_{CO_2} = 0.2$, the range of the equilibrium temperature is about 50°C at 1 kb, 40°C at 3 kb, and only 30°C at 5 kb. Therefore, paragenesis (11) may be of some petrogenetic value. However, the following common paragenesis (14) is much more significant:

Reaction

$$1\ \text{Di} + 3\ \text{Do} = 2\ \text{Fo} + 4\ \text{Cc} + 2\ CO_2 \qquad (14)$$

$$\text{Paragenesis Fo} + \text{Di} + \text{Do} + \text{Cc} \qquad (14)$$

Reaction (14) has been investigated at 1, 3, and 5 kb fluid pressure by Käse and Metz (1979). The small temperature stability range of para-

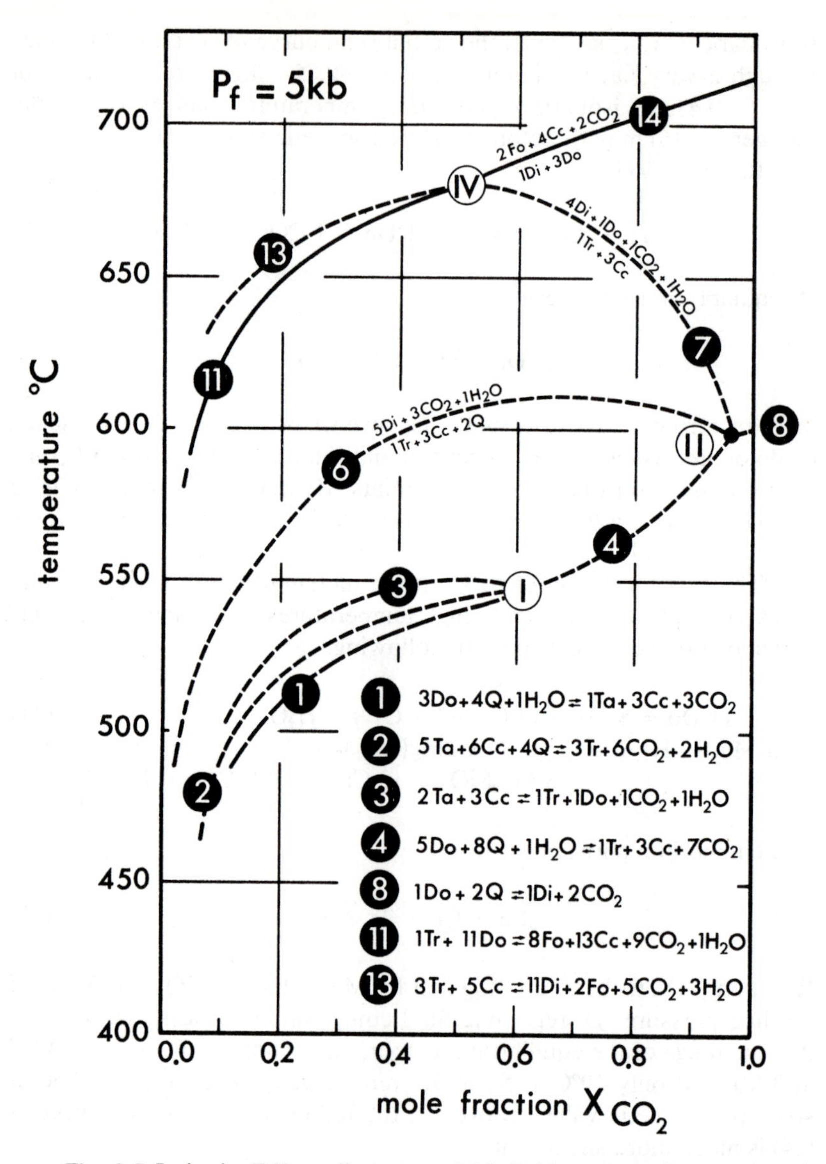

Fig. 9-5 Isobaric T-X_{CO_2} diagram at 5 kb fluid pressure for reactions in siliceous dolomites. Curve (11) from Metz (1976), curve (14) and position of (IV) from Käse and Metz (1979); other approximate data from Metz and Puhan (1978, personal communication).

genesis (14), at various fluid pressures, has been pointed out before and is shown in Figure 9-4.

The importance of the paragenesis of the isobaric invariant point (IV) was also pointed out earlier and demonstrated in Figure 9-4, *i.e.*, the paragenesis

$$\text{Fo} + \text{Di} + \text{Tr} + \text{Do} + \text{Cc} \qquad \text{(IV)}$$

Reaction (13) extends from point (IV), as shown in Figure 9-5:

$$3\ \text{Tr} + 5\ \text{Cc} = 11\ \text{Di} + 2\ \text{Fo} + 5\ CO_2 + 3\ H_2O \qquad (13)$$
$$\text{Paragenesis Fo} + \text{Di} + \text{Tr} + \text{Cc} \qquad (13)$$

At temperatures above reactions (13) and (14) the three-mineral paragenesis

$$\text{Di} + \text{Fo} + \text{Cc}$$

is stable over a large range of temperature and rock composition. In addition to equilibrium parageneses (13), (14), and (11), Trommsdorff (1972) has discovered the four-mineral parageneses of the reaction equilibria (2), (6), (3), (4), (8), and (7), shown in Figures 9-2 and 9-5 and in the table given earlier. This shows the importance of experimental work as a stimulus to field studies.

Concluding Remarks

In this section on the metamorphism of siliceous dolomitic limestones, some points of general significance have emerged:

1. Dolomite in sedimentary rocks is an important reactant only if quartz is present as well in the original rock.
2. Dolomite present in excess over the amount required for various reactions persists into high-grade regional metamorphism. Dolomite dissociates into $MgO + CO_2 + CaCO_3$ only at the high temperatures and low pressures of contact metamorphism. If X_{CO_2} is close to zero, dolomite + water will react to form brucite + $CO_2 + CaCO_3$ (Figure 9-6).
3. The presence of calcite in siliceous dolomitic sediments is not necessary in any of the reactions. Rather, all the calcite taking part in the reactions discussed here has been produced in previous reactions starting with dolomite + quartz only.
4. Similar to dolomite, any calcite present in excess over the

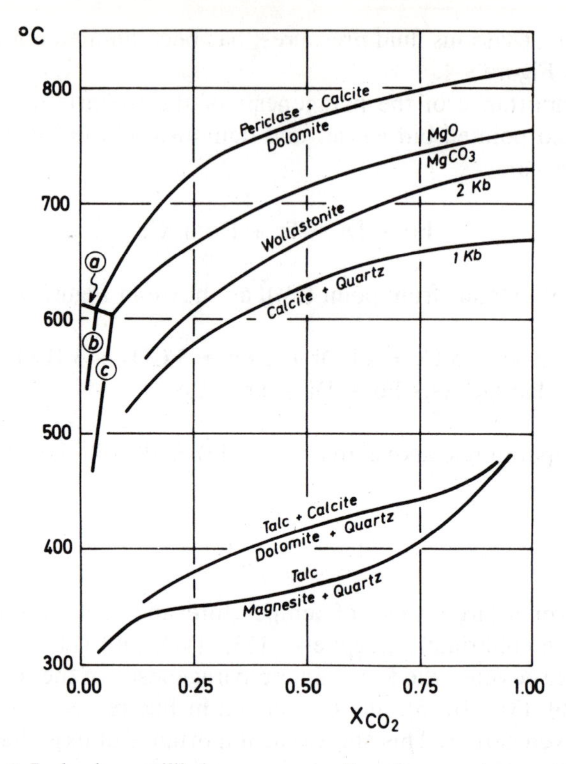

Fig. 9-6 Isobaric equilibrium curves of various metamorphic reactions. X_{CO_2} is the mole fraction of CO_2 in the fluid phase which consists of CO_2 + H_2O. The fluid pressure P_f is 1 kb, except where indicated otherwise.

amount necessary for the various reactions persists into high-grade regional metamorphism. Eventually, quartz + calcite may react to form wollastonite under conditions discussed in the following section.

5. Similar considerations apply to *magnesite* ($MgCO_3$) when present in excess amounts in siliceous dolomitic sediments. However, this is a very special case because the magnesite that is present in a siliceous dolomite commonly is completely consumed in metamorphic reactions. Reactions (5) and (15) can take place only if some magnesite is present as an additional constituent in siliceous dolomites. These parageneses are very rare compared to those resulting from metamorphism of siliceous dolomite having calcite (instead of magnesite) as an additional constituent.

Influence of FeO The numerical data given in this chapter refer to minerals composed only of the components CaO-MgO-SiO_2-CO_2-H_2O;

i.e., minerals devoid of FeO. Partial replacement of Mg by Fe^{2+} in dolomite is, however, common and this will influence the equilibrium data. This has to be investigated in each specific case and work on this problem would be highly desirable. Calculations by Thompson (1975) indicate that the equilibrium temperature of reactions occurring in siliceous dolomites may change by only 10–20°C if the FeO component is present in commonly observed concentrations.

Influence of Al_2O_3 Al_2O_3 in metamorphic siliceous dolomites (as well as in ultramafic rocks) gives rise to the formation of clinochlore, a MgAl-chlorite. In the absence of quartz, this chlorite will persist to high temperatures. Thus the paragenesis forsterite + calcite + dolomite + clinochlore has been recorded from medium- and even high-grade rocks (amphibolite facies, sillimanite zone) of the Central Alps by Trommsdorff (1966) and Trommsdorff and Schwander (1969). If potassium is present, phlogopite is formed instead of clinochlore (see below).

At high temperatures of metamorphism and probably at rather low pressures, clinochlore may be partially replaced by MgAl-spinel. This is a common constituent of the brucite marbles, the so-called predazzites, of the Bergell Alps. In these rocks forsterite may occur, and clinochlore [together with clinohumite, $Mg(OH,F)_2 \cdot 4\ Mg_2SiO_4$ and chondrodite, $Mg(OH,F)_2 \cdot 2\ Mg_2SiO_4$] is less common than the spinel (Trommsdorff and Schwander, 1969).

In metamorphosed siliceous carbonate rocks, some other Al-bearing minerals may occur in widely distributed but small amounts. These minerals are biotite (instead of phlogopite), muscovite, plagioclase, and (at medium- and high-grade) scapolite. The latter two minerals may coexist; however, replacement of plagioclase by scapolite has been observed very often.

It should be realized in this context that two of the several scapolite end members may be written as

$$3\ CaAl_2Si_2O_8 \cdot CaCO_3 \text{ and } 3\ NaAlSi_3O_8 \cdot CaCO_3,$$

i.e., as a combination of $CaCO_3$-component with anorthite- or albite-component, respectively. Although scapolite in metamorphic rocks is generally viewed as secondary, after plagioclase, it may be a stable reaction product formed from plagioclase and calcite (and/or $CaSO_4$) at higher metamorphic temperatures. This is strongly suggested by the experimental work of Newton and Goldsmith (1976) and of Orville (1975). Also, the stable coexistence of scapolite and plagioclase is shown by Goldsmith and Newton (1977).

The following phlogopite producing reaction has been investigated experimentally by Puhan and Johannes (1974) at 1 kb, and at higher pressures by Puhan (1978):

$$3 \text{ dolomite} + 1 \text{ K feldspar} + 1\ H_2O =$$
$$1 \text{ phlogopite} + 3 \text{ calcite} + 3\ CO_2$$

At any given X_{CO_2} and P_f, this reaction takes place at a temperature about 20–10°C lower than the talc producing reaction (1). The isobaric equilibrium curve of the phlogopite producing reaction runs approximately parallel to that of reaction (1); see Figure 9-5. This makes it clear that the equilibrium paragenesis

$$\text{Do} + \text{K spar} + \text{Phlog} + \text{Cc}$$

is not a suitable temperature indicator at a given fluid pressure.

At higher temperature, the following reaction takes place:

$$5 \text{ phlogopite} + 6 \text{ calcite} + 24 \text{ quartz} =$$
$$3 \text{ tremolite} + 5 \text{ K feldspar} + 2\ H_2O + 6\ CO_2$$

This bivariant reaction has been investigated by Hoschek (1973) and by Hewitt (1975).

Formation of Wollastonite

The carbonates dolomite and magnesite react whenever quartz is present at low-grade metamorphism. This is not true of calcite, the most widespread carbonate. As shown in Figure 9-7, much higher temperatures are needed for calcite to react with quartz; therefore, calcite + quartz are constituents of very many mineral parageneses from very-low-to high-grade. Even at the highest temperatures attained in regional metamorphism, calcite + quartz are generally stable. They only react when either the CO_2-rich fluid phase is considerably diluted by H_2O or the pressure of the CO_2-rich fluid is low as in shallow contact metamorphism.

The reaction

$$\text{calcite} + \text{quartz} = \text{wollastonite} + CO_2$$
$$CaCO_3 + SiO_2 = CaSiO_3 + CO_2$$

typically takes place in the higher temperature ranges of shallow contact metamorphism. In regional metamorphism, wollastonite is formed only in exceptional cases where water from surrounding rocks was introduced into thin layers of carbonate rocks and diluted the CO_2 concentration of the fluid phase. These conclusions are based on the following experimental results.

The reaction for the formation of wollastonite as written above is univariant. However, since water is present in the pore fluid of the rocks, this reaction is bivariant in nature. The reaction represents an equilibrium involving the four components CaO, SiO_2, CO_2, and H_2O, which constitute the four phases calcite, quartz, wollastonite, and vapor; therefore, there are two degrees of freedom. At any given fluid pressure, the equilibrium temperature increases with increasing X_{CO_2}. Greenwood (1967a) has determined the isobaric equilibrium curve at a fluid pressure of 1000 and 2000 bars. The data listed in Table 9-1 are shown graphically in Figure 9-6, p.126. At lower fluid pressures, the equilibrium curve lies at lower temperatures.

In Figure 9-7 the dependence of the equilibrium temperature on the fluid pressure P_f is shown for various constant compositions of the fluid phase (constant X_{CO_2}). Figure 9-7 and the two corresponding curves in Figure 9-6 provide a clear illustration of a bivariant equilibrium. It is apparent that the reaction of wollastonite formation is not a suitable temperature indicator if nothing is known about the composition of the fluid phase. In an attempt to estimate the composition of the fluid phase, petrographical observations are used.

Wollastonite is never found in rocks metamorphosed at relatively low temperatures of 400° to 500°C, not even in contact metamorphic rocks formed at shallow depths corresponding to pressures of only a few hundred bars. This fact indicates that the X_{CO_2} of the fluid present during metamorphism is never very small; presumably it is always rather large. It is probably correct that at the same given pressure, the temperature of wollastonite formation in nature is perhaps 10° to 30°C lower than the corresponding equilibrium temperature at X_{CO_2} = 1.0. Accordingly, the formation of wollastonite in contact aureoles surrounding shallow-seated intrusions at a depth of 2 km (500 bars) takes place at about 600°C. In the case of deeper intrusions at 4 km (1000 bars), these temperatures would be about 650° to 670°C and, at depths of 7 to 8 km (2000 bars), the temperatures would exceed 700°C. In regional metamorphism, where higher pressures of several kilobars prevail, even temperatures of 700°

Table 9-1

	Temperature, °C	
X_{CO_2}	P_f = 1000 bars	P_f = 2000 bars
0.25	580	610
0.50	630	670
0.75	660	715
1.0	670	730

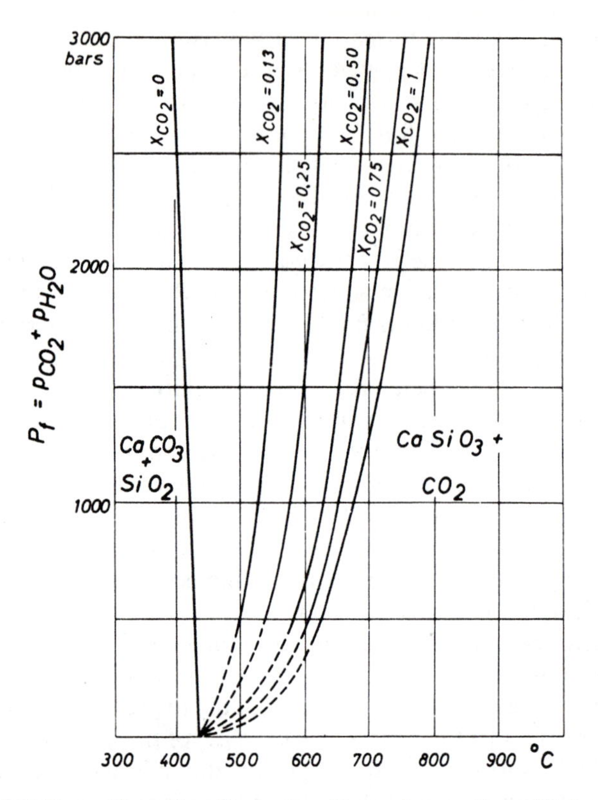

Fig. 9-7 Formation of wollastonite. Dependence of equilibrium temperatures on the fluid pressure P_f for various constant compositions of the fluid phase (const. X_{CO_2}). The diagram is based on experimental data taken from Greenwood (1967) and Harker and Tuttle (1956). Note: The curve designated $X_{CO_2} = 0$ is incorrectly labelled; it should be labelled partial pressure of CO_2 = 1 atm.

to 800°C are not high enough to allow the formation of wollastonite. Therefore, it is not surprising that wollastonite is not found in rocks formed even at the highest temperature of regional metamorphism; calcite and quartz remain a stable association. Only few examples of the occurrence of wollastonite in such rocks are known (Misch, 1964; Trommsdroff, 1968). In these cases layers of wollastonite-bearing rocks only a few centimeters thick are intercalated with other metamorphic rocks. In such thin layers, a dilution of the CO_2 liberated by the wollastonite reaction by water entering from surrounding rocks must have been possible. In this exceptional case the fluid phase in the calcsilicate rocks at the time of metamorphism had a low X_{CO_2}.On the other hand, the general absence of wollastonite indicates that this strong dilution of CO_2 does not take place if the original intercalctions of siliceous limestones have a greater thickness.

The reaction of wollastonite formation is bivariant and will proceed within a certain range of temperature. Thus, the equilibrium assemblage

wollastonite + calcite + quartz

is expected to have an areal distribution forming a reaction-isograd band or area rather than a line. This distribution has, in fact, been recorded by Trommsdorff (1968) in the western Bergell Alps.

In the discussion of the metamorphism of siliceous dolomitic limestones, wollastonite formation was not mentioned. However, in Figure 9-2 the reaction curve of wollastonite formation will be situated at temperatures markedly above those of reactions (13) and (14). At such conditions, the previous paragenesis

diopside + calcite + quartz

will be replaced, depending on bulk compositions, by

either diopside + wollastonite + calcite
or diopside + wollastonite + quartz

Also, the paragenesis stable along the isobaric equilibrium curve of the reaction wollastonite = calcite + quartz may be encountered.

diopside + wollastonite + calcite + quartz

The curves of *dissociation of magnesite and dolomite* are also shown in Figure 9-6, valid for $P_f = 1$ kb. Their dissociation requires even higher temperatures than that of calcite + quartz.

The curve for the reaction

$$\begin{array}{ccccc} \text{magnesite} & = & \text{periclase} & + & CO_2 \\ MgCO_3 & & MgO & & CO_2 \end{array}$$

has been calculated as well as experimentally determined by Johannes and Metz (1968). After metamorphism, the periclase that forms is easily altered to brucite, $Mg(OH)_2$, by circulating water. Therefore, most brucite marbles do not contain periclase.

The curve of dissociation of magnesite shown in Figure 9-6 is only significant in the metamorphism of rare deposits of pure magnesite or pure magnesite + dolomite. The introduction of CO_2 into serpentinites produces magnesite + quartz, and in the presence of quartz, magnesite will react and disappear at temperatures much lower than those of the dissociation of magnesite and of the formation of wollastonite. Periclase and brucite may also form at low temperatures by reactions other than the dissociation of magnesite. This will be discussed in Chapter 11 dealing with the metamorphism of ultramafic rocks.

At a temperature somewhat higher than that of the dissociation of magnesite, the following reaction takes place:

$$\begin{array}{lcl} \text{dolomite} & = & \text{periclase} + \text{calcite} + CO_2 \\ CaMg(CO_3)_2 & = & MgO + CaCO_3 + CO_2 \end{array}$$

Based on experimental data at $X_{CO_2} = 1.0$ determined by Harker and Tuttle (1955), Metz (personal communication) has calculated the equilibrium curve shown in Figure 9-6. It is evident that dolomite is stable to the highest temperatures of regional metamorphism unless the CO_2-rich fluid phase has been considerably diluted. Thus, either pure dolomite or the assemblage calcite + dolomite ± forsterite will be stable at high temperatures. The assemblage calcite + dolomite + forsterite has been formed in high-grade equivalents of dolomites which originally contained only a small amount of quartz.

The idea that dolomite is not stable in higher grade rocks is wrong. However, its presence or absence depends strongly on the nature of the associated minerals.

Dolomite will be decomposed to periclase + calcite in the innermost aureoles of hot intrusions and in xenoliths engulfed by such magmas. In shallow contact metamorphism the decomposition of dolomite to periclase + calcite becomes more probable at lower P_f. At $X_{CO_2} = 1.0$, the dissociation temperature is 820°C at 1000 bars, 760°C at 500 bars, and only about 700°C at 250 bars (corresponding to a depth of about 1 km).

The dissociation of dolomite can take place at appreciably lower temperatures only if X_{CO_2} of the fluid phase is very small, *i.e.*, if some mechanism has been effective locally in diluting the liberated CO_2 with H_2O. In this case, phase relations shown in Figure 9-6 at very low values of X_{CO_2} are valid. Here, the curves of dissociation of dolomite and magnesite are terminated by line (a) at about 600°C (1 kb). This line indicates the transition

$$\text{brucite} = \text{periclase} + H_2O \tag{a}$$

At temperatures below line (a) brucite is stable, and curves (b) and (c) refer to the following reactions:

$$\text{dolomite} + H_2O = \text{brucite} + \text{calcite} + H_2O \tag{b}$$
$$\text{magnesite} + H_2O = \text{brucite} + CO_2 \tag{c}$$

However, dilution of CO_2-rich fluid by H_2O seems to be a rare case restricted to narrow bands. The formation of parageneses which require a high X_{CO_2} during metamorphism of siliceous dolomites suggests that the formation of wollastonite and the dissociation of dolomite into periclase + calcite commonly take place at high X_{CO_2} as well. Therefore, intrusions that produced the assemblage periclase (or brucite pseudomorphs) + calcite in their contact zones must have risen to shallow depth where the fluid pressure was very low. Thus, it is apparent that wollastonite and the assemblage periclase + calcite cannot commonly occur in regionally metamorphosed rocks formed at greater pressures. This demonstrates how differences in the value of P_f may significantly determine the mineral assemblages produced by metamorphism.

Metamorphism of Carbonates at Very High Temperatures and Very Low Pressures

Special circumstances of very high temperature contact metamorphism may give rise to Ca and Ca-Mg silicates not yet mentioned. Such special conditions exist if, at very shallow depth, carbonate rocks are engulfed by magma or are heated at the immediate contact to a very high temperature. These conditions, different from those of the normal high-grade contact zone (K feldspar-cordierite-hornfels zone) are typical of the so-called sanidinite facies.

Apart form the term "facies," the designation "sanidinite facies" is poorly chosen. Although it is true that the occurrence of sanidine in hornfelses is an indication of high temperatures and rather rapid cooling, other minerals than sanidine are more typical of this type of contact metamorphism developed in the subvolcanic environment of intrusions and characterized by very high temperatures and very low pressures. The temperature range can be further subdivided on the bases of mineral assemblages. Fyfe *et al.* (1958) proposed a division into (a) monticellite-melilite subfacies and (b) larnite-merwinite-spurrite subfacies. On the other hand, Sobolev (cited in Reverdatto, 1964) drops the name "sanidinite facies" and recognizes (a) the monticellite-melilite facies and (b) the spurrite-merwinite facies. After substituting the term zone for facies, we shall partly follow these authors' proposal and use the designation

monticellite-melilite zone. (a)

For the higher temperature zone a new designation is suggested by recent experiments:

tilleyite/spurrite zone. (b)

In this zone (b) the association calcite + wollastonite is no longer stable; it is replaced by calcite + tilleyite *or* calcite + spurrite, depending on the relative amounts of wollastonite and calcite originally present.

It should be realized, however, that different sequences of zones can be distinguished according to various reactions taking place in wollastonite-bearing marbles. Experiments in such rocks have been carried out at high temperatures and very low CO_2 pressures ($X_{CO_2} = 1.0$) by Tuttle and Harker (1957), Harker (1959), and Zharikov and Shmulovich (1969). Starting out with the assemblage wollastonite + calcite, a number of minerals will form under the specific conditions considered here which are not ordinarily realized in metamorphism. The minerals consisting of components CaO, SiO_2, and CO_2 are

Tilleyite	$Ca_5[(CO_3)_2/Si_2O_7]$
Spurrite	$Ca_5[CO_3/(SiO_4)_2]$
Rankinite	$Ca_3(Si_2O_7)$
Larnite	β-Ca_2SiO_4

The following reactions lead to the formation of these minerals:

$$3 \text{ calcite} + 2 \text{ wollastonite} = 1 \text{ tilleyite} + CO_2 \quad (1)$$

$$1 \text{ tilleyite} = 1 \text{ spurrite} + CO_2 \quad (2)$$

$$1 \text{ tilleyite} + 4 \text{ wollastonite} = 3 \text{ rankinite} + 2\, CO_2; \text{ only at } CO_2 \text{ pressures greater than 150 bars.} \quad (3)$$

$$1 \text{ spurrite} + 4 \text{ wollastonite} = 3 \text{ rankinite} + CO_2; \text{ only at } CO_2 \text{ pressures smaller than 150 bars.} \quad (4)$$

$$1 \text{ spurrite} + 1 \text{ rankinite} = 4 \text{ larnite} + CO_2 \quad (5)$$

When the fluid phase consists of CO_2 only, all raction equilibria are univariant. The equilibrium curves as given by Zharikov and Shmulovich (1969) are shown in Figure 9-8. The mineral parageneses stable in the fields between curves are also indicated; any two minerals may coexist if they are adjacent on the composition line calcite-wollastonite. It should be noted that reaction curves (2), (3), and (4) intersect at an invariant point (double circle) at about 900°C and at a CO_2 pressure of about 150 bars. The invariant point is characterized by the paragenesis tilleyite + spurrite + rankinite + wollastonite.

The assemblage formed is dependent on reaction sequence and on bulk composition. The influence of bulk composition is very pronounced. When the ratio CaO:SiO_2 of the rock is smaller than 1.5 or larger than 2.5, the succession of parageneses with increasing temperature is simple:

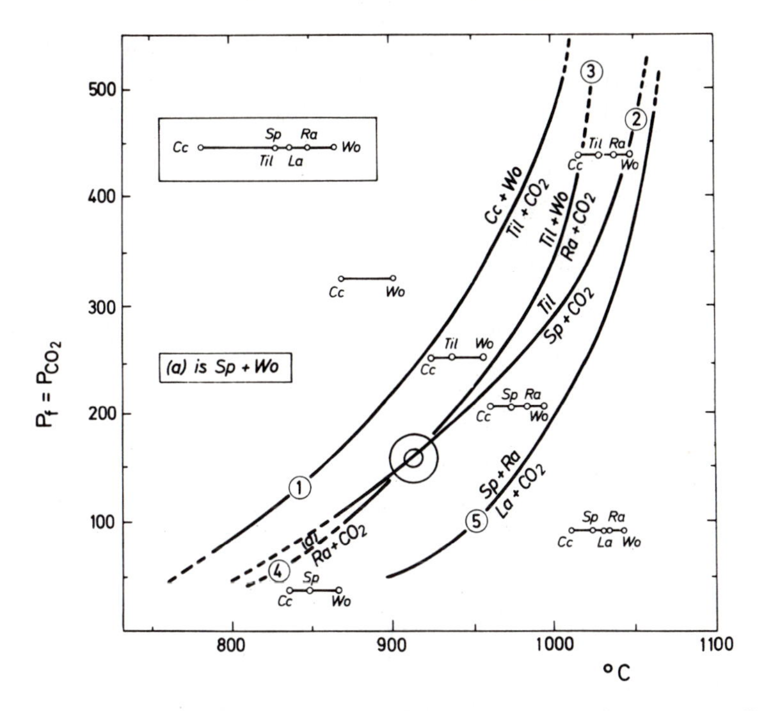

Fig. 9-8 P_{CO_2}-T diagram at very high temperatures of the system CaO-SiO_2-CO_2 in the composition range between calcite and wollastonite.

CaO:SiO_2 < 1.5 – Calcite + wollastonite
Calcite + tilleyite
Calcite + spurrite

CaO:SiO_2 > 2.5 – Calcite + wollastonite
Wollastonite + tilleyite
(Wollastonite + spurrite)
Wollastonite + rankinite

Only if the CaO:SiO_2 ratio of a rock is between 1.5 (the composition of rankinite) and 2.5 (the composition of tilleyite and spurrite), can parageneses form like

Spurrite + rankinite
Spurrite + larnite or
Larnite + rankinite

Reactions discussed so far do not involve Mg-bearing minerals. However, medium temperature metamorphism of siliceous dolomitic limestones will produce diopside and/or forsterite coexisting with calcite. At very high temperatures and low pressures, reactions among these minerals lead to the formation of the following Mg-bearing minerals:

Monticellite $CaMgSiO_4$
Åkermanite $Ca_2Mg(Si_2O_7)$
Melilite $Ca_2(Mg, Al)(Si, Al)_2O_7$
Merwinite $Ca_3Mg(SiO_4)_2$

These minerals, except merwinite, are formed within the stability field of wollastonite + calcite, *i.e.*, at conditions above curve (1) in Figure 9-8.

Monticellite may be formed by at least two reactions:

$$\begin{array}{l} \text{1 diopside} + \text{1 forsterite} + \text{2 calcite} = \text{3 monticellite} + 2\,CO_2 \\ 1\,CaMgSi_2O_6 + 1\,Mg_2SiO_4 + 2\,CaCO_3 = 3\,CaMgSiO_4 + 2\,CO_2 \end{array} \quad (6)$$

$$\begin{array}{l} \text{1 calcite} + \text{1 forsterite} = \text{1 monticellite} + \text{1 periclase} + 1\,CO_2 \\ 1\,CaCO_3 + 1\,Mg_2SiO_4 = 1\,CaMgSiO_4 + 1\,MgO + 1\,CO_2 \end{array} \quad (7)$$

For both reactions the equilibrium temperatures for various values of P_{CO_2}, at $X_{CO_2} = 1$, have been determined by Walter (1963). The two equilibrium curves, univariant at $X_{CO_2} = 1$, practically coincide. They show the same great dependence on CO_2 pressure as the curves in Figure 9-8. At the very low value of $P_{CO_2} = 100$ bars, the equilibrium temperature is already as high as 725°C. It may be concluded that at temperatures realized in nature, monticellite can form only at very low CO_2 pressures of, at most, a few hundred bars. The same is true of the formation of the other minerals. For example, akermanite may be formed according to the following reaction:

$$\text{1 calcite} + \text{1 diopside} = \text{1 akermanite} + 1\,CO_2 \quad (8)$$

According to Walter (1965), the equilibrium conditions of this reaction at $X_{CO_2} = 1.0$ are the same as those of reactions (6) and (7). Other reactions leading to the formation of monticellite or akermanite are also given by Walter (1963).

At a given CO_2 pressure, appreciably higher temperatures are required for the formation of merwinite. The following reaction takes place when the stability field of the assemblage calcite + wollastonite has been surpassed and spurrite has formed:

$$2 \text{ monticellite} + 1 \text{ spurrite} = 2 \text{ merwinite} + 1 \text{ calcite} \qquad (9)$$

The occurrence of merwinite alone does not permit an evaluation of temperature and pressure; rather, the assemblage merwinite + calcite is significant because this indicates that reaction (9) has proceeded to the right. According to the data of Walter (1965), the coexistence of merwinite + calcite indicates that the temperature of 820°C has been exceeded and that the CO_2 pressure was below 50 bars. However, this result may be somewhat different when it is realized that merwinite may as well have been formed by the following reaction (10):

$$\text{akermanite} + \text{calcite} = \text{merwinite} + CO_2 \qquad (10)$$

In rocks with excess calcite, the assemblage merwinite + calcite, which is also formed by reaction (9), will be present after reaction (10) has taken place.

Reaction (10) as well as reactions (8), (7), and (6) have been experimentally investigated by Zharikov and Bulatov (1977) at 1 kb fluid pressure and at a mole fraction of CO_2 smaller than 0.4. From their results, the equilibrium temperature at 1 kb fluid pressure and at $X_{CO_2} = 0.1$ is

for reaction (6) : 755°C
for reaction (8) : 790°C
for reaction (7) : 875°C
for reaction (10): 890°C

Note that at a very low mole fraction of CO_2, reaction (7) takes place at much higher temperature than reaction (6), in contrast to the situation at $X_{CO_2} = 1.0$ which was discussed above.

Contact metamorphism taking place at very shallow depth and at very high temperatures is a very unusual and rare phenomenon. Such metamorphism may be accompanied as, *e.g.*, at Crestmore, by intensive metasomatism, and gases forced into the country rock may greatly increase the heating effect of the intrusion.

References

Bickle, M. J. and Powell, R. 1977. *Contr. Mineral Petrol.* **59:** 281–292.
Bowen, N. L. 1940. *J. Geol.* **48:** 225–274.
Eskola, P. 1922. *J. Geol.* **30:** 265–294.
Fyfe, W. S., Turner, F. J., and Verhoogen, J. 1958. *Geol. Soc. Am. Memoir 73.*
Garde, A. A. 1977. *Contr. Mineral. Petrol.* **62:** 265–270.

Goldsmith, J. R. and Newton, R. C. 1969. *Am. J. Sci.* **267–A,** Schairer Volume: 160–190.

——— 1977. *Amer. Mineral.* **62:** 1063–1081.

Gordon, T. M. and Greenwood, H. J. 1970 *Am. J. Sci.* **268:** 225–242.

Greenwood, H. J. 1961/62. *Ann. Rept. Carnegie Inst. Geophys. Lab. 1961/62:* 82–85.

——— 1967a. *Am. Mineral.* **52:** 1169–1680.

——— 1967b. *In* P. H. Abelson, ed. *Researches in Geochemistry.* Vol. 2, pp. 542–567. John Wiley & Sons, New York.

Greenwood, H. J. 1975. *Amer. J. Sci.* **275:** 573–593.

Guitard, G. 1966. *Compt. Rend. Acad. Sci. Paris* **262:** 245–247.

Harker, R. I. 1959. *Am. J. Sci.* **257:** 656–667.

——— and Tuttle, O. 1955. *Am. J. Sci.* **253:** 209–244.

——— 1956. *Am. J. Sci.* **254:** 231–256.

Hewitt, D. A. 1975. *Amer. Miner.* **60:** 391–397.

Hoschek, G. 1973. *Contr. Mineral. Petrol.* **39:** 231–237.

Hutcheon, I., and Moore, J. M. 1973. *Can. J. Earth Sci.* **10:** 936–947.

Jansen, J. B. H., v. d. Kraats, A. H., v. d. Rijst, H., and Schuiling, R. D. 1978. *Contr. Mineral. Petrol.* **67:** 279–288.

Johannes, W. and Metz, P. 1968 *Neues Jahrb. Mineral. Mh.* **1968:** 15–26.

Käse, H. R. and Metz, P. 1979 *Contr. Mineral Petrol.* (in press).

Metz, P. 1967. *Geochim. Cosmochim. Acta* **31:** 1517–1532.

——— 1970, *Contr. Mineral. Petrol.* **28:** 221–250.

——— 1970. *Habilitationsschrift.* Göttinger 1970: 1–20.

——— and Puhan, D. 1970. *Contr. Mineral. Petrol.* **26:** 302–314.

——— 1971. *Contr. Mineral. Petrol.* **31:** 169–179.

——— 1976. *Contr. Mineral. Petrol.* **58:** 137–148.

——— 1977. *Tectonophysics.* **43:** 163–167.

Metz, P. and Trommsdorff. 1968. *Contr. Mineral. Petrol.* **18:** 305–209.

Metz, P., and Winkler, H. G. F. 1963. *Geochim. Cosmochim. Acta* **27:** 431–457.

Misch, P. 1964 *Contr. Mineral. Petrol.* **10:** 315–356.

Newton, R. C. and Goldsmith, J. R. 1976. *Z. Krist.* **143:** 333–353.

Orville, P. M. 1975. *Geochim. Cosmochim. Acta* **39:** 1091–1105.

Puhan, D. 1976. *Contr. Mineral. Petrol.* **58:** 23–28.

——— 1978. *Neues Jahrb. Mineral. Monatsh.* **1978:** 110–127.

Puhan, D. and Hoffer, E. 1973. *Contr. Mineral. Petrol.* **40:** 207–214.

Puhan, D. and Johannes, W. 1974. *Contr. Mineral. Petrol.* **48:** 23–31.

Puhan, D. and Metz, P. 1973 *In* D. Puhan and E. Hoffer. *Contr. Mineral. Petrol.* **40:** 207–214.

Reverdatto, V. V. 1964. *Geochem. Intern.* **1:** 1038–1053.

Skippen, G. B. 1971. *J. Geol.* **79:** 457–481.

——— 1974. *Am. J. Sci.* **274:** 487–509.

——— and Hutcheon, I. 1974. *Canadian Miner.* **12:** 327–333.

Slaughter, J., Kerrick, D. M., and Wall, V. J. 1975 *Am. J. Sci.* **275:** 143–162.

Steck, A. 1969. *Kaledonische Metamorphose. NE-Gronland,* Habilitationsschrift, Basel.

Thompson, A. B. 1975. *Contr. Mineral. Petrol.* **53:** 105–127.
Tilley, C. E. 1948. *Mineral Mag.* **28:** 272–276.
Trommsdorff, V. 1966. *Schweiz. Mineral. Petrog. Mitt.* **46:** 421–429.
——— 1968. *Schweiz. Mineral. Petrog. Mitt.* **48:** 655–666, 828–829.
——— 1972. *Schweiz. Mineral. Petrog. Mitt.* **52:** 567–571.
——— and Schwander, H. 1969. *Schweiz. Mineral. Petrog. Mitt.* **49:** 333–340.
Turner, F. J. 1968. *Metamorphic Petrology.* McGraw-Hill Book Company, New York.
Tuttle, O. F., and Harker, R. I. 1957, *Am. J. Sci.* **255:** 226–234.
Walter, L. S. 1963, *Am. J. Sci.* **261:** 488–500, 773–779.
——— 1965. *Am. J. Sci.* **263:** 64–77.
Zharikov, V. A., and Shmulovich, K. I. 1969. *Geochim. Intern.* **6:** 849–869.
Zharikov, V. A., and Bulatov, V. K. 1977. *Tectonophysics,* **43:** 145–162.

Chapter 10

Metamorphism of Marls

Marls constitute a group of sedimentary rocks with a very large range of composition. The clay constituents, predominantly dioctahedral micas (illite, phengite, muscovite), kaolinite, chlorite, and quartz, are mixed with calcite and/or dolomite in varying proportions. The chemical composition of such rocks can be expressed only by a very complex system: K_2O-CaO-Al_2O_3-MgO-FeO-SiO_2-CO_2-H_2O, and commonly Na_2O. The approach to an understanding of such a complex system involves the study of simpler subsystems. The system CaO-Al_2O_3-SiO_2-CO_2-H_2O has been investigated in some detail.

Thompson (1971) has made a theoretical study of reactions at temperatures of very-low-grade metamorphism. Minerals such as laumontite, lawsonite, prehnite, kaolinite, quartz, and calcite are stable. The CaAl silicates are generally derived from the alteration of mafic rocks and not from marls. However, metamorphism of marls could produce laumontite (or, at pressures above 3 kb, lawsonite) according to the reaction

$$\text{kaolinite} + \text{calcite} + 2\ \text{quartz} + 2\ H_2O = \text{laumontite} + CO_2$$

and, at higher temperature, prehnite, according to the reaction

$$\text{laumontite} + \text{calcite} = \text{prehnite} + \text{quartz} + 3\ H_2O + CO_2$$

The CaAl-minerals laumontite, prehnite, and lawsonite are stable only if the CO_2 content of the fluid phase does not exceed 1% (see Chapter 12). The assemblage lawsonite + quartz + dolomite, together with chlorite, phengite, and paragonite, has been found as a product of metamorphosed marl in southwestern Crete by Seidel and Okrusch (1976).

The following minerals may be present in marls subjected to very-low-grade metamorphism: pyrophyllite, paragonite, mixed-layer paragonite/muscovite and phengite, in addition to quartz, calcite, and/or dolomite (Frey, 1978). These metamorphites contain graphite which probably drastically reduced the activity of water in favour of CH_4 in the fluid phase; see Figure 3-2. Reduction of water activity shifts the lower stability of pyrophyllite to lower temperatures than those shown in Figure 14-1, curve 1. This effect may account for the presence of pyrophyllite in very-low-grade rocks.

Chlorite, paragonite, and phengite/muscovite, in addition to quartz and calcite, persist to higher grades. However, new minerals like chloritoid, margarite, and zoisite/clinozoisite are formed by low-grade metamorphism of marls.

In marls which originally contained an aluminum-rich clay constituent (kaolinite), margarite $CaAl_2[(OH)_2/Al_2O_{10}]$ will form at low-grade metamorphism (greenschist facies). This mineral was considered rare except in metamorphic emery deposits. However, Frey and Niggli (1972) and Frey (1978) recently discovered, by systematic x-ray diffraction studies of metamorphosed marly black shales, that margarite is an important rock-forming mineral and may even constitute the dominant sheet silicate. It is safe to predict that mineral identification by x-ray diffraction supplemented by microscopic work will prove margarite to be common in low-grade metamorphic areas. The stability of margarite in various mineral assemblages will be reviewed further on in this chapter.

The Ca-free mineral chloritoid occurs in low-grade metamorphic marls if chlorite together with kaolinite or pyrophyllite were present at an earlier stage. Also during low-grade metamorphism, chloritoid in metamarls reacts with calcite and quartz to produce chlorite and zoisite or margarite (Frey, 1978).

The CO_2-free system CaO-Al_2O_3-SiO_2-H_2O at high pressures and temperatures has been studied by Boettcher (1970), and the same system with CO_2 as an additional component has been investigated by Storre (1970), Kerrick (1970), Gordon and Greenwood (1971), and Storre and Nitsch (1972). The minerals in this system are graphically represented in Figure 10-1. From the data given by these authors an isobaric equilibrium diagram at $P_f = 2$ kb has been compiled which excludes margarite (Figure 10-2).

A selection of mineral assemblages is also shown in Figure 10-2. These will not be reviewed in detail. However, the stability of zoisite and grossularite and the assemblage wollastonite + anorthite are of special interest.

The appearance of *zoisite* (or iron-poor epidote, *i.e.*, clinozoisite) characterizes the beginning of low-grade metamorphism. Zoisite is formed between 350° and 400°C by reactions, some of which are given in Chapter 7 (see Figure 7-2). Subsequently, zoisite is a member of many

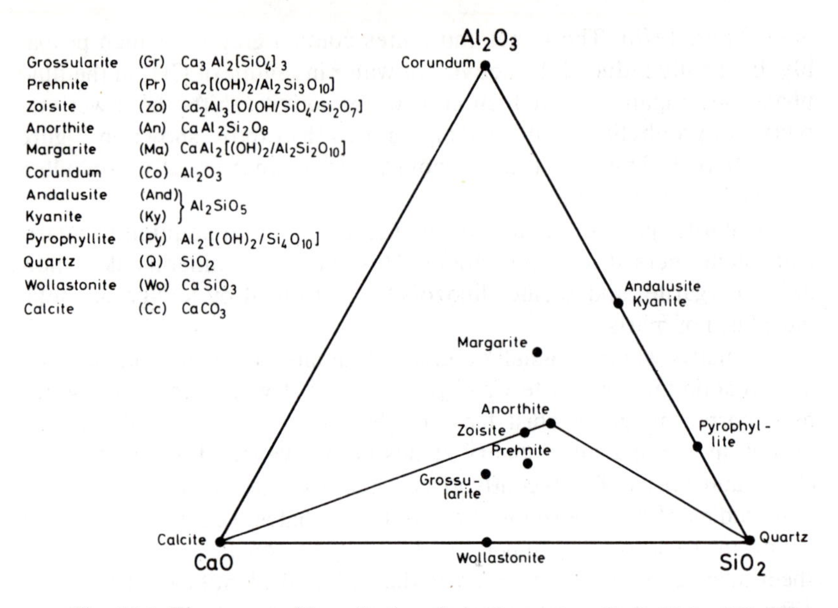

Fig. 10-1 The composition of minerals in the system CaO-Al_2O_3-SiO_2-CO_2-H_2O.

parageneses in a temperature range of about 200°C. The stability field of zoisite is different, depending on whether quartz is absent or present:

In the *absence of quartz,* the stability field of zoisite is restricted by reaction (9) with respect to X_{CO_2}:

$$2 \text{ zoisite} + 1\ CO_2 = 3 \text{ anorthite} + 1 \text{ calcite} + 1\ H_2O \quad (9)$$

and by reaction (10) with respect to temperature:

$$6 \text{ zoisite} = 6 \text{ anorthite} + 2 \text{ grossularite} + 1 \text{ corundum} + 1\ H_2O \quad (10)$$

In the *presence of quartz,* the existence of zoisite is restricted further with respect to temperature by reaction (8):

$$4 \text{ zoisite} + 1 \text{ quartz} = 1 \text{ grossularite} + 5 \text{ anorthite} + 2H_2O \quad (8)$$

and with respect to X_{CO_2} by reaction (9) which is also valid in the presence of quartz. The isobaric invariant points (III) and (IV) in Figure 10-2 are produced by the intersection of curves (8) and (9) and of (10) and (9), respectively. The corresponding mineral assemblages deserve special attention in fieldwork. Puhan (personal communication) found paragenesis (III) in his work in South West Africa. The P_f-T dependency

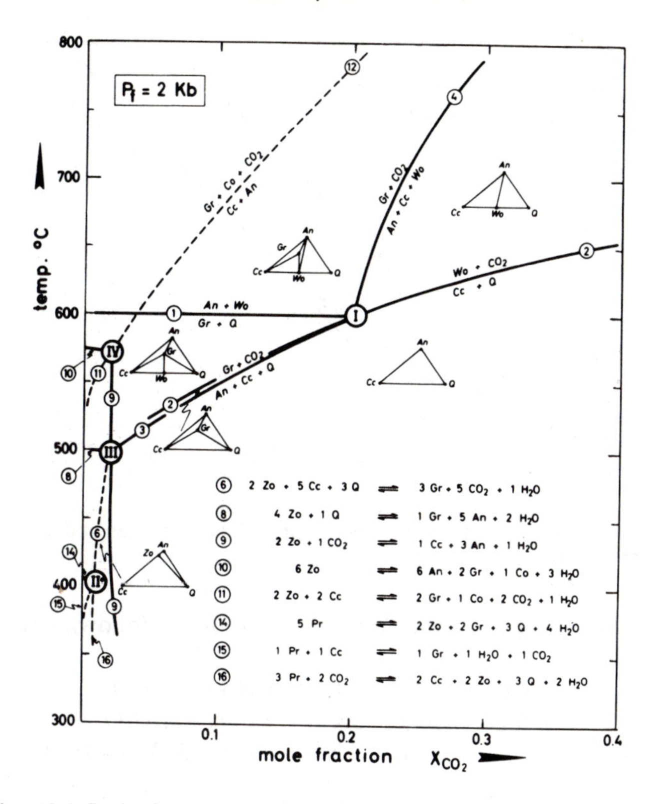

Fig. 10-2 Isobaric (P_f = 2 kb) phase relations in the system CaO-Al_2O_3-SiO_2-H_2O-CO_2 to 800°C. Reactions involving margarite are excluded. The parageneses are represented in the subtriangle Cc-An-Q, which is shown in Fig. 10-1. (Compiled by Storre and Nitsch, 1972.)

This figure takes into account all available experimental data; it differs from figures previously given by Storre (1970, Fig. 4 and 5) mainly in the following aspects pointed out by Gordon and Greenwood (1971): The formation of grossularite + quartz at low temperatures is due to the decomposition of prehnite; reaction (14) takes the place of a reaction which now is realized to be metastable. Consequently, the former isobaric invariant point (II) has been changed to (II*).

of (III) and (IV) is shown in Figure 10-3 (Storre, 1973, unpublished). The parageneses are:

$$\text{(III) Zo + Gr + Qz + An + Cc}$$
$$\text{(IV) Zo + Gr + Co + An + Cc}$$

In most rocks anorthite is present as a component of plagioclase. This reduction of the anorthite activity will displace the curves of Figure 10-3 to lower temperatures. The magnitude of this effect is still unknown.

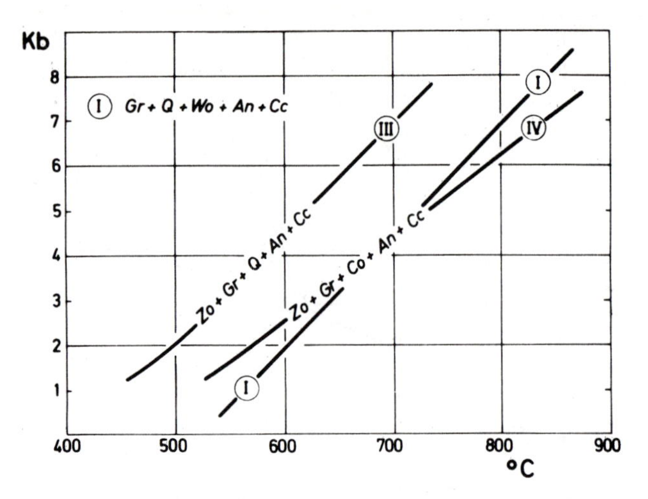

Fig. 10-3 Equilibrium conditions of mineral assemblages stable at isobaric invariant points (III), (IV), and (I) of Figure 10-2.

The *stability of grossularite* is restricted by the following reactions:

$$1\ \mathrm{Pr} + 1\ \mathrm{Cc} = 1\ \mathrm{Gr} + 1\ \mathrm{H_2O} + 1\ \mathrm{CO_2} \quad (15)$$

$$2\ \mathrm{Zo} + 5\ \mathrm{Cc} + 3\ \mathrm{Q} = 3\ \mathrm{Gr} + 5\ \mathrm{CO_2} + 1\ \mathrm{H_2O} \quad (6)$$

$$5\ \mathrm{Pr} = 2\ \mathrm{Zo} + 2\ \mathrm{Gr} + 3\ \mathrm{Q} + 4\ \mathrm{H_2O} \quad (14)$$

From approximately 400°C at 2 kb, grossularite is stable in the presence of quartz to about 600°C. At higher temperatures, quartz and grossularite react:

$$1\ \mathrm{Gr} + 1\ \mathrm{Q} = 2\ \mathrm{Wo} + 1\ \mathrm{An} \quad (1)$$

This reaction stabilizes the assemblage wollastonite + anorthite.

The wollastonite + anorthite assemblage has been reported in low pressure and high temperature contact metamorphism and in special cases of regional metamorphism. It may be rewarding to search for the isobaric invariant paragenesis (I) Gr + Q + Wo + An + Cc (see Figure 10-3). In the absence of quartz, grossularite occurs in various assemblages to very high temperatures [875°C at 2 kb (after Boettcher, 1970)]. With respect to X_{CO_2}, the stability of grossularite is limited by reaction (6) and the following:

$$1\ \mathrm{Gr} + 2\ \mathrm{CO_2} = 1\ \mathrm{An} + 2\ \mathrm{Cc} + 1\ \mathrm{Q} \quad (3)$$

$$1\ \mathrm{Gr} + 1\ \mathrm{CO_2} = 1\ \mathrm{An} + 1\ \mathrm{Cc} + 1\ \mathrm{Wo} \quad (4)$$

Previously, it was not known that zoisite and grossularite + quartz are restricted to fluid compositions of low CO_2 content.[1] Whenever wollastonite is associated with these minerals, it has been formed at low X_{CO_2} values and consequently at much lower temperatures than in massive beds of siliceous calcite marbles where X_{CO_2} was large. The formation of wollastonite at low X_{CO_2} takes place only in thin calcareous layers embedded with pelitic rocks, if water from the pelitic rocks has lowered X_{CO_2} in the calcareous layers. Such occurrences have been observed by Misch (1964) in the northwest Himalayas, and Gordon and Greenwood (1971) pointed out that the zonal sequence of calc-silicate assemblages, described by Misch, can be understood on the basis of phase relations in the system CaO-Al_2O_3-SiO_2-CO_2-H_2O.

The calc-silicate assemblages are embedded with crystalline schists and comprise a roughly circular area with a radius of about 20 km. The associated schists contain kyanite in zones 4 and 3 and sillimanite in the higher temperature zones 2 and 1. All zones represent medium-grade metamorphism (amphibolite facies), but the inner core of zone 1 extends into high-grade metamorphism. This indicates a total pressure of about 6 to 7 kb and a temperature of about 650°C in the core of zone 1.

A phase diagram at that total pressure is not yet available. However, the sequence of parageneses is probably the same as at 2 kb (Figure 10-2). The temperatures at 6 kb will of course be higher[2] by approximately 150°C, but the temperature differences between reactions will be almost the same as at 2 kb.

Gordon and Greenwood (1971) present an evaluation of metamorphic conditions under which the calc-silicate assemblages of the northwest Himalayas were formed. Figure 10-2, valid for 2 kb, is not suitable for a quantitative evaluation of conditions at very much higher pressure. However, it serves to demonstrate an important point: The petrogenetic significance of parageneses may be very different. Some parageneses may indicate the temperature and some may indicate the composition of the fluid phase, but others do not reflect definite metamorphic conditions, *i.e.*, they are of no petrogenetic significance. As an illustration, mineral assemblages from the northwest Himalayas will be interpreted on the basis of phase relations shown in Figure 10-2 as if they had been formed at 2 kb.

The sequence of zones from low to high temperatures is in the order 4 to 1:

---

[1]Grossularite without quartz is also restricted to low X_{CO_2} values at $P_f = 2$ kb, but at high pressure grossularite extends to larger X_{CO_2} values (Gordon and Greenwood, 1971).

[2]See Boettcher, 1970.

Zone 4: Paragenesis *Zo* + *Cc* + *Q*; this is possible to the right of curve (6) and left of (9).
Evaluation: Temperature below 500°C at 2 kb; X_{CO_2} very low (smaller than 0.02 at 2 kb).

Zone 3: Paragenesis *An* + *Cc* + *Q*; this is possible to the right of curve (9) and below (3).
Evaluation: Both temperature and X_{CO_2} undefined over a wide range.
Paragenesis *Gr* + *Cc* + *Q*; this is possible only in the narrow range between curves (2) and (3) or (6), respectively.
Evaluation: Temperature undefined over a wide range from 400° to 600°C at 2 kb; X_{CO_2} very low or low (between 0.02 and 0.2 at 2 kb).

Zone 2: Paragenesis *Gr* + *Cc* + *Wo*; this is possible above curve (2) and left of (4).
Evaluation: Temperature and X_{CO_2} undefined over a wide range.
Paragenesis: *Gr* + *An* + *Q*; this is possible below curve (1) and above (8) and (3).
Evaluation: Temperature ranges from 500 to 600°C at 2 kb; X_{CO_2} very low or low (between 0.02 and 0.2 at 2 kb).

Zone 1: Paragenesis *Wo* + *An* + *Q*; this is possible above curves (1) and (2).
Evaluation: Temperature greater than 600°C at 2 kb; X_{CO_2} undefined over a wide range.

Although the temperatures indicated by any one paragenesis lie within a wide range, the whole sequence of parageneses clearly gives the information that the four zones extend over a temperature range from less than 500° to greater than 600°C (at a pressure of 2 kb).

Ca, Al Mica Margarite [3]

Margarite has a higher Al_2O_3 content than zoisite and anorthite (see Figure 10-1). It occurs in low-grade calcareous metapelitic rocks in which kaolinite or a bauxite mineral was an original constituent. Accord-

---

[3]Perhaps margarite is more frequent than is supposed. It is very difficult to distinguish between margarite and muscovite in thin sections, especially when these two phases occur together. On the basis of the strong x-ray interference lines at d = 2.51 and 1.48 A, it can be easily identified from powder x-ray diffraction patterns. Most analysed margarites show that a considerable amount of the paragonite component is present in solid solution.

ing to observations by Frey and Niggli (1972), margarite may be associated with muscovite, paragonite, chlorite, chloritoid, quartz, graphite, calcite, and dolomite. Assemblages in the higher temperature part of low-grade metamorphism include muscovite, biotite, chlorite, clinozoisite, plagioclase (oligoclase-andesine), garnet, quartz, graphite, calcite, and dolomite. From petrographic observations it has been concluded that margarite is formed first at approximately the beginning of low-grade metamorphism (*i.e.*, greenschist facies) by the following reaction (Frey, 1978):

$$\{\text{pyrophyllite} + \text{calcite} \pm \text{paragonite}\} = \{\text{margarite} + \text{quartz} + H_2O + CO_2\}$$

The conditions for equilibrium are not yet known. However, experiments to determine the upper stability of margarite in the CO_2-free system have been carried out by Chatterjee (1971, 1974), Velde (1971), and Storre and Nitsch (1972). Nitsch and Storre (1972) have studied that part of the CO_2-bearing system $CaO\text{-}Al_2O_3\text{-}H_2O\text{-}CO_2$ in which margarite plays a role. From their work it is evident that margarite, like zoisite, can form only in the presence of a water-rich fluid. When the CO_2 content of the fluid exceeds a certain value (depending on total pressure and temperature), margarite breaks down to calcite, Al_2SiO_5 and H_2O. On the other hand, when at a given relatively low X_{CO_2}, temperature surpasses a certain (pressure dependent) limit, margarite breaks down to anorthite and an Al-rich phase; thus, the anorthite content of plagioclase increases abruptly.

Commonly, margarite occurs in quartz-bearing phyllites and low-grade micaschists. Therefore, the stability of the association margarite + quartz is of petrogenetic significance. At P_f larger than about 2 kb, the upper limit of margarite + quartz is governed by the following reaction:

$$1 \text{ margarite} + 1 \text{ quartz} = 1 \text{ anorthite} + 1 \text{ kyanite-andalusite} + 1\ H_2O \qquad \text{(a)}$$

According to experimental determinations by Storre and Nitsch (1974) the equilibrium conditions of this reaction, at $X_{CO_2} = 0$, are

515 ± 25°C at 4 kb H_2O pressure
545 ± 15°C at 5 kb H_2O pressure
590 ± 10°C at 7 kb H_2O pressure
650 ± 10°C at 9 kb H_2O pressure

This demonstrates a very strong influence of pressure on the stability of margarite + quartz and indicates that this association (and even more so margarite without quartz) is expected to occur not only in low-grade (greenschist facies) rocks but also in medium-grade rocks subjected to high pressures.

At pressures less than about 2 kb, the following reaction (b) takes place instead of reaction (a):

$$1 \text{ margarite} + 4 \text{ quartz} = 1 \text{ anorthite} + 1 \text{ pyrophyllite} \quad \text{(b)}$$

At pressures greater than approx. 9 kb, reaction (c) occurs:

$$4 \text{ margarite} + 3 \text{ quartz} = 2 \text{ zoisite} + 5 \text{ kyanite} + 3\ H_2O \quad \text{(c)}$$

Reaction (c) is of special interest because it provides a *pressure indicator: Wherever zoisite + kyanite are observed in stable association, the total pressure must have exceeded 9 kb* (Storre and Nitsch, 1974).

Reactions (a) and (c) are graphically shown in Figure 15-3 as curve (8) and above 9 kb, as its continuation having a different slope.[4]

Plagioclase + Calcite Assemblages

The assemblage plagioclase + calcite may occur in an interlayered sequence of rocks of diverse composition, *e.g.*, calcite-bearing pelites (Bündnerschiefer of the Swiss Alps), calc-silicate rocks, and siliceous marbles. Wenk (1962) has followed such rocks through a prograde succession of metamorphic zones in the Swiss Alps in order to test whether the anorthite content of plagioclase increases with increasing temperature of metamorphism. It would be ideal to have rocks with identical bulk composition that could be followed into higher grades. However, in practice this restriction is too severe, and Wenk chose rocks which had at least similar (not always identical) mineral associations.

Wenk found that the "anorthite content of plagioclase associated with calcite is almost uniform in a given region but varies from area to

---

[4]The data are valid for a naturally occurring margarite which contains 23 mole percent paragonite component. Pure $CaAl_2[(OH)_2/Al_2Si_2O_{10}]$ is unknown in rocks; the component $NaAl_2[(OH)_2/AlSi_3O_{10}]$ amounts to 10 to 50% with an average of about 30%. Reaction (a), *i.e.*, curve (8) in Figure 15-3, is shifted toward slightly lower temperatures if the reaction involves pure margarite and, above 9 kb, the curve of reaction (c) would show a more pronounced negative slope. Thus, depending on the variable composition of margarite, reaction (a) represents a narrow reaction-isograd band, whereas reaction (c) takes place within a larger temperature range.

area depending on the grade of metamorphism." Omitting some exceptions, the anorthite content increases with increasing temperature. He further observed that the plagioclase composition changes abruptly rather than continuously. He divided the Swiss Alps into metamorphic zones (Figure 8-5) based on ranges of plagioclase composition, *i.e.*, 0 to 5, 10 to 30, 30 to 70, 70 to 85, 85 to 100% anorthite component.

The stepwise increase of anorthite content in plagioclase must be due to a series of distinct reactions which produce anorthite component with rising temperature. Examples of such reactions are shown in Figure 10-2. However, it is also evident from this phase diagram that a number of other reactions also form anorthite component in response to an increase of X_{CO_2} at constant temperature and P_f. If X_{CO_2} is different in nearby layers of identical bulk composition, different plagioclase compositions must be expected, although P_f and T have been the same.

It is obvious that the observed increase of anorthite content in plagioclase with rising temperature is a complex phenomenon and that the "exceptions" deserve special attention. Of course, the stepwise change of anorthite content in plagioclase can take place only in those parts of a rock where appropriate reactants are available, while in the other areas plagioclase composition remains unchanged. Therefore, in compositionally nonhomogeneous rocks, different plagioclase compositions may be expected to lie side by side even in the area of a thin section.

Further experimental and theoretical work is needed for a rigorous correlation between plagioclase composition and metamorphic temperature in a complex rock system. To do this, it is essential to keep as many variables constant as possible. Compositions of plagioclase either from the same paragenesis or from several parageneses related by simple reactions must be compared.

This complex problem is being investigated by Orville (personal communication), who has pointed out that in metamorphism of marls the following assemblage may be an indicator of metamorphic temperature and fluid composition:

plagioclase + calcite + quartz + zoisite + muscovite + K feldspar

As an approximation this paragenesis may be represented in the NaO_2-free system K_2O-CaO-Al_2O_3-SiO_2-H_2O-CO_2. A preliminary study by Johannes and Orville (1972) deals with equilibria in this system at 7 kb.

More recently, Hewitt (1973) investigated the following reaction, which is relevant to the problem:

1 muscovite + 1 calcite + 2 quartz =
1 K feldspar + 1 anorthite + 1 CO_2 + 1 H_2O

At constant fluid pressure the equilibrium temperature has a maximum value at CO_2:H_2O = 1:1, *i.e.*, at $X_{CO_2} = 0.5$. Under this condition the equilibrium data are:

475 ± 20°C at 2 kb
533 ± 7°C at 4 kb
559 ± 9°C at 5 kb
584 ± 4°C at 6 kb
606 ± 5°C at 7 kb

Hewitt also showed that, at 6 kb fluid pressure, the isobaric univariant equilibrium curve T-X_{CO_2} is symmetrical about $X_{CO_2} = 0.5$ with points

for 570°C at $X_{CO_2} = 0.22 \pm 0.33$ and 0.78 ± 0.05
for 555°C at $X_{CO_2} = 0.11 \pm 0.02$ and 0.88 ± 0.02

It would be useful to know the T-X_{CO_2} data in the range of low X_{CO_2} values at different fluid pressures. This would enable the determination of isobaric invariant points such as those which arise from the intersection of the above equilibrium curve with that of the previously mentioned reaction (9):

$$2 \text{ zoisite} + 1\ CO_2 = 1 \text{ calcite} + 3 \text{ anorthite} + 1\ H_2O$$

The T-P_f curve of the isobaric invariant point would represent the conditions of formation of the assemblage

zoisite + K feldspar + anorthite + muscovite + calcite + quartz

which occurs in metamorphic marly rocks. However, the plagioclase is rarely anorthite. The effect of an addition of albite component to anorthite is to progressively decrease somewhat the equilibrium temperature.

Other assemblages found in metamorphosed marls have been studied by Frey and Orville (1974). They payed particular attention to the variation of plagioclase composition in response to metamorphic grade. Most of the assemblages included plagioclase, calcite, and margarite. They arrived at an understanding of the following surprising petrographic observations:

(a) Due to the presence of margarite, the first-appearing plagioclase in the lower temperature part of low-grade metamorphism is not albite as expected but has a composition of approximately An_{30}.

(b) Plagioclase in most assemblages, among others in the margarite-bearing assemblage

margarite + zoisite + quartz + calcite + plagioclase,

becomes richer in An content with increasing temperature. However, this is not so with the following paragonite- and margarite-bearing parageneses:

margarite + paragonite + quartz + calcite + plagioclase

and

margarite + paragonite + quartz + zoisite + plagioclase

In these associations the An content remains constant with increasing temperature. This has been observed in low-grade and in the lower temperature range of medium-grade metamorphism.

The assemblage margarite + plagioclase is stable to a temperature somewhat higher than that of the boundary between low-grade and medium-grade metamorphism. However, its stability is drastically limited when the water pressure exceeds about 7–9 kb. At such pressures, the following reaction takes place (Franz and Althaus, 1977):

plagioclase + margarite + H_2O = paragonite + zoisite + quartz

The high pressure assemblage paragonite + zoisite + quartz has been observed in low-grade terrains where glaucophane occurs in neighboring rocks of appropriate composition. This paragenesis is expected to be stable, at high pressures, in medium-grade rocks as well.

Vesuvianite

Vesuvianite (also called idocrase) is apparently a rather rare mineral formed in metamorphosed marls. Its complex composition may be represented by the following formula in which some Fe^{2+} replaces Mg and some F may replace OH:

$$Ca_{10}Mg_2Al_4[(OH)_4/(SiO_4)_5/(Si_2O_7)_2]$$

Vesuvianite is well known as a constituent in calcium-rich metamorphites of high level contact metamorphism, although this mineral is in no way restricted to contact metamorphism. Vesuvianite, together with diopside and/or grossularite, and vesuvianite + epidote, have been described from regional medium-grade (amphibolite facies) terrain by Tilley (1927) and by Trommsdorff (1968). Chatterjee (1962) demonstrated that it can also appear in low-grade metamorphism (greenschist facies) in the following assemblages: vesuvianite + chlorite, vesuvianite + actinolite/tremolite, and vesuvianite + epidote. This typically metamorphic mineral thus possesses a very large stability range.

Braitsch and Chatterjee (1963) suggested the following reactions in which vesuvianite takes part:

5 vesuvianite + 4 epidote + 11 quartz =
16 grossularite + 10 diopside + 12 H_2O

3 vesuvianite + 2 tremolite + 7 quartz =
16 diopside + 6 grossularite + 8 H_2O

1 vesuvianite + 1 clinochlore + 8 quartz =
3 anorthite + 7 diopside + 6 H_2O

A systematic study of vesuvianite assemblages is not yet available.

References

Boettcher, A. L. 1970 *J. Petrol.* **11:** 337–379.
Braitsch, O., and Chatterjee, N. D. 1963. *Contr. Mineral. Petrol.* **9:** 353–373.
Chatterjee, N. D. 1962. *Contr. Mineral. Petrol.* **8:** 432–439.
——— 1971. *Naturwiss.* **58:** 147.
——— 1974. *Schweiz. Mineral. Petrog. Mitt.* **54:** 753–767.
Franz, G., and Althaus, E. 1977. *Neues Jahrb. Mineral. Abhand.* **130:** 159–167.
Frey, M. 1974. *Schweiz. Mineral. Petrog. Mitt.* **54:** 753–767.
——— 1978. *J. Petrol.* **19:** 95–135.
Frey, M. and Niggli, E. 1972. *Naturwiss.* **59:** 214–215.
——— and Orville, Ph.M. 1974. *Am. J. Sci.* **274:** 31–47.
Gordon, T. M., and Greenwood, H. J. 1971. *Am. Mineral.* **56:** 1674–1688.
Hewitt, D. A. 1973. *Am. Mineral* **58:** 785–791.
Johannes, W., and Orville, P. M. 1972. *Fortschr. Mineral.* **50:** 46–47.
Kerrick, D. M. 1970. *Geol. Soc. Am. Bull.* **81:** 2913–2938.
Misch, P. 1964. *Contr. Mineral. Petrol.* **10:** 315–356.
Nitsch, K.-H and Storre, B. 1972. *Fortschr. Mineral.* **50**, Beiheft 1: 71–73.
Seidel, E., and Okrusch, M. 1976. *Bull. Soc. Geol. France* **18:** 151–154.
Storre, B. 1970. *Contr. Mineral. Petrol.* **29:** 145–172.

——— and K.-H. Nitsch. 1972. *Contr. Mineral. Petrol.* **35:** 1–10.
——— 1974. *Contr. Mineral. Petrol.* **43:** 1–24.
Thompson, A. B. 1971. *Contr. Mineral. Petrol.* **33:** 145–161.
Tilley, C. E. 1927. *Geol. Mag.* **64:** 372–376.
Trommsdorff, V. 1968. *Schweiz. Mineral. Petrog. Mitt.* **48:** 655–666, 828–829.
Velde, B. 1971. *Mineral. Mag.* **38:** 317–323.
Wenk, E. 1962. *Schweiz. Mineral. Petrog. Mitt.* **42:** 139–152.

Chapter 11

Metamorphism of Ultramafic Rocks: Systems MgO-SiO_2-CO_2-H_2O and MgO-CaO-SiO_2-H_2O

Two different rock types are predominantly composed of MgO and SiO_2: ultramafic rocks and the much rarer siliceous magnesite rocks. The major mineral constituents of ultramafic rocks are olivine, orthopyroxene, and/or clinopyroxene, and the MgO/FeO ratio of the rocks is very high. CaO is another component present in clinopyroxene.

Metamorphism of ultramafic rocks requires the access of H_2O and/or CO_2. Particularly H_2O is commonly available and the introduction of H_2O converts ultramafic rocks into serpentinites, consisting mainly of antigorite or lizardite/chrysotile (varieties of serpentine), minor amounts of talc, quartz, or brucite, and some magnetite. The transition from lizardite/chrysotile to antigorite in serpentinite rocks has been observed to take place in the Swiss Alps close to the upper limit of pumpellyite (Dietrich and Peters, 1971, *i.e.,* close to the beginning of low-grade (greenschist facies) metamorphism.

Access of H_2O *and* CO_2 to ultramafic rocks is demonstrated by the conversion of serpentinites to rocks consisting of talc + magnesite ± dolomite. Introduction of H_2O and a small amount of CO_2 often results in rims of talc + magnesite enclosing serpentinite which previously formed by the introduction of H_2O alone. Furthermore, there are rare cases where rocks consisting of magnesite + anthophyllite or of magnesite + enstatite (a rock called sagvandite) were formed from ultramafic rocks by introduction of a fluid having a *high* concentration of CO_2. Petrogenetic interpretations will be presented on the basis of experimental work in the system MgO-SiO_2-CO_2-H_2O. Further on, the effects of components CaO and FeO will be considered.

Experimental work prior to 1967 [reviewed in Turner (1968)] did not take into account the presence of CO_2 in addition to H_2O. Recent work has investigated the influence of X_{CO_2} and is of more general significance. Here, phase relations in the system $MgO-SiO_2-H_2O-CO_2$ are based on the investigations of Greenwood (1967) and Johannes (1969); in both papers earlier work by them and others are also taken into account.

Minerals in the system $MgO-SiO_2-H_2O-CO_2$ are arranged in order of increasing ratio $(Mg \times 100)/(MgO + SiO_2)$:

Quartz (Q)	0% MgO	Serpentine (S)	60% MgO
Talc (Ta)	43	Forsterite (Fo)	67
Mg-anthophyllite (Antho)	47	Brucite (B)	100
$Mg_7[(OH)_2/Si_8O_{22}]$		Periclase (P)	100
Enstatite (En)	50	Magnesite (M)	100

In the system SiO_2-MgO, the compositions of ultramafic rocks will lie between forsterite and enstatite because olivine and pyroxene are the predominant constituents. The presence of Ca-bearing clinopyroxene in addition to orthopyroxene makes it necessary to consider CaO as a component as well; this will be done later.

Figure 11-1 indicates the results of serpentinization: If the ultramafic rock is a dunite it plots at Fo. If it consists of olivine and a small amount of pyroxene, its bulk composition plots between Fo and En to the right of point S. On serpentinization, such a rock will yield a mixture of predominantly serpentine and some brucite. On the other hand, a rock consisting mainly of pyroxene and some olivine plots between Fo and En to the left of point S; in this case a lot of serpentine and some talc will be formed on serpentinization. Both types of serpentinite assemblage may form in different layers of a compositionally layered ultramafic body.

The bulk compositions of most serpentinites plot between MgO 60 and 67%, *i.e.*, between points S and Fo in Figure 11-1. Consequently, whenever serpentine is a reactant, the reaction products will be two minerals—one to the right of point S (forsterite, brucite, or magnesite) and the other to the left (enstatite, anthophyllite, or talc). Quartz is to be expected only at a temperature lower than that of the reaction

$$1 \text{ serpentine} + 2 \text{ quartz} = 1 \text{ talc} + 1 \text{ H}_2\text{O}$$

Fig. 11-1 Graphical representation of the ratios SiO_2/MgO in the stated minerals.

The following reactions have to be considered:

Reactions at extremely small values of X_{CO_2}:

1 serpentine + 1 magnesite = 2 forsterite + 2 H_2O + 1 CO_2 (1)
2 serpentine + 3 CO_2 = 1 talc + 3 magnesite + 3 H_2O (3)
1 serpentine + 3 CO_2 = 2 quartz + 3 magnesite + 2 H_2O (5)
1 serpentine + 1 brucite = 2 forsterite + 3 H_2O (6)
5 serpentine = 6 forsterite + 1 talc + 9 H_2O (7)
1 serpentine + 2 quartz = 1 talc + 1 H_2O (8)
1 brucite + 1 CO_2 = 1 magnesite + 1 H_2O (19)
1 brucite = 1 periclase + 1 H_2O (20)

Reactions within a large intermediate range of X_{CO_2} *between the extremes:*

1 talc + 5 magnesite = 4 forsterite + 1 H_2O + 5 CO_2 (2)
4 quartz + 3 magnesite + 1 H_2O = 1 talc + 3 CO_2 (4)
9 talc + 4 forsterite = 5 anthophyllite + 4 H_2O (9)
1 anthophyllite + 1 forsterite = 9 enstatite + 1 H_2O (10)
7 talc = 3 anthophyllite + 4 quartz + 4 H_2O (11)
1 anthophyllite = 7 enstatite + 1 quartz + 1 H_2O (12)

Reactions at extremely large values of X_{CO_2}:

1 anthophyllite + 9 magnesite = 8 forsterite + 1 H_2O + 9 CO_2 (13)
2 talc + 1 magnesite = 1 anthophyllite + 1 H_2O + 1 CO_2 (14)
7 magnesite + 8 quartz + 1 H_2O = 1 anthophyllite + 7 CO_2 (15)
1 anthophyllite + 1 magnesite = 4 enstatite + 1 H_2O + 1 CO_2 (16)
1 enstatite + 2 magnesite = 2 forsterite + 2 CO_2 (17)
2 magnesite + 2 quartz = 1 enstatite + 2 CO_2 (18)

The phase diagram for the system MgO-SiO_2-H_2O-CO_2 at $P_f = 2$ kb (Figure 11-2) shows the reasons for grouping reactions according to the value of X_{CO_2}. A number of reactions take place only if X_{CO_2} is very small. Especially serpentine—either alone or associated with another mineral—is stable only at very small values of X_{CO_2}. Indeed, the presence of serpentine in a rock is a very good indicator that the fluid phase present during rock alteration contained very little CO_2 or none at all; X_{CO_2} must have been less than 10 mole percent, otherwise serpentine would have been altered to magnesite + quartz [reaction (5)] or to magnesite + talc [reaction (3)]; see Figure 11-3.

On the other hand, at very high X_{CO_2}, reactions (14) and (15) lead to

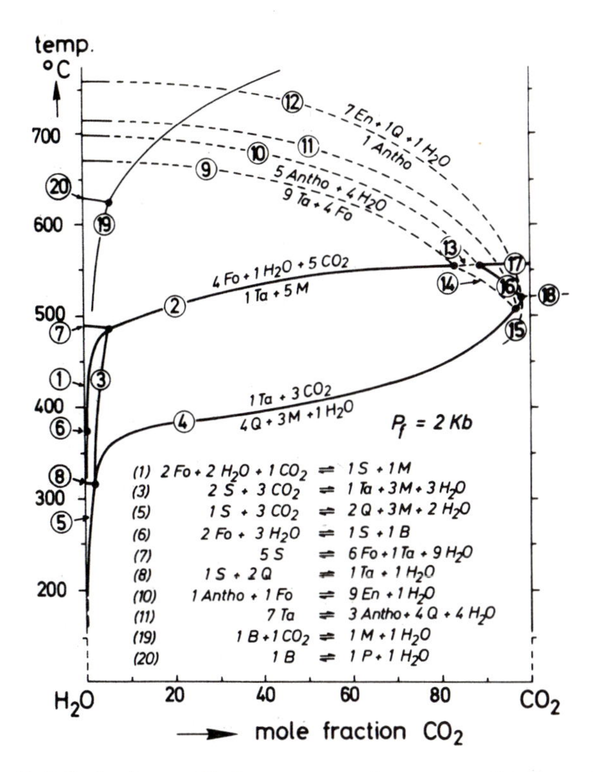

Fig. 11-2 Isobaric equilibrium curves of reactions (1) through (20) in the system MgO-SiO_2-H_2O-CO_2 (Johannes, 1969).
Remark: The experimental data for reactions (6) and (7) have been obtained using chrysotile as the serpentine species; it is known now that antigorite is the stable species at the experimental conditions. Using antigorite instead of chrysotile gives rise to somewhat higher equilibrium temperatures (Evans *et al.*, 1976).

the assemblage anthophyllite + magnesite and reactions (16) and (18) produce enstatite + magnesite (Figure 11-4). These assemblages are stable within a temperature range of about 50°C at 2 kb. Therefore, these parageneses, which are petrographically well known, would not be suitable as temperature indicators even if the dependence of the reactions on P_f were known. However, reactions (17) and (13) depend very little on X_{CO_2} and would serve that purpose if their positions at various P_f were known.

Figure 11-2 shows that reactions (1), (3), and (5), occurring at low values of X_{CO_2}, and reactions (9), (10), (11), (12), (2), and (4), occurring within a large range of X_{CO_2}, are stable over a large temperature range. On the other hand, reactions (6), (7), and (8) and the isobaric invariant

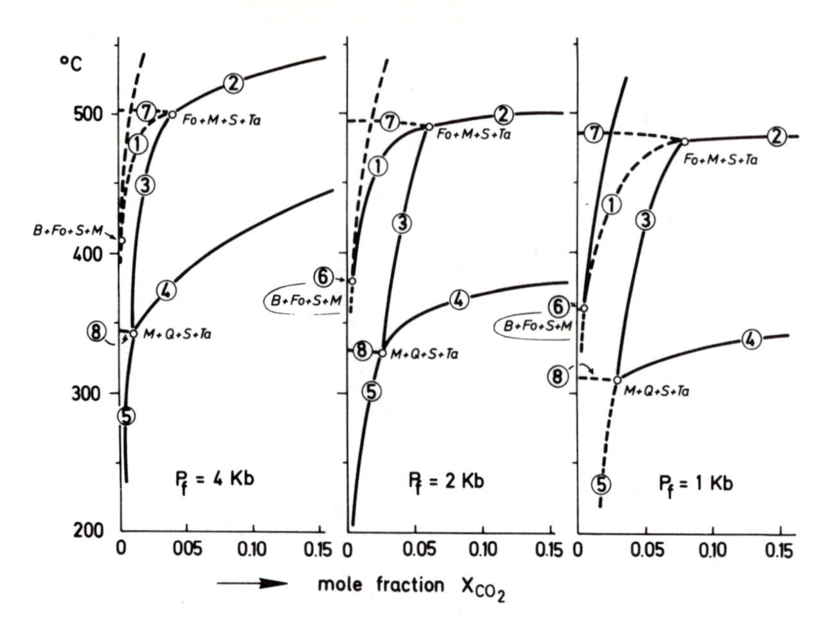

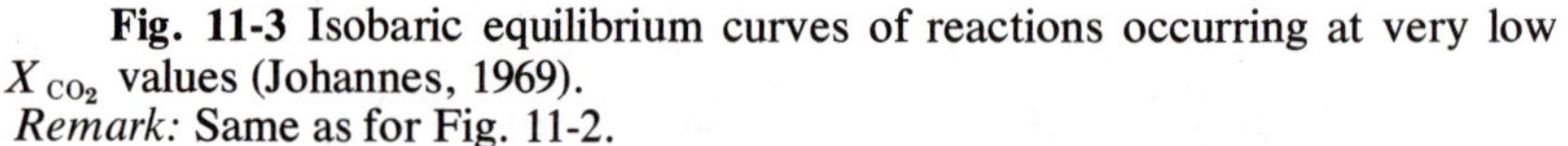

Fig. 11-3 Isobaric equilibrium curves of reactions occurring at very low X_{CO_2} values (Johannes, 1969).
Remark: Same as for Fig. 11-2.

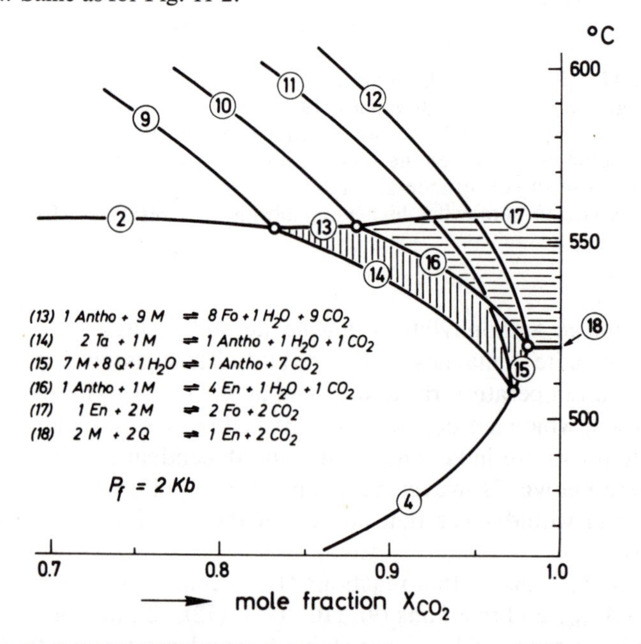

Fig. 11-4 Isobaric equilibrium curves of reactions occurring at large X_{CO_2} values. Horizontally ruled: enstatite + magnesite; vertically ruled: anthophyllite + magnesite (Johannes, 1969).

points, are valuable temperature indicators or indicators of P_f if the temperature is known from other observations. In particular, the following parageneses are petrogenetically significant:

(8) Serpentine + quartz + talc
(8/3) Serpentine + quartz + talc + magnesite, invariant point very close to (8)
(6) Serpentine + brucite + forsterite (first possible appearance of forsterite!)
(6/1) Serpentine + brucite + forsterite + magnesite, isobaric invariant point very close to (6)
(7) Serpentine + forsterite + talc
(7/3) Serpentine + forsterite + talc + magnesite, isobaric in-variant point; temperatures very close to those of (7).

Using chrysotile as the serpentine species, the conditions of reactions (6) and (7) have been determined by various authors. These data would be petrogenetically significant if they could be applied to nature. However, this is not the case. In the following reactions it is not chrysotile but antigorite which is the stable serpentine species:

$$1 \text{ antigorite} + 1 \text{ brucite} = 2 \text{ forsterite} + 3\, H_2O \qquad (6)$$
$$5 \text{ antigorite} = 6 \text{ forsterite} + 1 \text{ talc} + 9\, H_2O \qquad (7)$$

In discussing the reasons for this observation, Evans *et al.* (1976) point out that the lower stability boundary of forsterite, *i.e.*, reaction (6), lies outside the stability field of chrysotile at most metamorphic pressures.

Whereas the composition of chrysotile is close to the ideal formula $Mg_3Si_2O_5(OH)_4$, antigorite is somewhat poorer in Mg and OH. Consequently, the transition from chrysotile to antigorite takes place by the following reaction (Evans *et al.*, 1976):

$$17 \text{ chrysotile} = 17 \text{ antigorite} + 3 \text{ brucite}$$

Calculated equilibrium conditions of this reaction suggest that chrysotile is not stable above approximately 300°C. However, field observations, cited in the above paper, clearly show that chrysotile persists metastably somewhat into the field of antigorite stability. Because of a small entropy change, the conversion of chrysotile to antigorite is very sluggish. Therefore, it is very likely that in quartz-bearing serpentinites, at temperatures between 300 and 360°C, the following reaction (8) takes place, chrysotile being present as a metastable serpentine species:

$$1 \text{ chrysotile} + 2 \text{ quartz} = 1 \text{ talc} + 1\, H_2O \qquad (8)$$

The equilibrium curves of reactions (7), (6), and (8) are shown in Figure 11-5. The data for (7) have been experimentally determined by Johannes (in Evans *et al.*, 1976). The data for (6) have been calculated by Evans and since then also determined experimentally at high pressure by Johannes (personal communication, 1977). The data for (8) have been deduced by Johannes (1969) from other experimentally determined equilibria.

The association chrysotile + quartz and that of the reaction equilibrium of (8), namely chrysotile + quartz + talc, are to be expected in very-low-grade metamorphism. Equilibrium paragenesis (6) is present in low-grade metamorphism where indeed the first appearance of olivine has been observed by Trommsdorff and Evans (1974) at small distance above the biotite isograd. The same authors have observed paragenesis (7), *i.e.*, the disappearance of antigorite at some distance above the staurolite isograd, *i.e.*, somewhat above the beginning of medium-grade metamorphism. From this it is obvious that serpentinites when situated in such a metamorphic terrain deserve careful investigation.

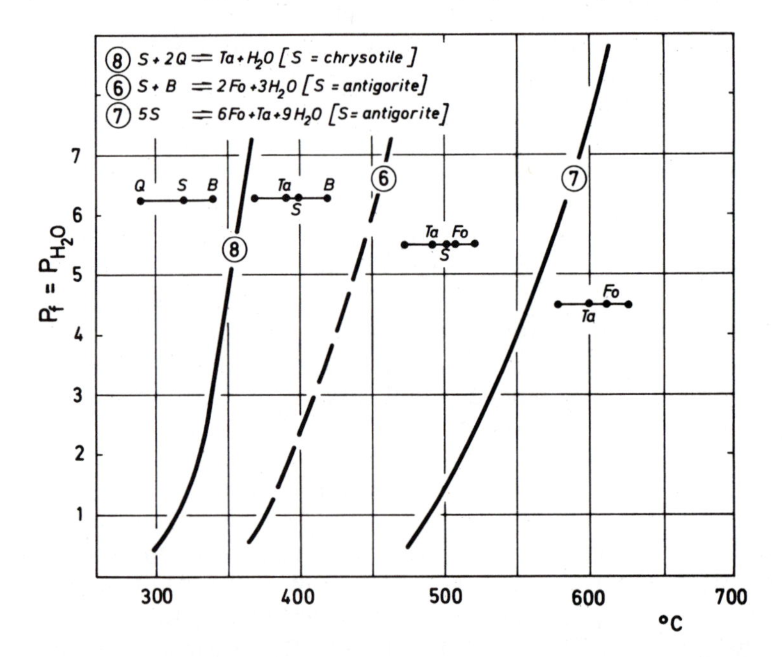

Fig. 11-5 Equilibrium curves of reactions (8), (6), and (7). The data for the stability of the three equilibrium parageneses are also valid if X_{CO_2} is not exactly zero. Any two minerals adjacent to each other on the line SiO_2-MgO are stable in the fields *between* reaction curves; the minerals designated by letters may occur in metaserpentinites. Reaction (8) involves chrysotile and reactions (6) and (7) involve antigorite as the stable serpentine species.

If X_{CO_2} is not known, mineral parageneses formed in serpentinites at higher temperatures do not constitute reliable (relative) temperature indicators, except those rare assemblages of reaction equilibria (13) and (17). The many possible parageneses which may be formed by metamorphism of serpentinites can be deduced easily from the various figures given in this chapter. However, it should be stressed that only the ones discussed in particular are useful petrogenetic indicators.

So far, most petrographic investigations have failed to consider the significance of certain three-mineral (or four-mineral) assemblages in the Ca-free system MgO-SiO_2-fluid. However, two-mineral parageneses in metamorphosed ultramafic rocks have been recorded. A sequence of parageneses with increasing temperature of metamorphism has been published by Evans and Trommsdorff (1970):

Talc or brucite + serpentine (species antigorite)
Forsterite + serpentine (species antigorite)
Forsterite + talc
Anthophyllite + talc and anthophyllite + forsterite
Enstatite + forsterite

Instead of orthorhombic anthophyllite, the monoclinic Mg-rich cummingtonite (magnesiocummingtonite) is often found as intergrown lamellae. Cummingtonite is believed to partially invert to anthophyllite during cooling (Rice *et al.*, 1974). This sequence agrees with the experimental results shown in Figures 11-2 and 11-5 and comprises reactions (8), (6), (7), (9), and (10). Also, the observed geographical overlap, with increasing temperature, of paragenesis talc + serpentine, formed by reaction (8), and paragenesis forsterite + serpentine, formed by reaction (6), is very well understood; it is due to the presence of different rock compositions in close proximity. The assemblage talc + serpentine is unaffected by reaction (6), as shown in Figure 11-5. On the other hand, the assemblage forsterite + serpentine disappears with increasing temperature and its place is taken by the assemblage talc + forsterite; there is no overlap of assemblages because all serpentine is decomposed by reaction (7). Furthermore, Evans and Trommsdorff have observed that the first appearance of the association forsterite + talc occurs somewhat above the beginning of medium-grade metamorphism marked by the formation of staurolite in metapelitic rocks—in agreement with experimental results; see Figure 15-2. At higher temperatures the expected sequence of reactions is the formation of anthophyllite + talc or forsterite by reaction (9), followed by the formation of enstatite + forsterite by reaction (10) (see Figure 11-7).

It should be noted that the formation of enstatite in ultramafic rocks from anthophyllite and forsterite takes place at an appreciably lower temperature than the breakdown of anthophyllite to enstatite + quartz [compare reactions (10) and (12) at the same X_{CO_2} values]. Furthermore, the first appearance of enstatite according to (10) perhaps takes place at lower temperature than the widespread first formation of orthopyroxene (mostly hypersthene) in metamorphic basic and acidic rocks (granulites) (to be discussed further on). These reactions are different from the ones in ultramafic rocks and therefore temperatures and parageneses will be different.

A sequence different from that observed by Evans and Trommsdorff may be inferred from Figure 11-2 if parts of a serpentinite body have been altered to quartz + magnesite at low temperatures by the addition of a fluid containing small amounts of CO_2. With increasing temperature, the sequence of parageneses will be

Quartz + magnesite
  Quartz + magnesite + talc [reaction (4)]
Talc + magnesite
  Talc + magnesite + forsterite [reaction (2)]
Forsterite + talc
  Forsterite + talc + anthophyllite [reaction (9)]
Anthophyllite + forsterite or talc
  Anthophyllite + forsterite + enstatite [reaction (10)]
Enstatite + forsterite

Although such a sequence is expected to occur in some environments, it has not yet been documented.

Presence of Al_2O_3, CaO, and FeO

So far, only parageneses in the system MgO-SiO_2-CO_2-H_2O have been considered. If Al_2O_3 and/or CaO are additional components of an ultramafic rock, additional minerals are encountered.

The presence of Al_2O_3 in serpentinites will commonly give rise to the formation of *Mg-rich chlorite*. It is stable into medium- and even high-grade metamorphism. At high temperature conditions Mg-chlorite typically occurs in association with enstatite and forsterite (and magnetite) (Trommsdorff and Evans, 1969). Thus, contrary to some views, chlorite is not restricted to very-low-grade and low-grade metamorphism. The stability of a mineral depends on the associated minerals. Thus, chlorite associated with muscovite reacts at the beginning of medium-grade metamorphism, commonly leading to the disappearance of chlorite. On the other hand, chlorite may remain stable as a member of other assemblages to much higher temperatures.

When, in rare cases, large amounts of Al_2O_3 and MgO are available in spinel-bearing layers of ultramafic bodies, *sapphirine* $Mg_2Al_4SiO_{10}$ is formed in high-grade ultramafic rocks. Sapphirine, which contains very little FeO, may be associated with enstatite, spinel, pyrope, hornblende of the variety pargasite $NaCa_2Mg_4Al[(OH)_2/Al_2Si_6O_{22}]$, anthophyllite (gedrite), diopside, bytownite, or phlogopite. Although sapphirine is stable over a wide range of pressure and high temperature conditions,[1] its occurrence depends crucially upon the appropriate bulk composition; especially the MgO/FeO ratio must be high enough to stabilize sapphirine. This conclusion is based on field and petrographic studies (Herd *et al.*, 1969; Lensch, 1971) as well as on experimental work (Seifert, 1974). The latter author has also shown that the assemblage

$$\text{sapphirine} + \text{enstatite} \pm \text{spinel}$$

is stable above about 3.5 kb water pressure when temperatures are in excess of 765°C.

CaO as an additional component of ultramafic rocks will be considered next. The main mineral constituent with CaO as a major component is diopsidic clinopyroxene. Therefore, bulk compositions of ultramafic rocks plot, in terms of CaO:MgO:SiO_2 ratios, in the area of the subtriangle diopside (Di)-forsterite (Fo)-enstatite (En) within the CaO-MgO-SiO_2 composition triangle (see Figure 11-6). In the metamorphism of ultramafic rocks—in particular, serpentinites—two more minerals, diopside and tremolite, have to be considered in addition to the minerals in the CaO-free system. All minerals of interest belong to the subsystem SiO_2-MgO-$CaMgSi_2O_6$ (diopside); their compositions are plotted in Figure 11-6.

Two types of metamorphic processes have to be distinguished:

1. By addition of CO_2 to an H_2O-rich fluid, a previous serpentinite has been converted to a rock consisting of magnesite + quartz + some dolomite. In this case, the series of reactions that were outlined earlier in the system CaO-MgO-SiO_2-CO_2-H_2O is pertinent, especially those at high values of X_{CO_2} (Chapter 9).
2. A serpentinite has been subjected to metamorphism without addition of CO_2, which is proved by the very presence of serpen-

[1]Attention is drawn to the fact that the rare association sapphirine + quartz, which cannot form in metamorphic ultramafic rocks, has a much more restricted field of stability in granulite terrains.

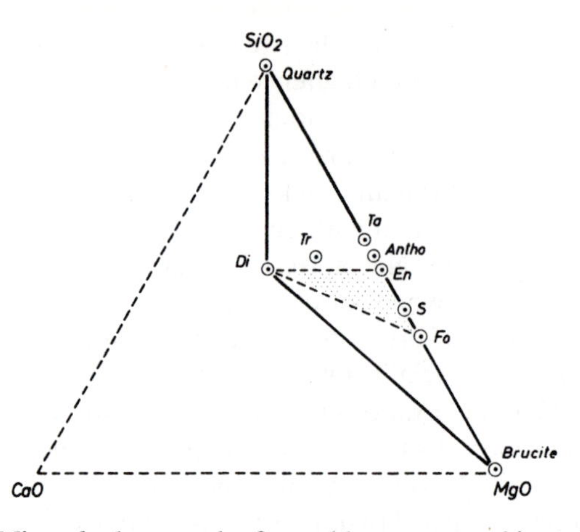

Fig. 11-6 Minerals that may be formed by metamorphism of serpentinites. Bulk compositions of ultramafic rocks plot in the stippled area. Possible mineral assemblages are shown in Figure 11-7.

tine. In this case, magnesite and dolomite are not stable and X_{CO_2} is negligible. This situation, with X_{CO_2} being zero or very small, has not been discussed in Chapter 9. It is of importance only in the metamorphism of ultramafic rocks and will, therefore, be treated here. This review is based mainly on publications by Evans and Trommsdorff (1970) and by Trommsdorff and Evans (1972).

In metamorphosed serpentintites, diopside forms at relatively low temperatures, in contrast to its appearance in the metamorphism of siliceous dolomites at rather high temperatures. In the absence of CO_2, diopside is stable together with serpentine (and brucite) at very-low-grade and low-grade metamorphism, and in association with other minerals it persists to higher temperatures. This is very similar to the behavior of forsterite which may form by reaction (6) at rather low temperatures. The parageneses

diopside + serpentine + brucite, followed by
diopside + serpentine + forsterite

occur in the temperature range of very-low-grade to low-grade metamorphism (see Figure 11-7).

With increasing temperature, reaction (21) takes place:

$$5 \text{ serpentine} + 2 \text{ diopside} = 1 \text{ tremolite} + 6 \text{ forsterite} + 9\, H_2O \quad (21)$$

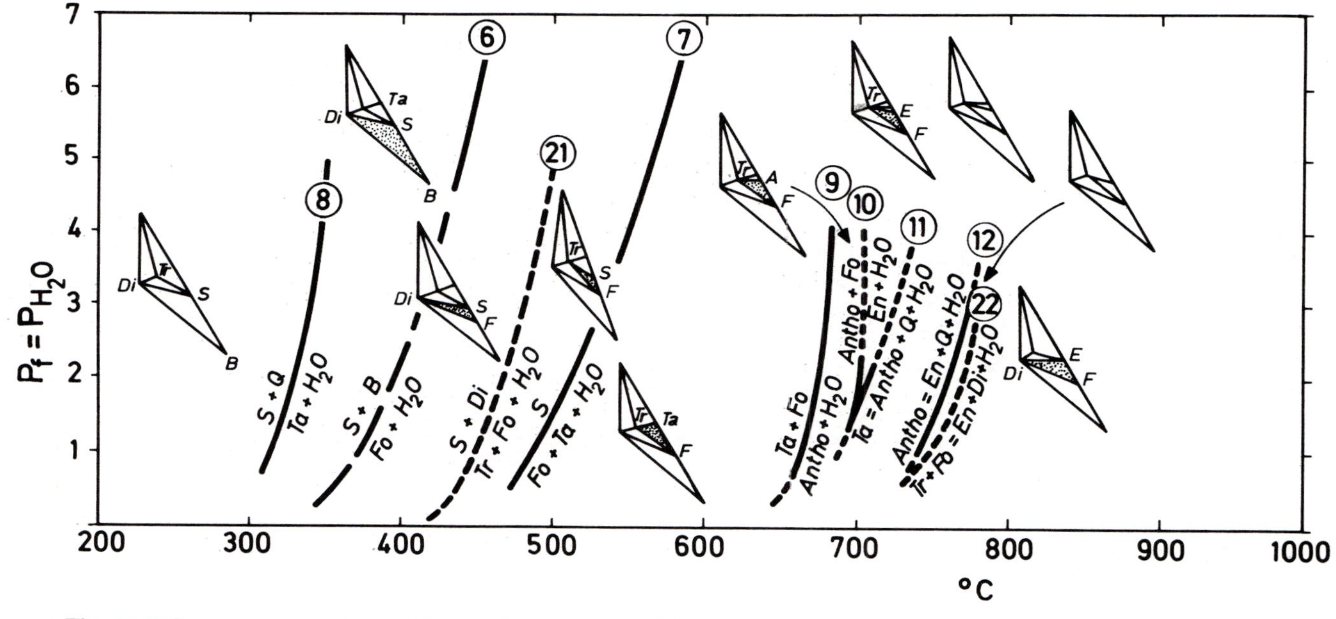

Fig. 11-7 Sequence of metamorphic reactions in serpentinites. Stippled area shows parageneses observed in the Central Alps and most likely to be encountered because of the compositions of serpentinites. [Modified from Evans and Trommsdorff (1970)]. Note that reactions (6), (21), and (7) involve antigorite as the serpentine species.

This is the only tremolite-producing reaction in metaserpentinites, and it first stabilizes tremolite at a *higher* temperature than diopside (!). Tremolite persists in various assemblages over a very large temperature range until reaction (22) takes place:

$$1 \text{ tremolite} + 1 \text{ forsterite} = 5 \text{ enstatite} + 2 \text{ diopside} + 1 \text{ H}_2\text{O} \quad (22)$$

CaO as an additional component in serpentinites makes it necessary to consider only two new reactions, (21) and (22), in addition to the ones discussed previously.

The sequence of reactions in prograde metamorphism of serpentinites in the absence of CO_2 is shown in Figure 11-7. The possible mineral associations between reaction curves and the parageneses actually observed in metaserpentinites by Evans and Trommsdorff (1970) in the Central Alps are also shown; the four-mineral equilibrium paragenesis of the reactions (21) and (22) is also indicated. The data for reactions (8), (6), and (7) are the same as in Figure 11-5; reactions (21) and (22) were calculated by Evans and Trommsdorff (1970) and reactions (9) to (12) by Greenwood (1963).

FeO as additional component. Ferrous iron substitutes for Mg in crystal structures. In ultramafic rocks, the ratio of FeO to MgO is very small and becomes even smaller through serpentinization because some oxidation causes the formation of magnetite. Little ferrous iron remains available for Mg substitution in the various minerals. The influence of the FeO component on equilibrium data is believed to be minimal. Trommsdorff and Evans (1972, 1974) have determined by microprobe analyses of coexisting minerals the following order of preference for Fe relative to Mg: anthophyllite > olivine > antigorite > tremolite > diopside > talc. Using these partitioning data, the authors deduced that shifts of equilibrium temperatures are less than 10°C. Only in the case of reaction (9) is the influence of Fe substitution somewhat larger.

References

Dietrich, V. and Peters, T. 1971. *Schweiz. Mineral. Petrogr. Mitt.* **51:** 329–348.

Evans, B. W., Johannes, W., Oterdoom, H., and Trommsdorff, V. 1976. *Schweiz. Mineral. Petrog. Mitt.* **56:** 79–93.

Evans, B. W. and Trommsdorff, V. 1970. *Schweiz. Mineral. Petrog. Mitt.* **50:** 481–492.

Greenwood, H. J. 1963. *J. Petrol.* **4:** 317–351.

——— 1967. *In* P. H. Abelson, ed. *Researches in Geochemistry*. Vol. 2 pp. 542–567. John Wiley & Sons, New York.

Herd, R. K., Windley, B. F., and Ghisler, M. 1969. *Geol. Surv. Greenland Rept. No. 24.*

Johannes, W. 1968. *Contr. Mineral. Petrol.* **19:** 309–315.

——— 1969, *Am. J. Sci.* **267:** 1083–1104.

Lensch, G. 1971. *Contr. Mineral. Petrol.* **31:** 145–153.

Rice, J. M., Evans, B. W., and Trommsdorff, V. 1974. *Contr. Mineral. Petrol.* **43:** 245–251.

Seifert, F. 1974, *J. Geol.* **82:** 173–204.

Trommsdorff, V. and Evans, B. W. 1969. *Schweiz. Mineral. Petrog. Mitt.* **49:** 325–332.

——— 1972. *Am. J. Sci.* **272:** 423–437.

——— 1974. *Schweiz. Mineral. Petrogr. Mitt.* **54:** 333–352.

Turner, F. J. 1968 *Metamorphic Petrology.* McGraw-Hill Book Company, New York.

Chapter 12

Metamorphism of Mafic Rocks

Basalts and pyroxene-andesites comprise by far the largest amount of mafic rocks and greatly predominate over their plutonic equivalents. Basalts and andesites are widespread in most geosynclines in the form of lava flows, pillow lavas, tuff layers, sills, and dykes. At least three fundamentally different cases of metamorphism of mafic igneous rocks must be distinguished:

a. Water has had access to the rock.
b. Water has not had access and load pressure was not very high.
c. Water has not had access but load pressure was very high.

In case (a) the metamorphic products, formed with decreasing temperature of metamorphism, are amphibolites, greenschists, and glaucophane schists, lawsonite-albite-chlorite schists, or laumontite-prehnite-chlorite schists and similar rocks, depending on pressure and temperature (schistosity may be absent). In case (b) nothing happens; the mafic rock persists as it is. This is well documented by the existence of an unchanged *inner* part of gabbro plutons in metamorphic terrains. In case (c) eclogite is formed in a wide range of temperature and mafic granulite is formed only at high temperatures. This will be discussed in later chapters. We are concerned here with case (a) where water has had access to originally anhydrous rock.

Transformations Except Those of Very-Low-Grade Metamorphism at Low Pressures

An investigation of 700 amphibolites, all derived from mafic rocks and distributed over an area of 5000 sq km in the Central Alps, was reported by Wenk and Keller (1969). This enabled a relationship to be established between the mineral composition of amphibolites and metamorphic grade. All amphibolites have plagioclase + hornblende as major constituents, and in some rocks these two minerals account for 95% of the mineral content. Some quartz is always present and commonly some of the following minerals occur in various parageneses: biotite, almandine-garnet, sphene, rutile. Clinozoisite or epidote may be an important constituent as is chlorite in low-grade amphibolites. Diopsidic pyroxene may be formed in amphibolites at high temperatures.

The authors have shown that the anorthite content of the plagioclase in amphibolites increases with increasing temperature of metamorphism. They were able to distinguish the following zones:

Albite-amphibolites
Oligoclase-amphibolites
Andesine-amphibolites
Labradorite-amphibolites

Such zonal division, however, can be made only on a statistical basis; individual amphibolites may have a plagioclase composition outside the range considered as typical for a given zone. Furthermore, when clinozoisite or epidote is a member of the plagioclase + hornblende + quartz paragenesis, the anorthite content is often higher than it is in their absence. This and other still unknown factors markedly influence the plagioclase composition. The extent of this influence is surprising in view of the rather narrow range of bulk compositions of basaltic-andesitic rocks. Considerable work will be required before this can be fully understood.

At present, the plagioclase composition of amphibolites does not furnish a reliable indicator of metamorphic grade, as had been expected. However, one feature seems to be reliable, namely, the *sudden change from albite (An $\leqslant$ 5)* to oligoclase of about An_{17}. This abrupt compositional change has also been observed in the association plagioclase + calcite (see Chapter 10). Wenk and Keller (1969) made the important observation that in the Central Alps the isograd plagioclase An_{17} + hornblende (in amphibolites) practically coincides with the isograd plagioclase An_{17} + calcite (in metamorphosed marls, etc.). This isograd is observed within the garnet zone and at a somewhat lower temperature than the appearance of staurolite in metapelites. The abrupt change in plagioclase composition from An_5 to An_{17} is universal and is estimated to lie 20° to 30°C below the boundary between low-grade and medium-grade metamorphism. It constitutes a readily detectable feature and is well suited to subdivide the temperature range of low-grade metamorphism.

The main constituents of basalts and andesites are plagioclase (from about An_{70} to An_{40} and clinopyroxene, commonly augite, with MgO dominating over FeO. In addition, some hypersthene and/or olivine may be present. Other than SiO_2, the main components are CaO, Na_2O, and Al_2O_3 from plagioclase; CaO, MgO, FeO, and some Al_2O_3 from augite; and MgO and FeO from olivine and/or hypersthene. Basaltic rocks have significantly more Al_2O_3 and CaO than ultramafic rocks; K_2O is subordinate.

After metamorphism, Na_2O is a major component only in sodium amphibole (glaucophane or crossite) and in albite. Albite may occur as a mineral or as a component of plagioclase solid solution. Taking this into account, it is possible to omit Na_2O in a graphical representation of parageneses. Furthermore, since MgO-to-FeO ratios do not vary much in this group of rocks, these components may be united to (Mg,Fe)O. Therefore, as an approximation, the ACF diagram can be used to show a number of parageneses that form under different conditions of metamorphism.

When reviewing the metamorphism of mafic rocks, it is interesting to start with the magmatic paragenesis of basaltic–andesitic rocks and to observe the metamorphic changes that take place at progressively lower temperatures of metamorphism. The following sequence of rocks is observed:

a. Amphibolites
b. Greenschists
c. Pumpellyite- or lawsonite-glaucophane-bearing rocks and related lower pressure equivalents

A wealth of petrographic information about such rocks is available. Of special interest are observations of progressive metamorphic changes along a geothermal gradient. For mafic metavolcanics formed at medium and high pressures, the publications of Wiseman (1934), van der Kamp (1970), Wenk and Keller (1969), Ernst *et al.* (1970), and Seki *et al.* (1969a) are major sources of information.

The sequence of metamorphic changes, beginning with the basaltic-andesitic precursor, will be shown with ACF diagrams. Figure 12-1 gives the average composition of the commonest basalt type, tholeiite (square, after Wedepohl, 1967); the ruled area within the triangle plagioclase-augite-(hypersthene, olivine) indicates the common range of basalts and andesites.

The main changes of basaltic-andesitic rocks during high-grade, medium grade, and the higher temperature part of low-grade metamorphism, where *amphibolites* are formed, are:

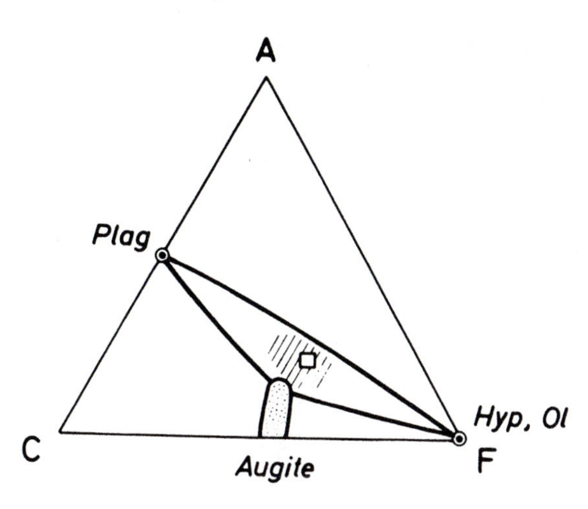

Fig. 12-1 *Basaltic-andesitic rocks;* mean value of tholeiites indicated. Plagioclase commonly between An_{65} and An_{40}.

1. Formation of *hornblende* and some almandine garnet (if pressure is high enough) mainly from clinopyroxene and from hypersthene, pigeonite, and/or olivine.
2. Formation of less anorthite-rich plagioclase and commonly clinozoisite or epidote from that amount of anorthite component which was removed from plagioclase. With decreasing temperature, the An content of plagioclase decreases and the amount of clinozoiste/epidote increases.
3. In low-grade metamorphism, chlorite + quartz are additional constituents of oligoclase- and albite-amphibolites. At still lower temperature, hornblende becomes unstable and decomposes to actinolite, chlorite, and clinozoisite/epidote. Thus, amphibolites are changed into greenschists in which albite is a major constituent.

Figure 12-2 represents the *labradorite/bytownite-amphibolites* in which the composition of plagioclase, as compared to that of the original basaltic rock, has hardly changes; consequently, the amount of clinozoisite/epidote is generally very small or nonexistent. However, at high temperatures and in the presence of water, hornblende is formed at the expense of pyroxenes and olivine. These amphibolites persist to the highest temperatures of metamorphism in the earth's crust, provided that water is present. Depending on composition, some diopsidic pyroxene may be present instead of almandine-garnet. However, at lower tem-

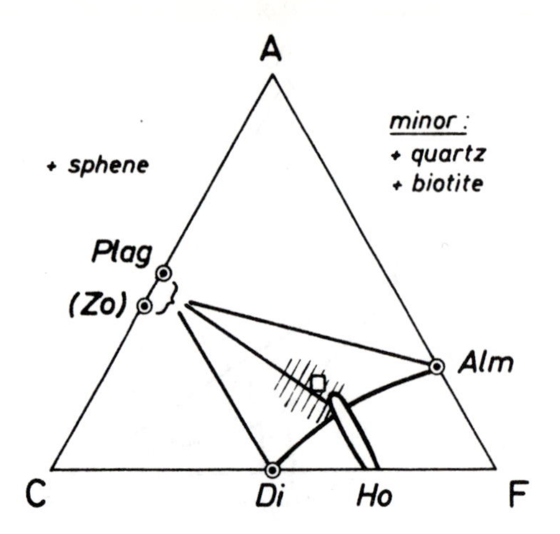

Fig. 12-2 *Labradorite/bytownite-amphibolite*. High-grade or higher temperature part of medium-grade and high-grade, depending on pressure. Little garnet is present and clinozoisite/epidote may be lacking. Besides hornblende, Ca-free amphibole (mostly cummingtonite) is a common additional constituent; this is not obvious from the diagram.

peratures pyroxene is absent, as is shown in Figure 12-3 for the *andesine- and oligoclase-amphibolites* of medium-grade.

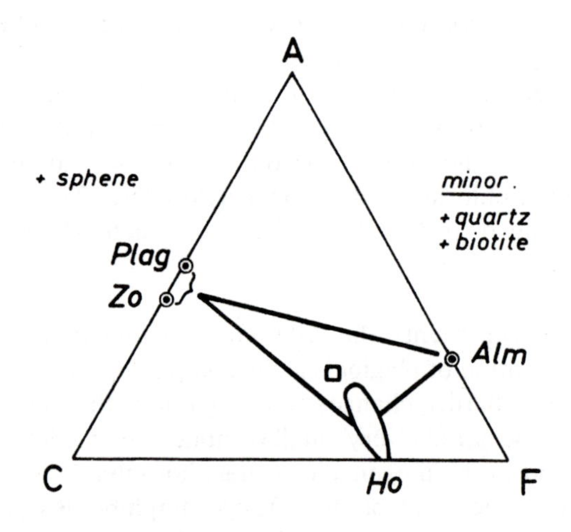

Fig. 12-3 *Andesine- and oligoclase-amphibolite*. Medium-grade. Garnet and clinozoisite/epidote may be lacking. Besides hornblende, Ca-free amphibole (mostly cummingtonite) is a common additional constituent; this is not obvious from the diagram.

Low-Grade

The *low-grade amphibolites* are shown in Figure 12-4; here, either albite or oligoclase coexists with hornblende, clinozoisite/epidote, garnet, quartz, *and chlorite*. The presence of chlorite in the paragenesis is diagnostic of low-grade. There are low-grade oligoclase-amphibolites with chlorite and medium-grade oligoclase-amphibolites without chlorite, *i.e.,* oligoclase-amphibolites are stable on both sides of the low-grade–medium-grade metamorphic boundary. With decreasing temperature, oligoclase is replaced by albite (+ clinozoisite); the plagioclase "jump" is encountered and *albite-amphibolites* are produced. At still lower temperatures hornblende becomes unstable; amphibolites no longer exist since they are defined as metamorphic rocks consisting predominantly of plagioclase and *hornblende*. The stability range of *greenschists or greenstones* has been reached.

Figure 12-5 refers to *greenschists* derived from mafic volcanics and formed in the lower temperature part of the range of metamorphic conditions which have been defined as low-grade. Hornblende has given *way to actinolite, chlorite, and clinozoisite/epidote; these minerals coexist with much albite and some quartz.* biotite *or* stilpnomelane, a little muscovite (phengite), and calcite may also be present. Greenschists are also formed from pelitic rocks, but they are distinguished from those derived from mafic rocks by the relative amounts of minerals: greenschists derived from pelites have much more quartz and phengite, less chlorite, albite, and epidote, and generally no actinolite.

Field observations suggest that actinolite changes to hornblende at

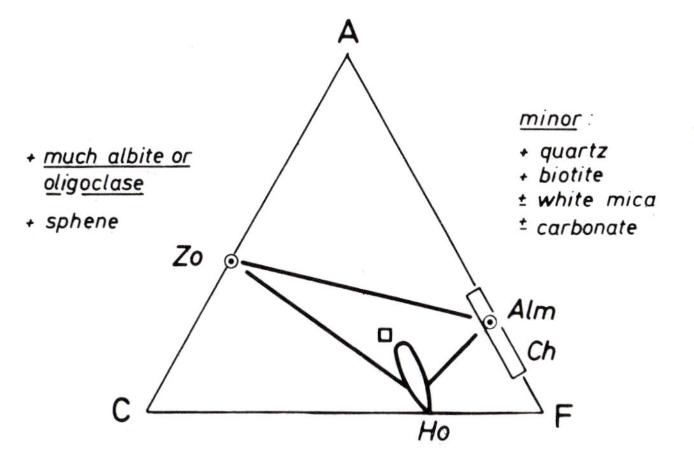

Fig. 12-4 *Albite/oligoclase-amphibolite = albite/oligoclase-hornblende-chlorite zone*. Low-grade; higher temperature part.

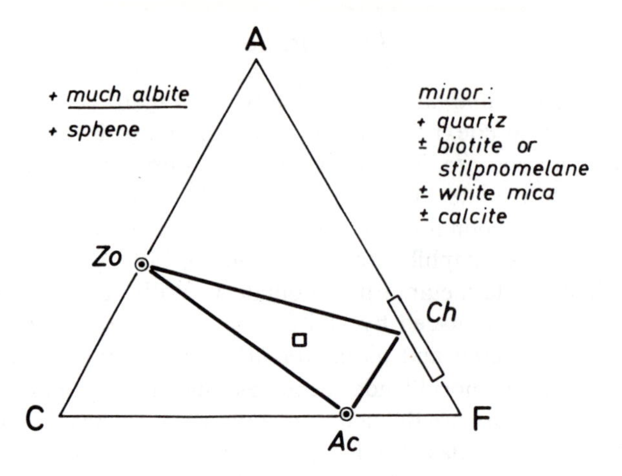

Fig. 12-5 *Albite-actinolite-chlorite zone*. Low-grade; lower temperature part: mafic greenschist.

about the same P,T conditions as for the appearance of almandine-garnet in metapelitic rocks at medium and high pressures. No data concerning these complex metamorphic changes are available. At present, it is only possible to estimate from the sequence of various metamorphic changes in different metamorphic terrains that the change from actinolite to hornblende will probably take place at about 500°C, rising only slightly with increasing pressure.

The distinction between hornblende and actinolite is easy only if iron-poor, colorless or pale-green actinolite has been formed, which contrasts with deeper-green hornblende. Only in such a case is it possible to separate an

albite-hornblende-chlorite zone

from an

albite-actinolite-chlorite zone

within the temperature range of low-grade metamorphism.

Very-Low-Grade

Mafic rocks of very-low metamorphic grade (roughly between 200° and 400°C) *exhibit a great variety of mineral parageneses* which are very sensitive to changes in temperature and pressure. In order to demonstrate this, the main differences between low-grade greenschists and various rocks of very-low-grade will be outlined briefly.

The typical assemblage of metabasaltic greenschists or greenstones is diagnostic of a specific metamorphic zone designated as

albite + actinolite + chlorite + zoisite/clinozoisite (epidote) zone

At lower temperatures zoisite/clinozoisite, in contrast to iron-rich epidote, is no longer stable. Consequently, clinozoisite or the iron-free clinozoisite component of epidote breaks down and forms new Ca-Al silicates like *lawsonite* or *laumontite* (at pressures below 3 kb) or the Ca-Al-Mg mineral *pumpellyite*.

Pumpellyite may be formed by various reactions, of which the following is probably common in a very large range (except at pressures lower than 2 to 3 kb):

$$\{\text{clinozoisite} + \text{actinolite} + H_2O\} = \{\text{pumpellyite} + \text{chlorite} + \text{quartz}\}$$

Lawsonite in metabasalts is probably formed by reactions mentioned earlier on p. 70, namely

$$\{\text{zoisite/clinozoisite} + \text{chlorite} + \text{quartz} + H_2O\} = \{\text{lawsonite} + \text{Al-poorer chlorite}\}$$

where chlorite supplies the required small amount of Al to form lawsonite from clinozoisite/zoisite, and the chlorite after the reaction will contain somewhat less Al. However, this suggestion still needs verification.

Whenever CO_2 is available in small amounts in a water-rich fluid, the following reaction leads to the formation of lawsonite:

$$1\ \text{zoisite/clinozoisite} + 1\ CO_2 + 5\ H_2O = 3\ \text{lawsonite} + 1\ \text{calcite}$$

In mafic rocks of very-low metamorphic grade, pumpellyite or lawsonite or both are diagnostic; they are present instead of iron-poor epidote, clinozoisite, or zoisite. At lower pressures *prehnite* $Ca_2[(OH)_2/Al_2Si_3O_{10}]$ occurs. At pressures below about 3 kb, laumontite is stable instead of lawsonite. Parageneses with these minerals will be considered on p. 179ff.

The formation of lawsonite, laumontite, pumpellyite, and prehnite from low-grade rocks is mainly caused by a decrease of temperature, whereas other reactions are caused by increasing pressure. These reactions are formation of lawsonite instead of laumontite, transformation of calcite into aragonite, and formation of the Na-amphiboles glaucophane or crossite. Glaucophane or crossite may also be formed in the lower temperature range of *low-grade* metamorphism, but the association with lawsonite and/or pumpellyite, instead of zoisite/clinozoisite, is diagnostic of *very-low-grade* metamorphism.

Glaucophane/crossite is formed by a reaction between chlorite and albite; actinolite may be an additional reactant:

$$\{\text{chlorite} + \text{albite} \pm \text{actinolite}\} = \{\text{glaucophane/crossite} + H_2O\}$$

A simplified equation, based on iron-free minerals, demonstrates the relationship between reactants and products:

$$\underset{\text{albite}}{4\ NaAlSi_3O_8} + \underset{\text{chlorite}}{Mg_6[(OH)_8/Si_4O_{10}]} = \underset{\text{glaucophane}}{2\ Na_2Mg_3Al_2[(OH)_2/Si_8O_{22}]} + 2\ H_2O$$

If the pressure has been sufficiently high to allow the formation of glaucophane, the amount of chlorite and albite in the metabasalts is greatly reduced or disappears entirely. Consequently, the rocks have a blue color. Generally, either chlorite or albite is used up completely in the reaction. Therefore, the following parageneses are common in very-low-grade mafic rocks within the *lawsonite-glaucophane zone:*

Glaucophane/crossite + lawsonite and/or pumpellyite
+ sphene + small amounts of albite or chlorite
Some specimens may contain actinolite, and minor amounts of white mica, stilpnomelane, quartz, calcite, or aragonite may be present.

This is indicated in Figure 12-6.

In addition to the reactions given above, more complex reactions leading to the formation of crossite have been suggested by Brown (1977) on the basis of petrographic studies:

$$\{\text{epidote} + \text{actinolite} + \text{albite} + \text{chlorite} + H_2O\} = \{\text{crossite} + \text{pumpellyite} + \text{quartz}\}$$

$$\{\text{actinolite} + \text{hematite} + \text{albite} + \text{chlorite} + H_2O = \{\text{crossite} + \text{epidote} + \text{quartz}\}$$

$$\{\text{epidote} + \text{chlorite} + \text{albite} + \text{quartz} + H_2O\} = \{\text{lawsonite} + \text{crossite} + \text{paragonite}\}$$

If the bulk composition of the rock was such that some albite remains after the formation of glaucophane, a complex reaction may be possible, forming jadeite-bearing pyroxene + quartz. Only rarely and in

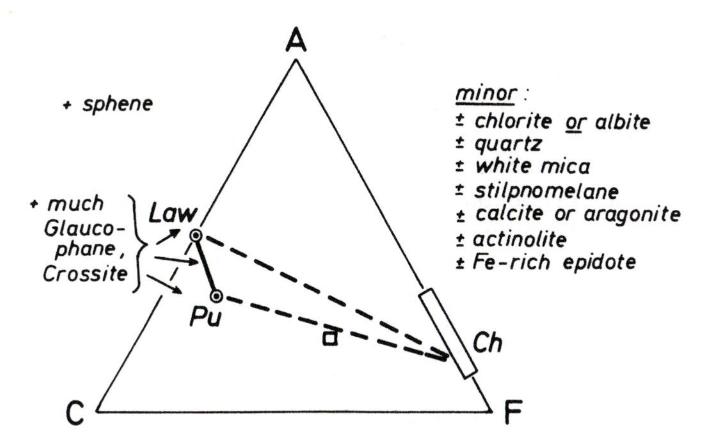

Fig. 12-6 *Lawsonite/pumpellyite-glaucophane zone*. Very-low-grade; high pressure.

small amounts have *Na pyroxenes* been formed in metabasalts. Aegirine-augite (with only 10% jadeite component) has been recorded from Calabria by E. de Roever (1972) and green omphacite or acmite with only a moderate content of jadeite component has been reported from California by Coleman and Clark (1968). These Na pyroxenes have less jadeite component than the jadeite pyroxenes which are typical of metagraywackes formed at very high pressures and very-low-grade metamorphism. It is necessary to make a clear distinction between Na pyroxenes in metabasalts and jadeitic pyroxenes in graywackes, because the simple breakdown reaction of albite to jadeite + quartz is relevant only in metagraywackes where the ratio of albite-to-mafic minerals is very high.

After the initial discovery of metamorphic *aragonite* by Coleman and Lee (1962) it has been reported from several places outside the American Pacific region—in Calabria (southern Italy) by Hoffmann (1970) and E. de Roever (1972) and in New Caledonia (island east of Australia) by Brothers (1970). These publications have contributed to the general knowledge about very-low-grade rocks.

The *P-T* range in which the discussed glaucophane/crossite-bearing rocks are formed will be designated as the

lawsonite/pumpellyite-glaucophane zone.

This is followed at lower pressures where glaucophane/crossite cannot form by the

lawsonite/pumpellyite-albite-chlorite zone.

Glaucophane-Free Mafic Rocks

At pressures lower than necessary for the formation of glaucophane/crossite, the minerals albite and chlorite are, of course, major constituents of metabasalts. However, the minerals lawsonite and/or pumpellyite are major constituents as well, as in the glaucophane rocks just discussed. Therefore, the commonest parageneses have as major constituents lawsonite and/or pumpellyite and albite and chlorite. Accordingly, the stability field of these rocks will be designated as the

lawsonite/pumpellyite-albite-chlorite zone.

The parageneses lawsonite and/or pumpellyite (depending on bulk composition), together with chlorite and much albite, have been indicated in Figure 12-7. Sphene is always an additional phase in variable amounts from a few percent up to 10 to 15%. The minor minerals are indicated as well in Figure 12-7.

Iron-rich epidote is usually rare but it has been observed together with lawsonite, occasionally in appreciable amounts (Hoffmann, 1970). Commonly, actinolite is also rare but in the Sanbagawa belt a distinct

pumpellyite + actinolite + albite + chlorite + lawsonite zone

has been mapped (Seki, 1969).

The association albite + chlorite + *epidote* + *actinolite* ± pumpel-

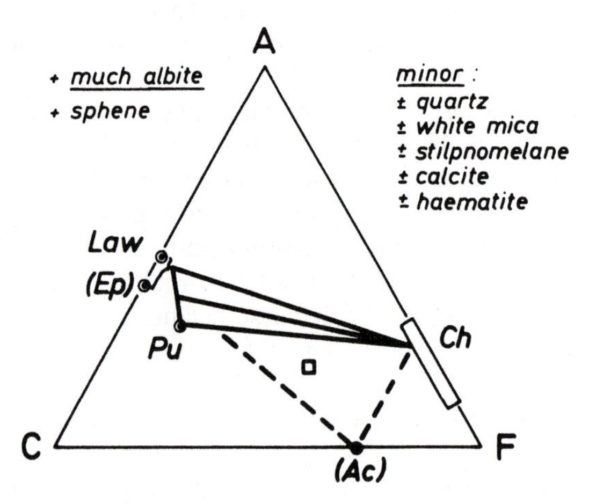

Fig. 12-7 *Lawsonite/pumpellyite-albite-chlorite zone.* Very-low-grade; medium to high pressure but lower than for glaucophane formation.

lyite recorded in California is exceptional, because the paragenesis albite + chlorite + actinolite + epidote is diagnostic of low-grade rather than very-low-grade metamorphism, provided epidote is iron-poor. In very-low-grade metamorphism actinolite coexists only with iron-rich epidote.

The following should be noted: The absence of glaucophane/crossite in this zone need not imply, as E. de Roever (1972) pointed out, an absence of blue amphiboles in general. The blue, iron-rich amphibole magnesioriebeckite[1] (as well as the nonjadeitic Na pyroxenes aergirine and aegirine-augite) may well be present in rocks of this zone of very-low-grade metamorphism. Within the range of bulk compositions of mafic rocks, paragenesis (a) or (b) is found close to (c):

magnesioriebeckite + epidote + albite + chlorite (a)
magnesioriebeckite + aegirine-augite (or aegirine) + albite + chlorite (b)
lawsonite ± epidote + albite + chlorite (c)

Are Prehnite-Bearing Rocks Formed Only at Low Pressures?

This section deals with the metamorphism of mafic rocks at pressures greater than 3 kb, the approximate pressure below which lawsonite is not stable. On p. 180ff we shall review the different transformations that may take place at pressures below approximately 3 kb. There, the occurrence of prehnite and of prehnite + pumpellyite plays a prominent role. However, the question arises as to whether prehnite with or without pumpellyite is invariably a criterion of pressures lower than about 3 kb or whether it may be formed at higher pressures as well.

Only one petrographic example is known which may cast doubt on the low pressure nature of prehnite. Metamorphic aragonite marble is associated with metasedimentary and metavolcanic rocks that consist of prehnite + pumpellyite + albite + chlorite ± quartz in northwest Washington. Vance (1968), who reported this interesting occurrence, drew from it the conclusion that the pressures for aragonite formation may have been much lower than experimental determinations demand. Newton *et al.* (1969) suggested that strain energy in calcite would lower the transition pressure by more than 3 kb and that this might have been the case in the investigated region. However, this problem remains unsolved. It is very likely that aragonite occurring in metamorphic rocks is an indicator of high pressures.

---

[1]Riebeckite is $Na_2Fe_3^{2+}Fe_2^{3+}$ $[(OH)_2/Si_8O_{22}]$; in the investigated magnesioriebeckites between 25 and 55% of the Fe^{2+} is replaced by Mg.

Experimental investigations at pressures between 5 and 9.5 kb of the reactions

$$\text{pumpellyite} + \text{quartz} = \text{prehnite} + \text{chlorite} + H_2O$$
$$\text{pumpellyite} + \text{actinolite} + \text{quartz} = \text{prehnite} + \text{chlorite} + H_2O$$

by Hinrichsen and Schürmann (1969) and Nitsch (1971) show that prehnite (with or without pumpellyite) may exist at high and medium pressures. Prehnite + pumpellyite is not limited to low pressure metamorphism, and the association of prehnite with (high pressure, low temperature) aragonite requires no special explanation. It is expected that in a rock sequence formed at medium or high pressures, rocks of appropriate composition will grade, with decreasing temperature, from pumpellyite (± lawsonite) + albite + chlorite + quartz or pumpellyite + actinolite + albite + chlorite + quartz rocks to albite + chlorite rocks containing prehnite ± pumpellyite ± lawsonite. This is suggested from the experimental data shown in Figure 12-11 on p. 190.

Summary of Metamorphic Zones in Mafic Rocks

lawsonite/pumpellyite-albite-chlorite zone. At higher pressure: lawsonite/pumpellyite-glaucophane zone (calcite- or aragonite-bearing).	very-low-grade
clinozoisite-albite-actinolite-chlorite zone (greenschist) At higher temperature: albite/oligoclase-hornblende-chlorite zone (low-grade amphibolite)	low-grade
amphibolites	medium- and high-grade

Very-Low-Grade Metamorphism at Low Pressures

Here, reactions below about 3 kb will be considered; at these pressures laumontite instead of lawsonite is stable. In this pressure range and at temperatures of very-low-grade metamorphism, the following minerals are significant:

Laumontite: $Ca[Al_2Si_4O_{12}] \cdot 4\ H_2O$

Wairakite: $Ca[Al_2Si_4O_{12}] \cdot 2\ H_2O$
Prehnite: $Ca_2[(OH)_2/Al_2Si_3O_{10}]$
Pumpellyite: $Ca_4(Mg,Fe)(Al,Fe)_5O(OH)_3[Si_2O_7]_2[SiO_4]_2 \cdot 2\ H_2O$

We shall continue to review the metamorphic changes in mafic rocks with decreasing temperature.

At low-grade metamorphism mafic metamorphic rocks (greenschists) are characterized by the assemblage shown in Figure 12-5:

Zoisite/clinozoisite (or relatively iron-poor epidote)
+ Actinolite + albite
+ Chlorite
± Quartz

At conditions of very-low-grade metamorphism, this paragenesis changes to the assemblage shown in Figure 12-8:

Pumpellyite + albite
+ Actinolite
+ Chlorite
± Quartz

Iron-rich epidote and some stilpnomelane may also occur. In certain cases, clinozoisite coexists with pumpellyite and chlorite; this assemblage may possibly represent the actual reaction equilibrium (Bishop,

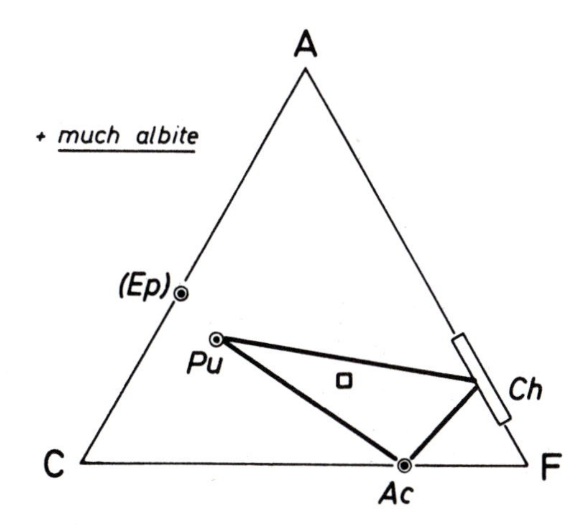

Fig. 12-8 *Pumpellyite-actinolite-chlorite zone.* Very-low grade; low but not very low pressure.

1972). The assemblage shown in Figure 12-8 is diagnostic of the so-called pumpellyite-actinolite facies of Hashimoto (1965) [see also Coombs *et al.* (1970)], which here is called the

pumpellyite-actinolite-chlorite zone.

The following reaction (No. 2 in Figure 12-11) leads to this paragenesis:

$$\{\text{clinozoisite} + \text{actinolite} + H_2O\} = \{\text{pumpellyite} + \text{chlorite} + \text{quartz}\}$$

This reaction was suggested by Seki (1969), and has been experimentally investigated by Nitsch (1971). This study indicated that the reaction cannot proceed at very low pressures but only at pressures above the invariant point shown in Figure 12-11. The exact position of this point, between about 1 and 3 kb, is unknown. As shown in Figure 12-7, pumpellyite + actinolite + chlorite occur at higher pressures but probably at pressures sufficiently low to prevent the formation of lawsonite in accompanying rocks, *i.e.,* at pressures below about 3 kb. This would be consistent with the common field observation that with decreasing temperature, rocks with pumpellyite + actinolite + chlorite are followed by rocks with prehnite and finally with laumontite.

Prehnite-pumpellyite assemblages have now been recognized in many areas; see Seki (1969), Jolly (1970), Coombs *et al.* (1970), and Martini and Vuagnat (1970). A so-called prehnite-pumpellyite facies has been recognized (Coombs, 1961); we prefer the designation

prehnite-pumpellyite-chlorite zone

Figure 12-9 shows, among others, the commonest assemblage in mafic rocks:

pumpellyite + prehnite + chlorite + albite ± quartz

and the less common assemblage:

prehnite or pumpellyite + chlorite + calcite + albite

Sphene is present in all these mafic rocks.

Transitional between the parageneses of Figures 12-9 and 12-8 is the paragenesis observed by Coombs *et al.* (1970):

prehnite + chlorite + actinolite + albite ± epidote

A number of reactions may produce the association pumpellyite + prehnite + chlorite. Therefore, this three-mineral assemblage by itself is a poor indicator of metamorphic conditions. However, if prehnite-pumpellyite assemblages are associated with metamorphic rocks containing laumontite or wairakite rather than lawsonite and/or glaucophane, it is clear that pressures have been rather low, probably less than 3 kb. In such case and in a reaction sequence reflecting decreasing temperature, the appearance of prehnite + pumpellyite + chlorite + albite indicates, according to the investigations by Nitsch (1971), that the temperature has been between about 400° and 300°C (see Figure 12-11). At these temperatures, laumontite will not form but wairakite may be present.

The following reactions produce either prehnite or pumpellyite; they are designated by the numbers used in Figure 12-11, p. 190:

{pumpellyite + actinolite + quartz} = {prehnite + chlorite} (1)
{pumpellyite + actinolite + epidote} = {prehnite + chlorite} (4)
{actinolite + epidote} = {prehnite + chlorite + quartz} (5)
{prehnite + chlorite + epidote} = {pumpellyite + quartz} (6)

According to these reactions, specific parageneses arise which may produce a zonation in areas of very-low-grade metamorphism. Some of the parageneses are indicated in Figure 12-11.

In detailed studies of very-low-grade metamorphic areas, the following sequence, with decreasing temperature, commonly has been found:

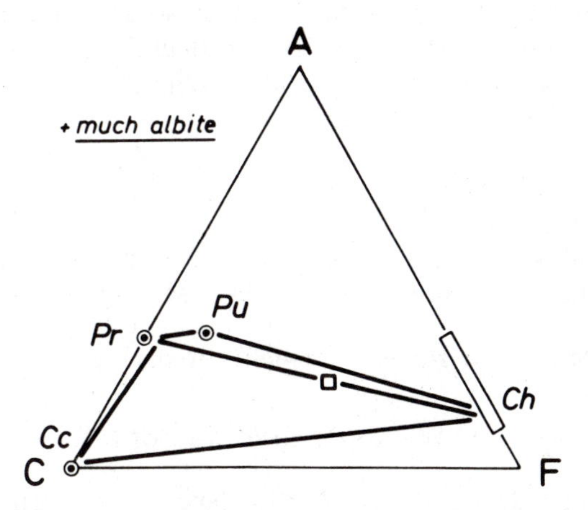

Fig. 12-9 *Prehnite-pumpellyite-chlorite zone.* Very-low-grade; low pressure. Note that this zone need not be restricted to low pressures.

Low-grade rocks belonging to the
clinozoisite + actinolite + chlorite + albite zone.
(mafic greenschists) pass into the following sequence
of very-low-grade rocks:

Pumpellyite + actinolite + chlorite + albite zone,
Prehnite + pumpellyite + chlorite + albite zone,
Laumontite + chlorite zone;

In some areas the laumontite zone is preceded by a
wairakite + chlorite zone

The *laumontite-chlorite zone* is characteristically developed in mafic rocks. This assemblage is very widespread in the Coombs *et al.* (1959) *zeolite facies,* loosely defined "to include at least all those assemblages produced under conditions in which quartz-analcime, quartz-heulandite, and quartz-laumontite commonly are formed." However, recently Coombs (1971) suggested the following, more restricted, definition: The zeolite facies is characterized by the association calcium-zeolite + chlorite + quartz in rocks of favorable bulk composition. Mafic rocks and many volcanogenic sediments have the required composition. This definition is very similar to Winkler's (1965) laumontite-prehnite-quartz facies; however, Coombs' recent designation of a certain range of metamorphic conditions by the assemblage Ca-zeolite + chlorite is admittedly superior in view of petrographic observations now available.

Whereas facies are based on associations of mineral assemblages from rocks of diverse bulk compositions, metamorphic zones are designated by a characteristic paragenesis (or, in some cases, by a mineral). Following Coombs' (1971) suggestion, we shall distinguish the following zones, all of which contain a Ca zeolite + chlorite and are developed in mafic rocks and derived sediments.

Wairakite-chlorite zone
Laumontite-chlorite zone
and possibly, if established as a stable association,
heulandite + prehnite + chlorite zone (see Figure 12-10b)

The sequence of these zones corresponds to decreasing temperature.

The Wairakite-Chlorite Zone

This zone is less common than the laumontite-bearing zone. It has been observed only in areas with very high geothermal gradients, *i.e.,* in active volcanic terrains. In the Wairakei district in New Zealand (Stei-

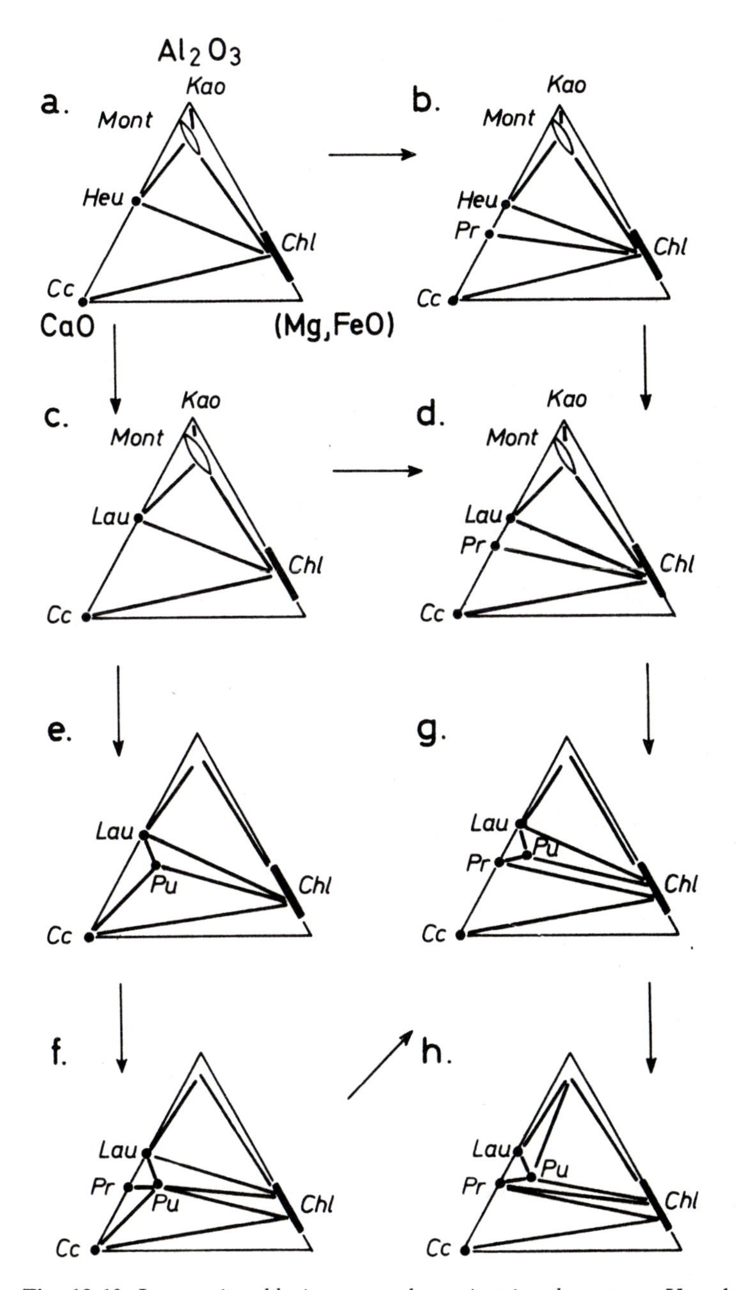

Fig. 12-10 *Laumonite-chlorite zone, shown in triangles c to g.* Very-low-grade; low pressure, h represents the *prehnite-pumpellyite-chlorite zone*. (Compare Figure 12-9 for mafic compositions.) (Compilation of some zeolite facies assemblages by Coombs, 1971.)

ner, 1955; Coombs *et al.*, 1959) and at several places in Japan (see Seki, 1969), cores from boreholes drilled in areas of geothermal activity showed zones of zeolites formed by the action of steam and hot solutions. The zones have the following sequence with increasing temperature:

Mordenite zone or mordenite + heulandite zone
Laumontite zone
Wairakite zone

At Tansawa Mountain in Japan this sequence can be followed further into higher temperature ranges; there, the

Wairakite zone is succeeded by the
Prehnite + pumpellyite + chlorite zone and by the
Clinozoisite + actinolite + chlorite zone

From field observations it is clear that wairakite occurs instead of laumontite at higher temperatures if pressures remain rather small (probably below 3 kb). The entire sequence may be expressed by the following reactions:

mordenite = {laumontite + albite + quartz + water}
heulandite = {laumontite + albite + quartz + water}
laumontite = {wairakite + water}

The compositions of the minerals are:

Mordenite: $(Ca,Na_2)[Al_2Si_{10}O_{24}] \cdot 7\ H_2O$
Heulandite: $(Ca,Na_2)[Al_2Si_7O_{18}] \cdot 6\ H_2O$
Laumontite: $Ca[Al_2Si_4O_{12}] \cdot 4\ H_2O$
Wairakite: $Ca[Al_2Si_4O_{12}] \cdot 2\ H_2O$

The Laumontite-Chlorite Zone

It is well documented that in response to decreasing temperature laumontite appears in place of pumpellyite and/or prehnite; eventually all pumpellyite and then all prehnite is replaced by laumontite. The reactions involving pumpellyite, prehnite, laumontite, and chlorite are not yet known[2]. There must be several reactions because, in rocks of mafic compositions, several different assemblages have been observed. This is shown by the compilation of "zeolite facies, laumontite subfacies" assemblages by Coombs (1971); see Figure 12-10. Mafic compositions plot on or somewhat below the join laumontite-chlorite. All parageneses of mafic bulk composition in triangles c to g contain laumontite. They probably have formed at a temperature not exceeding 300°C. Nevertheless, parageneses including laumontite (all having additional albite as a major constituent) may differ markedly:

Laumontite + chlorite + pumpellyite
Laumontite + chlorite + prehnite
Laumontite + chlorite + calcite

Note, in Figure 12-10 g, that the laumontite + chlorite zone may overlap with the prehnite + pumpellyite + chlorite zone. It is also evident from Figure 12-10 that my earlier (1965) designation "laumontite-prehnite-quartz" for the whole "zeolite facies" can no longer be upheld; this assemblage corresponds to the very restricted conditions of triangle d within the laumontite-chlorite zone.

At temperatures lower than those of stages b and c in Figure 12-10, prehnite and laumontite are not stable; the Ca-zeolite heulandite or its variety clinoptilolite takes their place. Probably the following reactions proceed:

$$\underset{\text{laumontite}}{CaAl_2Si_4O_{12} \cdot 4\,H_2O} + \underset{\text{quartz}}{3\,SiO_2} + 2\,H_2O = \underset{\text{heulandite}}{CaAl_2Si_7O_{18} \cdot 6\,H_2O}$$

{heulandite + albite and/or adularia + water} = clinoptilolite

Starting with mafic rocks and water, a sequence of reactions has been discussed here that is observed with decreasing temperature. It may be more familiar to treat metamorphic changes with increasing temperature. In this case, rocks consisting of heulandite + chlorite + albite (or analcime + quartz) as major constituents are often used to begin

[2]The following reaction, suggested by Coombs (1961) may be one of various possible reactions:

[pumpellyite + quartz + H_2O] = [laumontite + prehnite + chlorite]

Another one has been given by Seki *et al.* (1969b).

[prehnite + pumpellyite + chlorite + quartz + water] = [laumontite + saponite]

discussion of prograde metamorphism. These rocks have previously been formed by alteration of mafic rocks in the *P-T* range designated as diagenetic in this treatise.

At about the same conditions, with increasing temperature, where laumontite is formed from heulandite, albite is formed from analcine + quartz:

$$\underset{\text{albite}}{NaAlSi_3O_8} + H_2O = \underset{\text{analcime}}{NaAlSi_2O_6 \cdot H_2O} + \underset{\text{quartz}}{SiO_2}$$

A typical field occurrence was described by Coombs (1961) as follows: "In the upper members of the Triassic Taringatura section, Southland, New Zealand, heulandite or its relative clinoptilolite is widespread. . . . Sedimentary beds consisting essentially of analcime and quartz also occur high in the Taringatura section. . . . At depths below about 17,000 ft (ca. 6 km) in the present stratigraphic section (which formerly had an overburden of an additional 5 km sedimentary succession), the analcime beds are represented by quartz-albite, sometimes with adularia, and heulandite gives way to a less hydrated lime zeolite, laumontite."

With the first appearance of laumontite (or lawsonite at higher pressures) mineral assemblages are encountered which are not stable in the sedimentary environment. For this reason we have defined this change as the beginning of metamorphism (Chapter 2); the field of metamorphism is bordered at lower temperatures by the field of diagenetic changes. During diagenesis many zeolites and other minerals are formed, but all of them are also stable under sedimentary conditions. By far the commonest zeolites are heulandite (or clinoptilolite) and analcime, and they can occur even as the chief constituents of some sediments. Other zeolites of sedimentary-diagenetic origin (some of which may persist somewhat into very-low metamorphic grade) are

Ca zeolites:	scolecite, gismondine, chabazite, levyne
Ca,Na zeolites:	mesolite, thomsonite, gmelinite, stilbite, mordenite
Na zeolite:	natrolite
K,Ca,Na zeolite:	phillipsite

For further information see Hay (1966) on zeolites and zeolitic reactions in sedimentary rocks and Miyashiro and Shido (1970) on "progressive metamorphism in zeolite assemblages."

Summary of Very-Low-Grade Metamorphic Zones with Increasing Temperature, Below the Stability of Lawsonite

Laumontite-chlorite zone
Wairakite-chlorite zone
Prehnite/pumpellyite-chlorite zone
Pumpellyite-actinolite-chlorite zone

Evaluation of Metamorphic Changes at Very-low-Grade

Phase relations relevant to very-low-grade metamorphism have been compiled in Figure 12-11. Much experimental work has been done recently and is summarized below:

(a) The reaction

analcime + quartz = albite + water

is taken as the boundary between diagenesis and very-low-grade metamorphism, *i.e.*, as an indication of the beginning of metamorphism. The earlier data for this reaction by Campbell and Fyfe (1965) have been superseded by those of Liou (1971a). The curve designated analcime + Q = albite refers to the experimentally determined equilibrium. Albite of intermediate structural state grew in the runs, whereas in nature low albite is formed. Taking this into account, Liou calculated the probable equilibrium conditions (broken line to the left of curve) involving low albite.

(b) A further indication of the beginning of metamorphism is the appearance of lawsonite at the expense of heulandite. The reaction

heulandite = lawsonite + quartz + water

has been investigated by Nitsch (1968).

(c) Reactions involving prehnite, pumpellyite, chlorite, clinozoisite/zoisite, actinolite, and quartz have been studied by Nitsch (1971). The following reactions are relevant:

$$\{\text{prehnite} + \text{chlorite} \pm \text{quartz}\} = \{\text{pumpellyite} + \text{actinolite} \pm \text{quartz}\} \quad (1)$$

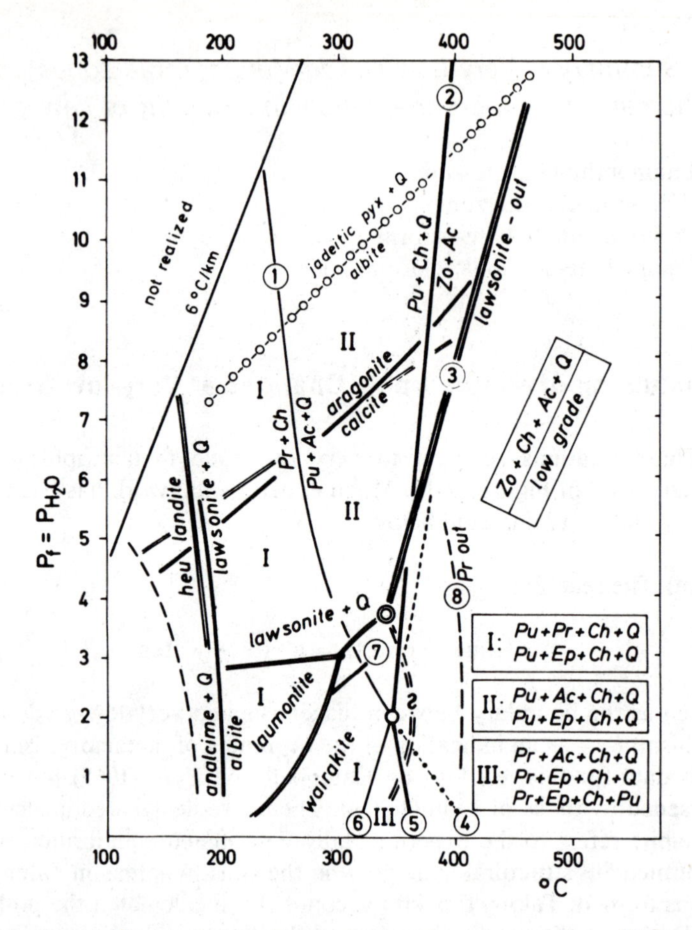

Fig. 12-11 Phase relations in very-low-grade metamorphism. H_2O pressure is equal to total pressure. Note that number 3 refers to the broken line further down.

{pumpellyite + chlorite + quartz} = {clinozoisite + actinolite} (2)

{pumpellyite + quartz} = {prehnite + clinozoisite + actinolite} (3)

{prehnite + chlorite} = {pumpellyite + clinozoisite + actinolite} (4)

{prehnite + chlorite + quartz} = {diopside + actinolite} (5)

{pumpellyite + quartz} = {prehnite + clinozoisite + chlorite} (6)

All reactions have been investigated in the iron-free system and therefore the univariant equilibria are correctly represented by a curve. With iron present, however, a band (probably narrow) will replace the curve in each of the six reactions and epidote instead of clinozoisite may be formed. The six equilibrium curves meet at an invariant point; its position within a pressure range of about 1 to 4 kb is still uncertain. But sequences of reactions suggest a location of the invariant point near 2 kb and at 345° ± 20°C. All equilibrium temperatures given are within the limit of ± 20°C. The location of reactions (3) and (4) in Figure 12-11 has been estimated.

(d) What was believed to be the upper stability of lawsonite has been investigated by Newton and Kennedy (1963) and by Nitsch (1972), but only recently Nitsch (1974) found that the previously studied reactions are metastable. The stable upper stability of lawsonite is represented by the following reaction (mentioned on page 79):

5 lawsonite = 2 zoisite-clinozoisite + 1 margarite + 2 quartz + 8 H_2O

For this reaction Nitsch gives the following equilibrium data:

345° ± 10°C at 4kb
385° ± 10°C at 7 kb
430° ± 10°C at 10 kb

These data for the upper stability of lawsonite are shown by a double line in Figure 12-11, although the specific margarite-producing reaction is not relevant when dealing with metamorphism of mafic rocks. Rather, the reactions mentioned on page 70 will be valid:

{lawsonite + chlorite} = {zoisite-clinozoisite + Al-richer chlorite + quartz + H_2O

and

3 lawsonite + 1 calcite = 1 zoisite-clinozoisite + 1 CO_2 + 5 H_2O

No equilibrium data for these reactions are available, but it is estimated that they will deviate only slightly from the maximum stability values of lawsonite.

(e) The reaction

1 lawsonite + 2 quartz + 2 water = 1 laumontite

studied by Crawford and Fyfe (1965) has been reinvestigated recently by Nitsch (1968), Thompson (1970a), and Liou (1971b). The results of the three investigations are in good agreement, yielding the following mean values:

200°C and 2.8 ± 0.2 kb	lawsonite is stable on
250°C and 3.0 ± 0.2 kb	the higher pressure side,
300°C and 3.1 ± 0.2 kb	*i.e.*, above ca. 3 kb

(f) Laumontite dehydrates with increasing temperature to wairakite:

$$1 \text{ laumontite} = 1 \text{ wairakite} + 2\ H_2O.$$

Equilibrium data determined by Liou (1971b) are

255° ± 5°C at 1 kb
282° ± 5°C at 2 kb
297° ± 5°C at 3 kb

(g) Above 3 kb laumontite is not stable and wairakite is formed from lawsonite and quartz in response to increasing temperature; this reaction is, however, restricted to a narrow pressure range:

$$1 \text{ lawsonite} + 2 \text{ quartz} = 1 \text{ wairakite}$$

Equilibrium data have also been determined by Liou (1971b):

315°C at 3.5 kb
360°C at 4.0 kb

The wairakite field may be restricted in its higher pressure range. This follows from Liou's (1971c) investigation of the reaction, designated as (7) in Figure 12-11:

$$2 \text{ wairakite} + n H_2O = \text{prehnite} + \text{montmorillonite} + 3 \text{ quartz} \quad (7)$$

(h) Wairakite may be stable within a temperature range of almost 100°C, according to experimental studies of Liou (1970). The dehydration reaction

$$\text{wairakite} = \text{anorthite} + \text{quartz} + \text{water}$$

takes place at the following conditions:

330°C at 0.5 kb	
350°C at 1 kb	(Temperatures are stated to be
370°C at 2 kb	correct within ± 5°C)
385°C at 3 kb	
390°C at 4 kb	

The temperature values at the higher pressures seem to be too high by about 30° when compared with the upper stability curve of lawsonite. These two reaction curves must meet in an invariant point (where laumontite is absent). Consequently, a slight adjustment of temperatures has been made so that the dehydration of wairakite has been plotted in Figure 12-11 at 350°C (instead of 385°C) and 3 kb.

(i) The upper stability of prehnite, given by the reaction

$$5 \text{ prehnite} = 2 \text{ zoisite} + 2 \text{ grossular} + 3 \text{ quartz} + 4 \text{ H}_2\text{O} \qquad (8)$$

has also been determined by Liou (1971c). The data which are shown in Figure 12-11 as reaction (8) are

403° ± 5°C at 3 kb
399° ± 10°C at 4 kb
393° ± 5°C at 5 kb

(j) Since aragonite occurs as a metamorphic mineral in very-low-grade rocks, it is appropriate to include the data of the calcite/aragonite inversion in Figure 12-11. The calcite/aragonite transformation has been reinvestigated by Johannes and Puhan (1971)[3]. It is of special value that they determined the transformation between 200°C and 480°C in a hydrothermal apparatus as well as in a piston-cylinder device. More recent work has been carried out by Crawford and Hoersch (1972) from 50° to 150°C. The resulting slope of that part of the equilibrium curve agrees very well with that obtained previously at higher temperatures by Boettcher and Wyllie (1968a), but it differs from that determined by Johannes and Puhan.

[3]Somewhat before Johannes and Puhan (1971) carried out their experiments, Zimmermann (1971) had investigated the calcite/aragonite transition between 150 and 350°C. He determined the same slope of the equilibrium curve as Boettcher and Wyllie (1968) but his curve is displaced towards 2 kb higher pressure values parallel to the relevant double line indicated in Figure 12-11. The reason for this shift is not yet understood.

Both curves are shown in Figure 12-11. The equilibrium data for the calcite-aragonite inversion of the two sets of determination are

5 kb at 180°C or 150°C
7 kb at 300°C or 300°C
9 kb at 400°C or 450°C

(k) In very-low-grade rocks, *jadeitic pyroxene* is a characteristic mineral formed at very high pressures in albite-rich and chlorite-poor rocks, *i.e.*, in *graywackes*. Generally, it is not the reaction albite = jadeite + quartz but a more complicated reaction that takes place, so that a jadeitic pyroxene is formed at somewhat lower pressures than would be necessary to form pure jadeite. Therefore, jadeitic pyroxene occurs together with quartz (but not pure jadeite) (E. de Roever, 1972). Newton and Smith (1967) have determined that the formation of a jadeitic pyroxene with the composition jadeite (82), acmite (14), and diopside (4) requires 7.5 kb at 200°C and 9.5 kb at 300°C. This boundary is also shown in Figure 12-11.

(l) The conditions leading to the formation of *glaucophane* (mainly from chlorite and albite) are not yet known. From field observation it is obvious that it forms at pressures lower than those necessary for the calcite/aragonite inversion. The estimated conditions given by E. de Roever (1972) are

200°C at 4 to 4.5 kb
300°C at 7 to 7.5 kb
400°C at 10 to 10.5 kb

However, the curve based on these values crosses the calcite-aragonite boundary between 250° and 300°C. A crossing of that line at about 300°C has also been suggested by Brothers (1970), but the slope appears to be wrong. At low temperatures glaucophane should not form in the calcite field but only at pressures somewhat higher than the calcite/aragonite inversion, whereas at higher temperatures glaucophane should form at pressures lower than the calcite-aragonite inversion. This deduction is based on many field observations, especially on the occurrence of glaucophane in low-grade greenschists, *i.e.*, from the higher pressure, low-grade "glaucophanitic greenschist zone," where aragonite has never been encountered.

This discussion shows how little is known about the conditions leading to the formation of glaucophane [and crossite, which E. de Roever (1972) believes to form at markedly higher pressures]. The estimate suggested on p. 89—200°C at 5 kb and 350°C at 7 kb—gives the approximate conditions indicated by the appearance of glaucophane. This line would be situated at slightly lower pressures than the calcite-aragonite inversion. The glaucophane field is not shown in Figure 12-11, although it has been sketched in Figure 7-4. The appearance of glaucophane in metamorphic rocks is very prominent but unfortunately little is known about its petrogenesis.

Summary of Phase Relations

Figure 12-11 represents a number of relevant phase relations in very-low-grade metamorphism, but the number of actual reactions during metamorphism is certainly greater and they cannot all be treated here. All data given are valid only for the condition that H_2O pressure is equal to total pressure. Very small amounts of CO_2 in the fluid will cause the decomposition of minerals like lawsonite, laumontite, etc., producing calcite in addition to other silicates. Therefore, when the minerals represented in Figure 12-11 are encountered in nature, it is certain that the fluid consisted predominantly of water.

Previously, on pp. 89 and 92, four *pressure divisions of very-low-grade metamorphism* were suggested. These are, with increasing pressure:

[laumontite]- or [wairakite]-very-low-grade.
[lawsonite/pumpellyite]-very-low-grade, with albite + chlorite, but no glaucophane in mafic metamorphic rocks. *Pumpellyite* may occur together with or instead of lawsonite.
[glaucophane + lawsonite/pumpellyite]-very-low-grade.
[jadeitic pyroxene + quartz]-very-low-grade, commonly developed in metagraywackes.

The glaucophane field occupying part of the large lawsonite field is not shown in Figure 12-11 because precise data are lacking. On the other hand, it is obvious from Figure 12-11 that a prominent additional pressure division can be based on the presence of aragonite (in addition to calcite).

From Figure 12-11 it is further evident that, within the large lawsonite field, a number of different parageneses may coexist with lawsonite. Characteristic parageneses are:

(a) To the right of reaction (2) and up to the stability limit of lawsonite, the paragenesis clinozoisite + actinolite + chlorite + quartz

may coexist with lawsonite.

The paragenesis *without* lawsonite is characteristic of low-grade rocks; however, in association with lawsonite, this paragenesis characterizes the highest temperature range of very-low-grade metamorphism near the boundary of low-grade and very-low-grade (according to the definitions given earlier in Chapter 7). Field occurrences of this particular paragenesis apparently have not yet been recorded, but they are expected to be found in further studies.

(b) The paragenesis

pumpellyite + actinolite + chlorite + quartz
(with or without lawsonite or wairakite)

may be formed between curves 2 and 1 of Figure 12-11 within a very large pressure range. This assemblage, together with lawsonite, has often been observed. However, depending on bulk chemistry, the following pumpellyite-bearing association may form within the same P,T range:

pumpellyite + epidote + chlorite + quartz

and this actinolite-free paragenesis may persist to lower temperatures to the left of curve 1. It is important to distinguish between the parageneses pumpellyite + actinolite and pumpellyite + epidote; the former is restricted to a smaller range of physical conditions. On the other hand, the absence of pumpellyite + actinolite does not necessarily mean that its stability field has not been reached; bulk composition may have prevented its formation, giving rise instead to the alternative association pumpellyite + epidote in field II of Figure 12-11. In field II, prehnite (together with pumpellyite and chlorite) does not occur.

(c) Parageneses with *prehnite and pumpellyite* are stable to the left of curves 1 and 5 in the fields I and III. The paragenesis

prehnite + pumpellyite + chlorite + quartz

covers the entire field, from very low to very high pressures.

From Figure 12-11 it seems possible that this paragenesis may occur together with rocks containing lawsonite, laumontite, or wairakite. But observations in nature (see Smith, 1969) strongly indicate that, with increasing temperature, the place of laumontite commonly (but not invariably) is taken by prehnite and/or pumpellyite within the low pressure range; thus a laumontite zone is succeeded by a prehnite + pumpellyite zone without laumontite (see Figure 12-10).

The other association with prehnite + pumpellyite, occurring in field III, is

prehnite + pumpellyite + epidote + chlorite

This assemblage is restricted to the narrow field between reactions (5) and (6). In the same field, prehnite-bearing assemblages without pumpellyite may be formed in rocks of appropriate bulk composition; this is indicated in field III in the inset of Figure 12-11.

Sequences of Metamorphic Zones

At very low pressures and with increasing temperature, the following sequence of zones has been recorded from the Tanzawa Mountains in Japan by Seiki *et al.* (1969b). In each zone albite and quartz are major constituents:

Clinoptilolite, stilbite
Laumontite + chlorite
Wairakite + chlorite (overlapping laumontite + chlorite)
Prehnite + pumpellyite + chlorite + epidote
Actinolite + epidote + chlorite (no pumpellyite or prehnite; low grade has been reached)

This sequence traverses the low pressure field III of Figure 12-11. This is only one example of a sequence observed in many localities.

At presumably normal geothermal gradients, the following sequence is observed in New Zealand (Coombs, 1960), in the western European Alps (Martini and Vuagnat 1970), in the Kii Peninsula, Sanbagawa belt of Japan (Seki, 1969), and in other places:

Laumontite + chlorite ± prehnite
Prehnite + pumpellyite + chlorite
Pumpellyite + actinolite + chlorite (no prehnite)
Actinolite + epidote + chlorite (neither prehnite nor pumpellyite; low grade has been reached).

This sequence apparently formed under a geothermal gradient of about 30°C/km, judging from Figure 12-11. These conditions must prevail in all areas where "normal" burial takes place. Mafic lava flows would then be converted to rocks displaying the above characteristic parageneses. Recently, these parageneses have also been discovered in the mafic lavas called *spilites*. Although a primary, magmatic origin of spilites is still upheld by some, such an origin is ruled out because of mineral parageneses found in many spilites. Although the common association chlorite + albite + calcite of many spilites is not very diagnostic, other assemblages mentioned above are. Thus, Coombs (1974) concludes that "spilites may occasionally be ascribed to the zeolite facies, and more commonly to the prehnite-pumpellyite, pumpellyite-actinolite, and greenschist facies. The prehnite-pumpellyite facies is particularly commonly represented, especially in New Zealand."

Significantly lower geothermal gradients are required in order to form *lawsonite*. Judging from Figure 12-11, a temperature of 200°C has not been exceeded at a depth of 10 km (*i.e.*, at a little less than 3 kb), if lawsonite is not preceded by laumontite, *i.e.*, the gradient was smaller than 20°C/km. If laumontite precedes the lawsonite zone, the thermal gradient was between 30° and 20°C/km.

A metamorphic series where lawsonite is encountered has been described from the Sanbagawa belt in Japan and in the Franciscan belt in California. In the Kanto Mountains, Sanbagawa belt, the sequence is

Pumpellyite ± lawsonite + chlorite + albite
Pumpellyite ± lawsonite + actinolite + chlorite
  in this zone sodic amphibole is common, and jadeitic pyroxene occurs in a narrow part of this zone
Epidote + actinolite + chlorite + albite;
  low grade has been reached; however, the presence of crossite in some units indicates the high pressure variety of low-grade metamorphism, *i.e.*, the glaucophanitic greenschist zone.

The zonal sequence in California is similar in general but shows significant differences in detail:

Laumontite
Lawsonite/pumpellyite + chlorite + albite
Lawsonite/pumpellyite + glaucophane (+ surplus chlorite)
  in this zone jadeitic pyroxene is very common in rocks of appropriate composition.[4]

In the Stoneyford-Goat Mountain area, *aragonite* formed at a grade somewhat lower than that of the glaucophane-crossite zone, whereas in

[4]This zone (but with surplus albite) has also been recognized in the European Western Alps by Bearth (1966) and in other places.

the Pacheco Pass area aragonite formed at a somewhat higher grade (Ernst *et al.*, 1970).

The occurrence of aragonite in California and its absence in the Japanese localities is one of the differences. Another is the fact that the pumpellyite + glaucophane association is practically absent in the Sanbagawa metamorphic areas (Seki, 1969). Instead, the pumpellyite + actinolite + chlorite zone is a prominent feature in the Sanbagawa areas, whereas such rocks are practically absent in the Franciscan belt of California. It seems that pumpellyite + glaucophane and pumpellyite + actinolite + chlorite are mutually exclusive within a certain range of physical conditions; the coexistence of the two amphiboles, actinolite and glaucophane, over a larger P,T field would not be expected.

A zonal sequence starting with laumontite, followed by a lawsonite + albite + chlorite zone, then by a glaucophane + lawsonite zone, and finally by a zone with jadeitic pyroxene (in metagraywackes) cannot have been formed under an approximately linear geothermal gradient. Rather, pressure must have increased more rapidly with increase of temperature than would be the case in a linear relationship. To give a possible example: P,T conditions may have changed from 2 to 3 kb at 200°C to 3 to 5 kb at 250°C and to 10 to 12 kb at 300° to 350°C. Such pressure-temperature distribution within a sector of the earth's crust gives rise to metamorphic belts characterized by the combination of high pressure and low temperature. The creation of these belts requires specific geologic and tectonic conditions, namely, a supply of "large amounts of debris which could rapidly be deposited to a sufficient thickness in a foundering trough and augmented by tectonic thickening to generate an abnormally low geothermal gradient." After metamorphism, rapid uplift must have taken place. [The quotation is from Ernst *et al.* (1970); in the memoir the authors review and propose models to explain the generation of the specific conditions necessary for high pressure, low temperature metamorphism. See also Ernst (1971a,b, 1973).]

Glaucophane-bearing rocks and related rocks are confined to Phanerozoic metamorphic belts, as E. de Roever (1956) pointed out. In a study on the "occurrence and mineralogic evolution of blueschist belts with time," Ernst (1972a) showed that such rocks are progressively more abundant in younger rocks. He concludes: "Although widespread in younger glaucophane terranes, lawsonite is rather uncommon in Paleozoic blueschists; metamorphic aragonite and jadeitic pyroxene + quartz are strictly confined to blueschist belts of Mesozoic and Cenozoic age. These observations are compatible with a suggested systematic decrease with time in the Earth's geothermal gradient (at least adjacent to convergent lithospheric plate junctions, the subduction zone locale where glaucophane schists appear to be generated)."

The Role of CO_2 in Very-Low-Grade Metamorphism

When dealing with Ca-Al silicates in metamorphosed marls (Chapter 10), it was shown that the stability of such minerals as margarite, grossularite, zoisite, and zeolites is very sensitive to X_{CO_2} of the fluid phase; a low value is essential for their preservation. The significance of CO_2 content in an H_2O-CO_2 fluid in governing the stability of zeolite and clay + carbonate assemblages was stressed by Zen (1961). Further theoretical considerations and calculations have been undertaken by Coombs *et al.* (1970) and Thompson (1971), who also refer to the previous literature. These authors have shown that heulandite-, laumontite-, and prehnite-bearing assemblages can be obtained isothermally and isobarically from marls, *i.e.,* from calcite + quartz + clay (kaolinite, montmorillonite) assemblages by increasing the chemical potential of H_2O relative to that of CO_2. Conversely, when the CO_2 content (or salinity) in the fluid increases beyond a certain value, the Ca-Al zeolites will be replaced by clay + carbonate or pyrophyllite + carbonate assemblages. It is known now that this holds in an analogous manner for wairakite, lawsonite, and very probably pumpellyite.

Calculations by Thompson (1971) indicate that, at $P_f = P_{total} = 2$ kb, the reactions

$$\text{laumontite} + CO_2 = \text{calcite} + \text{kaolinite} + 2\ \text{quartz} + 2\ H_2O$$
$$\text{prehnite} + \text{quartz} + 3\ H_2O + CO_2 = \text{laumontite} + \text{calcite}$$

are in equilibrium with an H_2O-CO_2 fluid of composition $X_{CO_2} \approx 0.01$. Both reactions proceed to the right within their respective temperature ranges when the CO_2 concentration increases.

Reconnaissance experiments by Liou (1970) suggest that

wairakite or laumontite will be replaced by calcite + montmorillonite

in the presence of a fluid containing approximately 1 mole% CO_2. Montmorillonite is stable only at low temperatures; therefore, the reaction involving wairakite is probably metastable.

At higher pressures, lawsonite is stable and Nitsch (1972) determined the equilibrium conditions, at 4 and 7 kb, of the reaction

$$\text{lawsonite} + 2\ \text{quartz} + CO_2 = \text{calcite} + \text{pyrophyllite} + H_2O\ .$$

The equilibrium composition of the fluid is $X_{CO_2} = 0.03 \pm 0.02$ and depends very little on fluid pressure and temperature. It should be noted that the above reaction does not take place in the presence of albite because, in such a case, paragonite will be formed instead of pyrophyllite.

All experimental and calculated values agree that very low CO_2 concentrations are required to keep the Ca-Al minerals stable. Very likely this also applies to the following reaction:

$$\{\text{pumpellyite} + \text{actinolite} + H_2O + CO_2\} = \{\text{chlorite} + \text{calcite} + \text{quartz}\}$$

An equilibrium assemblage consisting of calcite, pyrophyllite or a clay mineral, and one of the Ca-Al minerals ± quartz demands that the fluid phase contains only one to a few mole percent CO_2. And whenever the very narrow X_{CO_2} range, valid for a certain temperature and pressure range, is transgressed, the slightly higher CO_2 content will cause the disappearance of all Ca-Al minerals which are diagnostic of certain metamorphic zones in very-low-grade metamorphism. The products of such reactions will be such minerals as calcite, pyrophyllite, clay minerals, and quartz, which are not diagnostic of very-low-grade metamorphism; they occur within a very large range of *P-T* conditions. In fact, in such cases it is not possible to distinguish low-grade (greenschist facies) conditions from very-low-grade conditions. When pyrophyllite is missing as well, not even the boundary between metamorphism and diagenesis can be identified.

On the other hand, parageneses which include Ca-Al minerals prove conclusively that the fluid consisted predominantly of H_2O. This conclusion is corroborated by the observation by Ernst (1972b) that sphene rather than rutile coexists with quartz and $CaCO_3$. He concludes that in the Sanbagawa areas X_{CO_2} had a maximum value of 0.03 and in the Franciscan terrain only 0.01. The metamorphic conditions in these areas were those of the pumpellyite, lawsonite-bearing zones (see p. 198).

Because of the very low concentration of CO_2 in the fluid the positions of the relevant equilibrium curves in Figure 12-11 are not affected; only if the fluid contains an appreciable amount of CO_2 will there be a noticeable shift toward lower temperatures.

Yet another consequence deserves attention: If it can be shown that a Ca-Al mineral has been formed by the reaction of clay + $CaCO_3$, as inferred by Ernst (1971c) in the case of pumpellyite and lawsonite in the metagraywackes from the central California coast ranges, the liberated CO_2 must have been considerably diluted by water that either was present in the rocks or had access to them.

References

Bearth, P. 1966. *Schweiz. Mineral. Petrog. Mitt.* **46:** 12–23.

Bishop, D. G. 1972. *Geol. Soc. Am. Bull.* **83:** 3177–3198.

Boettcher, A. L. and Wyllie, P. J. 1968. *J. Geol.* **76:** 314–330.

Brothers, R. N. 1970. *Contr. Mineral. Petrol.* **25:** 185–202.

Brown, E. H. 1977. *Contr. Mineral. Petrol.* **64:** 123–136.

Campbell, A. S. and Fyfe, W. S. 1965. *Am. J. Sci.* **263:** 807–816.

Coleman, R. G. and Clark, J. R. 1968. *Am. J. Sci.* **266:** 43–59.

Coleman, R. G. and Lee, D. E. 1962. *Am. J. Sci.* **260:** 577–595.

Coombs, D. S. 1960. *Intern. Geol. Congr. Copenhagen,* Pt. 13, 339–351.

——— 1961, *Australian J. Sci.* **24:** 203–215.

——— 1971. *Advan. Chem. Ser. No. 101,* 317–327.

——— 1974. *In* C. Amstutz, ed. *Spilite-Rocks.* Springer-Verlag, Heidelberg.

———, Ellis, A. J., Fyfe, W. S., and Taylor, A. M. 1959. *Geochim. Cosmochim. Acta* **17:** 53–107.

Coombs, D. S., Horodyski, R. J., and Naylor R. S. 1970. *Am. J. Sci.* **268:** 142–156.

Crawford, W. and Fyfe, W. S. 1965. *Am. J. Sci.* **263:** 262–270.

Crawford, W. and Hoersch, A. L. 1972. *Am. Mineral.* **57:** 995–998.

de Roever, E. W. F. 1972. *Lawsonite-albite facies metamorphism.* Academisch Proefschrift, GUA, Amsterdam.

de Roever, W. P. 1956. *Geol. Mihnbouw* **18:** 123–127.

——— 1972. *Tschermaks Mineral. Petrog. Mitt.* **18:** 64–75.

Ernst, W. G. 1971a. *Am. J. Sci.* **270:** 81–108.

——— 1971b. *Contr. Mineral. Petrol.* **34:** 43–59.

——— 1971c. *J. Petrol.* **12:** 413–437.

——— 1972a. *Am. J. Sci.* **272:** 657–668.

——— 1972b. *Geochim Cosmochim. Acta* **36:** 497–504.

——— 1973. *Tectonophysics* **17:** 255–272.

———, Seki, Y., Onuki, H., and Gilbert, M. C. 1970 *Geol. Soc. Am. Memoir.* **124:** 1–276.

Hashimoto, M. 1965. *Geol. Soc. Japan J.* **72:** 253–265.

Hay, R. L. 1966 *Geol. Soc. Am. Special Paper No. 85.*

Hinrichsen, T. and Schürmann, K. 1969. *Neues. Jahrb. Mineral. Monatsh.* **1969:** 441–445.

Hoffman, C. 1970. *Contr. Mineral Petrol.* **27:** 283–320.

Johannes, W. and Puhan, D. 1971. *Contr. Mineral. Petrol.* **31:** 28–38.

Jolly, W. T. 1970. *Contr. Mineral. Petrol.* **27:** 204–224.

van de Kamp, P. C. 1970. *J. Geol.* **78:** 281–303.

Liou, J. G. 1970. *Contr. Mineral. Petrol.* **27:** 259–282.

——— 1971a. *Lithos* **4:** 389–402.

——— 1971b. *J. Petrol.* **12:** 379–411.

——— 1971c. *Am. Mineral* **56:** 507–531.

Martini, J. and Vuagnat, M. 1970. *Fortschr. Mineral.* **47:** 52–64.

Miyashiro, A. and Shido, F. 1970. *Lithos* **3:** 251–260.

Newton, R. C. and Kennedy, G. C. 1963. *J. Geophys. Res.* **68:** 2967–2983.
Newton, R. C., Goldsmith, J. R., and Smith, J. V. 1969. *Contr. Mineral. Petrol.* **22:** 335–348.
Newton, R. C. and Smith, J. V. 1967. *J. Geol.* **75:** 268–286.
Nitsch, K.-H. 1968. *Naturwiss.* **55:** 388.
——— 1971. *Contr. Mineral. Petrol.* **30:** 240–260.
——— 1972. *Contr. Mineral. Petrol.* **34:** 116–134.
——— 1974. *Fortschr. Mineral.* **51.** (Abstract). Beiheft 1, p. 34.
Seki, Y. 1969. *J. Geol. Soc. Japan* **75:** 225–266.
———, Ernst, W. G., and Onuki, H. 1969a. *Suppl. Geol. Soc. Am. Memoir 124.*
Seki, Y., *et al.* 1969b. *J. Japan. Assoc. Mineral. Petrol. Econ. Geol.* **61:** 1–75.
Smith, R. E. 1969. *J. Petrol.* **10:** 144–164.
Steiner, A. 1955. *Mineral. Mag.* **30:** 691–698.
Thompson, A. B. 1970a. *Am. J. Sci.* **268:** 267–275.
——— 1970b. *Am. J. Sci.* **268:** 454–458.
——— 1971. *Contr. Mineral. Petrol.* **33:** 145–161.
Vance, J. A. 1968. *Am. J. Sci.* **266:** 299–315.
Wedepohl, K. H. 1967. *Geochemie* (Sammlung Göschen), Berlin.
Wenk, W. and Keller, F. 1969. *Schweiz. Mineral. Petrog. Mitt.* **49:** 157–198.
Winkler, H. G. F. 1965. *Petrogenesis of Metamorphic Rocks.* Springer-Verlag, New York-Berlin.
Wiseman, J. D. H. 1934. *Geol. Soc. London. Quart. J.* **90:** 354–417.
Zen, E-An. 1961. *Am. J. Sci.* **259:** 401–409.
Zimmermann, H. D. 1971. *Nature Physical Sci.* **231:** 203–204.

Chapter 13

Very-Low-Grade Metamorphism of Graywackes

Graywackes are sedimentary clastic rocks which consist of quartz, feldspar, and sheet silicates (micas, chlorite, clay minerals, etc.) and commonly contain rock fragments. If the debris are derived from mafic rocks, metamorphic graywackes show qualitatively the same mineral parageneses as have been discussed in Chapter 12.[1] Quantitatively, however, mafic minerals and Ca-Al minerals are commonly present only in minor amounts, whereas white mica and especially quartz occur in larger amounts, The amount of feldspar in metagraywackes may be very different, depending on the initial quartz-to-feldspar ratio in the clastic sediment; feldspar may be very conspicuous.

If graywackes do not contain mafic rock fragments but pelitic constituents in addition to quartz and feldspar, the metamorphic changes are essentially those of pelitic rocks; these are reviewed in Chapter 14. Generally graywackes do not furnish metamorphic parageneses which cannot be observed as well or better in metapelites. However, this is not the case in very-low-grade metamorphism. Due to the fact that many graywackes do not contain albite (as most pelites do, if it is present at all) but rather anorthite-bearing plagioclase, minerals like pumpellyite and lawsonite may be formed in small amounts from detrital plagioclase which then becomes albitized. (Also, the formation of these Ca-Al minerals by the interaction of interstitial clay + $CaCO_3$ seems possible; see p. 200f.)

At very large pressures, albite, which is commonly much more abundant than in pelites, is the major reactant forming jadeitic pyroxene. This mineral is very prominent in metagraywackes produced in low temperature, very high pressure metamorphism.

---

[1]The laumontite → prehnite + pumpellyite → pumpellyite + actinolite sequence described from the Western Alps developed in "sandstones" which really are graywackes containing up to 90% andesitic fragments. The prehnite + pumpellyite "facies" from New Zealand also formed in metagraywackes.

A classic occurrence of very-low-grade metagraywackes, which in their highest pressure zone have developed jadeitic pyroxene, is in California. Franciscan metagraywackes from Diablo Range, central California coast ranges, have been studied by Ernst (1971). All metagraywackes contain quartz, albite, and/or jadeitic pyroxene as main constituents. White mica, chlorite, and minor amounts of sphene and carbonaceous matter are always present, whereas $CaCO_3$, stilpnomelane, iron oxide, and/or pyrite occur in some samples; rock fragments are prominent. Newly formed lawsonite and/or pumpellyite commonly constitute only 1 to 5% of the rocks.

Ernst has deduced the following sequence of characteristic mineral associations:

Pumpellyite + albite ± calcite
Lawsonite + albite ± calcite
Lawsonite + albite ± aragonite
Lawsonite + jadeitic pyroxene[2] ± aragonite

In a few of the latter (highest pressure) rocks, glaucophane occurs as a minor constituent; the amount of glaucophane is small because graywackes contain only a small amount of chorite, which, at sufficiently high pressures, reacts with albite to produce glaucophane.

Probable conditions for the formation of jadeitic pyroxene in metagraywackes are indicated in Figure 12-11. The values shown are lower by 0.5 kb at any given temperature compared to the equilibrium curve of pure jadeite + quartz = albite. This is realistic because the main components acmite $NaFe^{3+}\ Si_2O_6$ and diopside $CaMgSi_2O_6$ enter into solid solution with the jadeite structure and lower the pressure necessary for formation of jadeitic pyroxene. Thus, a jadeite-rich pyroxene $Jd_{82}Ac_{14}Di_4$ similar in composition to that found in the Franciscan formation is formed at pressures lower by 0.8 kb as compared with pure jadeite + quartz = albite (Newton and Smith, 1967; Boettcher and Wyllie, 1968b, Johannes *et al.*, 1971).

References

Boettcher, A. L. and Wyllie, P. J. 1968. *Geochim. Cosmochim. Acta* **32:** 999–1012.
Ernst, W. G. 1971. *J. Petrol.* **12:** 413–437.
Johannes, W., *et al.* 1971. *Contr. Mineral. Petrol.* **32:** 24–38.
Newton, R. C., and Smith, J. V. 1967. *J. Geol.* **75:** 268–286.

---

[2]The jadeitic pyroxene has a significant content of Fe^{+3}, *i.e.*, of acmite (aegirine).

Chapter 14

Metamorphism of Pelites

General Statement

Slates, phyllites, and mica schists are derived from clay. They belong to the large group of metamorphic rocks known as metapelites. In addition to quartz, the following minerals are common constituents of these rocks:

- Pyrophyllite, andalusite, kyanite, sillimanite
- Staurolite
- Chloritoid, cordierite
- Chlorite, almandine-rich garnet
- The K-bearing micas phengite, muscovite, and biotite
- The Na-bearing mica paragonite and solid solutions of paragonite and phengite/muscovite
- The Ca-bearing mica margarite
- Stilpnomelane
- Alkali feldspar is present in some very-low- and low-grade rocks and is widespread in high-grade rocks where it commonly takes the place of muscovite.

Most of these minerals may be graphically presented in an AFM diagram; see Figure 5-10 in Chapter 5.

Metamorphism of Pelitic Rocks at Very-Low- and Low-Grade

Pelites consist of clay minerals (illite, montmorillonite, kaolinite), chlorite, detrital muscovite, occasionally some feldspar, and quartz, which is a major constituent. The quantities of these minerals may vary considerably. Calcite commonly is also present; if it occurs in major amounts the rocks are known as marly clays and marls. During diagenesis various changes of the clay minerals take place in response to circulating fluids and slightly increased temperature. A review of transformations in clays and shales is given by Dunoyer de Segonzac (1970) and by Frey (1970, 1978).

Montmorillonites and irregular mixed-layer clay minerals are decomposed during advanced diagenesis; therefore, they are absent in pelitic rocks at the beginning of metamorphism. Illite or phengite (with crystallinity 7.5 on the Kubler scale), chlorite, and quartz are dominant and often are the only minerals in slates. This assemblage is not diagnostic and may persist with increasing metamorphic grade; only the crystallinity of mica is improved.

The typical metamorphic mineral pyrophyllite may, however, be formed in very-low-grade slates of appropriate composition. Pyrophyllite, once believed to be very rare, is a widespread mineral in slates and phyllites devoid of albite and K feldspar. In some pelitic rocks pyrophyllite and phengite (sericite) are the only sheet silicates (Weber, 1972), but more commonly chlorite is also present and the amount of pyrophyllite is subordinate. Any albite and K feldspar in contact with pyrophyllite will react to form paragonite and muscovite (phengite), respectively; commonly a mixed-layer paragonite/muscovite is also formed.

An earlier statement (Winkler, 1967) that pyrophyllite appears at the beginning of the greenschist facies, *i.e.*, at the beginning of low-grade metamorphism, is now known to be wrong. Due to the wide application of x-ray diffraction in mineral identification, petrographic studies have proven that pyrophyllite[1] is formed during very-low-grade metamorphism in pelitic rocks (Frey, 1970, 1978; Weber, 1972). It has been found in slates adjacent to metabasalts which contain prehnite + pumpellyite + epidote (restricted to field III of Figure 12-11, about 325° to 350°C) and thus belongs to the prehnite-pumpellyite-chlorite zone of very-low-grade metamorphism. It is not known yet whether pyrophyllite may also occur at still lower temperature in the laumontite-chlorite zone, although Frey suggested that kaolinite + quartz react to pyrophyllite + water near the beginning of that zone, *i.e.*, close to the very beginning of metamorphism, as defined in this treatise.

This uncertainty and the effect of pressure would be resolved if equilibrium data were known with certainty for the reaction

$$\underset{\text{kaolinite}}{Al_2[(OH)_4/Si_2O_5]} + \underset{\text{quartz}}{2SiO_2} = \underset{\text{pyrophyllite}}{Al_2[(OH)_2/Si_4O_{10}]} + H_2O \qquad (1)$$

[1]Pyrophyllite is easily detected by its basal reflections at 9.2, 4.6, and 3.05A; it is distinguished from talc by its (060) reflection at 1.49 A.

Sometime ago it was believed that accurate data were available, but subsequent experiments have increased the uncertainty to about ± 50°C! The most recent experimental results by Thompson (1970), who also reviews the earlier investigations, are approximately midway between the extremes. They are given below for $P_{H_2O} = P_{total}$:

1 kb and 325° ± 20°C
2 kb and 345° ± 10°C
4 kb and 375° ± 15°C

If these data, shown in Figure 14-1 as curve 1, were equilibrium data and thus applicable to nature, pyrophyllite would not form near the

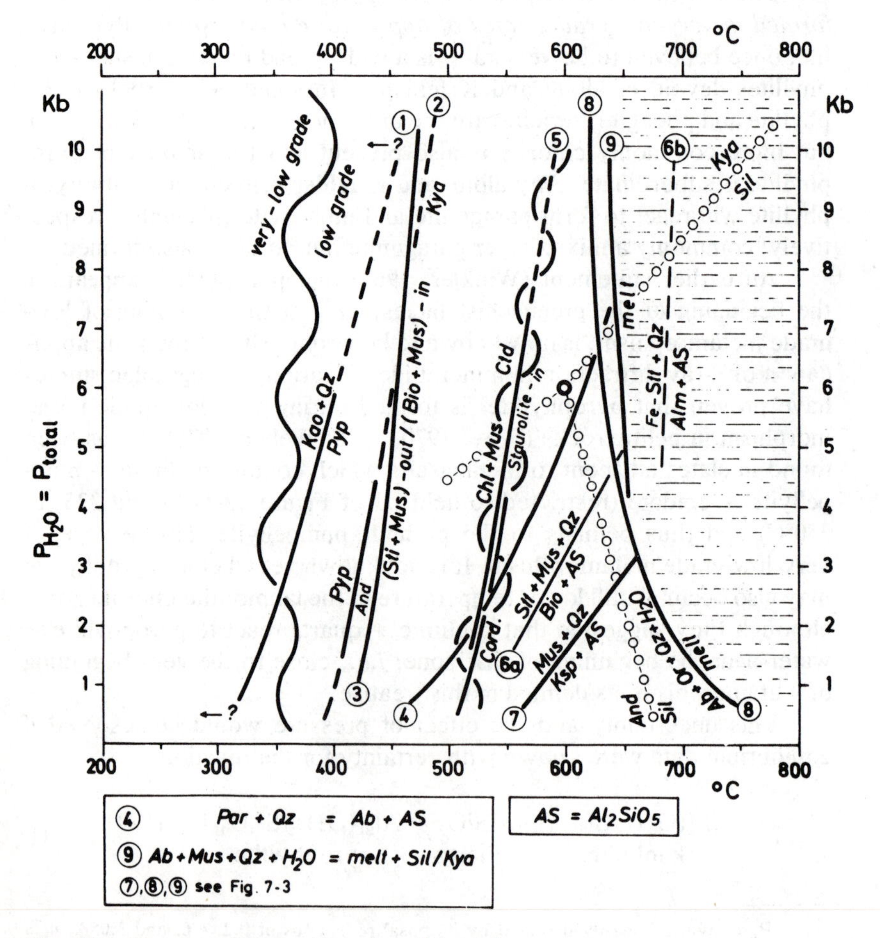

Fig. 14-1 Metamorphic reactions in pelitic rocks.

beginning of metamorphism, *i.e.*, at about 200°C and low pressure, but at appreciably higher temperature. Only at H_2O pressure lower than about 2 to 3 kb would pyrophyllite be formed within the range of very-low-grade metamorphism (corresponding to the upper part of the prehnite-pumpellyite-chlorite zone), while at higher pressures pyrophyllite would first appear in low-grade metamorphism. Consequently, at these higher pressures, kaolinite + quartz would persist into the lower range of low-grade metamorphism. Such an occurrence has, to my knowledge, not been observed. Nevertheless, the data given for reaction (1) may be equilibrium data valid for a water pressure equal to the fluid pressure. However, this condition may not be realized during the metamorphism of pelites if they contain graphite. As was pointed out at the beginning of Chapter 10, the presence of graphite may generate CH_4 during the metamorphism of black shales and marls, thereby drastically decreasing the activity of water at temperatures below about 400°C. This effect is more pronounced at lower temperatures (see Figure 3-2). Therefore, the formation of pyrophyllite by reaction (1) may take place at considerably lower temperatures than shown by curve 1 in Figure 14-1. This could account for the presence of pyrophyllite in very-low-grade, graphite-bearing black shales (Frey, 1978).

The *upper stability of pyrophyllite* is important because the first appearance of either andalusite or kyanite in quartz-bearing rocks requires that the upper stability of pyrophyllite has been reached or exceeded. This is true even if the Al_2SiO_5 mineral did not form from pyrophyllite.

A number of investigators have studies the breakdown reaction of pyrophyllite:

$$\underset{\text{pyrophyllite}}{Al_2[(OH)_2/Si_4O_{10}]} = \underset{\substack{\text{andalusite-}\\\text{kyanite}}}{Al_2SiO_5} + \underset{\text{quartz}}{3\ SiO_2} + H_2O \qquad (2)$$

The experimental data differ considerably, but the lowest temperature values obtained by Hemley (1967) and Kerrick (1968) appear to be equilibrium data. The values are given below; they are also plotted in Figure 14-1 as curve 2.

1.0 kb and 400° ± 15°C
1.8 kb and 410° ± 15°C
3.9 kg and 430° ± 15°C

At all these conditions andalusite is produced. (Kerrick gives references to all previous investigations.) The new determinations restrict the stability field of pyrophyllite to temperatures approximately 80°C lower than those of earlier experiments and thus provide an explanation for the appearance of andalusite or kyanite in low-grade (greenschist facies) metapelites.

Paragonite is formed in very-low-grade metamorphism. Frey (1969, 1978), in a study of progressive changes in pelites (at rather low pressures), demonstrated that the first appearance of paragonite takes place at very-low-grade metamorphic conditions, certainly at a grade no higher than that of the prehnite/pumpellyite zone. The very-low-grade formation of paragonite at high pressure, *i.e.*, in the lawsonite-glaucophane zone is also known (Chatterjee, 1971).

The following reactions, suggested by Chatterjee (1973) and Zen (1960), may lead to the formation of paragonite in response to very-low-grade metamorphism:

{Na-montmorillonite + albite} = {paragonite + quartz}
{kaolinite (or pyrophyllite) + albite} =
{paragonite + quartz + H_2O}

Paragonite $NaAl_2[(OH)_2/AlSi_3O_{10}]$ as a separate mineral commonly occurs together with phengite, chlorite, and quartz; calcite and/or dolomite and pyrophyllite may also be present. Coexisting phengite contains a few mole percent paragonite component in solid solution. Also, a mixed-layer paragonite/phengite, coexisting with the above minerals has been found by Frey (1969, 1978) in very-low-grade metamorphic rocks. This complex phase may represent an intermediate stage in the transformation of mixed-layer illite-montmorillonite to paragonite + phengite + chlorite. Paragonite is first formed by very-low-grade metamorphism and persists to low- and medium-grade. In relatively Al-rich rocks, paragonite may be formed first by very-low-grade metamorphism and persist to low- and medium-grade. However, in less Al-rich metapelites of common composition, paragonite does not form before the beginning of medium-grade metamorphism (Hoffer, 1978). In such a case, the paragonite-forming reactions are, of course, different from those given above. Furthermore, paragonite when formed by very-low-grade metamorphism need not always persist to medium-grade. According to observations by Frey (1978), paragonite reacts in the presence of carbonates and quartz to form plagioclase and margarite or zoisite/clinozoisite close to the low-grade/medium-grade boundary. Assuming a water pressure of about 2 kb, this upper stability limit of paragonite would agree with the upper stability of the paragenesis paragonite + quartz; if, however, these rocks were metamorphosed at a higher pressure, the breakdown of paragonite in the presence of carbonate and quartz would take place at lower temperatures than the reaction

$$\{\text{paragonite} + \text{quartz}\} = \{Al_2SiO_5 + \text{albite} + H_2O\} \tag{4}$$

This reaction has been investigated by Chatterjee (1972); his data are shown as curve (4) in Figure 14-1.

Paragonite is not compatible with K feldspar (Zen, 1960; Albee, 1968). This is well understood since Hemley and Jones (1964) have shown that

K feldspar + paragonite react to form muscovite + albite.

However, in fine grained slates and phyllites, where minerals can be detected only by x-ray diffraction, K feldspar and paragonite have been noted as constituents. In such cases it must be assumed that the two minerals are not adjacent to each other and thus do not coexist in the strict sense. At low-grade metamorphic conditions, paragonite does not occur together with biotite in common metapelites (Albee, 1968). However, in medium-grade metapelites, paragonite is known to coexist with biotite in the presence of staurolite and/or kyanite (Hoffer, 1978).

Stilpnomelane in metasediments. Stilpnomelane, common in very-low-grade mafic rocks, may be formed also in very-low-grade metasediments having a particular chemical composition. Frey (1970, 1973) observed that stilpnomelane occurs in metamorphic iron-oolitic or glauconitic horizons. Stilpnomelane appears at approximately those conditions at which pyrophyllite and paragonite are first formed in sediments of other compositions. It is clear now that stilpnomealne is not restricted to high pressures but is in fact formed within a very large pressure range. However, it is not yet known whether a minimum pressure of about 1 kb is required to stabilize this mineral.

Stilpnomelane can easily be mistaken for biotite or iron-rich oxichlorite under the microscope. Also, it may occur as very small needles only 20 μ long and 1 to 2 μ wide. It is best identified by its x-ray diffraction pattern characterized by a strong basal reflection at 12 Å.

Frey (1973) showed that in rocks consisting of glauconite + quartz + calcite ± iron-rich chlorite, the paragenesis stilpnomelane + K feldspar is produced at conditions of the prehnite/pumpellyite-chlorite zone of very-low-grade metamorphism. From microscopic observation, the following reaction has been inferred:

$$\{\text{glauconite} + \text{quartz} \pm \text{chlorite}\} =$$
$$\{\text{stilpnomelane} + \text{K feldspar} + H_2O + O_2\}$$

In addition to quartz and calcite, chlorite commonly accompanies stilpnomelane + K feldspar, but phengite is absent in the rocks investigated by Frey.

At slightly higher temperature, the following reaction also has been inferred by Frey:

$$\{\text{chlorite} + \text{K feldspar}\} = \{\text{biotite} + \text{stiplnomelane} + \text{quartz} + H_2O\}$$

Above the reaction temperature, rocks consist of (commonly brown) biotite + stilpnomelane + K feldspar + quartz + calcite.[2] This biotite-forming reaction takes place at conditions where pumpellyite is still stable in mafic rocks, *i.e.*, during very-low-grade metamorphism (Frey, 1973; Seki *et al.*, 1969). Therefore, a clear distinction must be made with regard to the first appearance of biotite:

a. Biotite in *very-low-grade* metamorphism and coexisting with *stilpnomelane*, quartz, and K feldspar or chlorite; *phengite absent*.
b. Biotite in *low-grade* metamorphism and coexisting with *phengite* ± chlorite, and quartz.

Case (b) corresponds to the beginning of Barrow's biotite zone in metapelites, whereas case (a) has a different petrogenetic significance and is observed only in sediments of restricted composition originally containing glauconite. Experimental data for these reactions are not yet available. Another significant reaction in pelitic rocks leads to the coexistence of stilpnomelane with phengite, quartz, chlorite, and, in some instances, K feldspar. This assemblage is common in the lower temperature part of low-grade metamorphism; it is expected that it will also be found in very-low-grade terrains.

The upper stability of stilpnomelane. No data are available concerning the upper stability of stilpnomelane by itself. However, the following reactions lead to the disappearance of stilpnomelane:

$$\{\text{stilpnomelane} + \text{phengite}\} = \{\text{biotite} + \text{chlorite} + \text{quartz} + H_2O\}$$

$$\{\text{stilpnomelane} + \text{phengite} + \text{actinolite}\} = \{\text{biotite} + \text{chlorite} + \text{epidote} + H_2O \qquad (3)$$

The latter reaction has been suggested by Brown (1971) on the basis of petrographic observation. The former reaction has been investigated

[2]Riebeckite may also be present in these as well as in the lower grade and even unmetamorphosed glauconitic rocks.

experimentally by Nitsch (1970). He gives the following data, including reversed runs at 4 and 7 kb H_2O pressure:

1 kb lower than 430°C
4 kb and 445° ± 10°C
7 kb and 460° ± 10°C

These data are shown in Figure 14-1 as curve 3 with the remark stilpnomelane + muscovite-out/biotite + muscovite-in, indicating the diagnostic significance of this reaction. It is not appropriate to use the designation "biotite-in" because this could refer to at least two different assemblages:

a. Biotite + stilpnomelane (Figure 14-3) or
b. Biotite + phengite, no stilpnomelane (Figure 14-5).

Reaction (3) constitutes a reaction-isograd which is designated biotite + chlorite + quartz + stilpnomelane + phengite. This equilibrium is not univariant; it is a band which allows the paragenesis biotite + stilpnomelane + muscovite ± chlorite + quartz to occur in a temperature range until, in the lower biotite zone, stilpnomelane eventually disappears in the presence of muscovite. Without muscovite, the assemblage stilpnomelane + biotite may persist into the upper biotite zone of low-grade metamorphism (Brown, 1971). Therefore, for mapping purposes, the reaction-isograd may be designated by the characteristic paragenesis which has just disappeared and by the paragenesis which has just appeared:

(stilpnomelane + muscovite)-out,
(biotite + muscovite)-in.

The parageneses of metamorphosed pelitic rocks are best represented in AFM diagrams. The conventional method is illustrated by an example in Figure 14-2a. Two coexisting minerals are joined by tie lines, each tie line corresponding to a different bulk composition. However, the configuration of tie lines in Figure 14-2a is schematic only and does not reflect a specific combination of metamorphic conditions. Instead of drawing the complete bundle of tie lines, the coexistence of two minerals may be schematically represented by showing one tie line only. Following Albee (1968), this method has been accepted because it allows a very clear illustration of mineral associations. It is of particular advantage in displaying three-mineral associations; because of their petrogenetic significance they are emphasized by ruled areas. In Figure 14-2b the para-

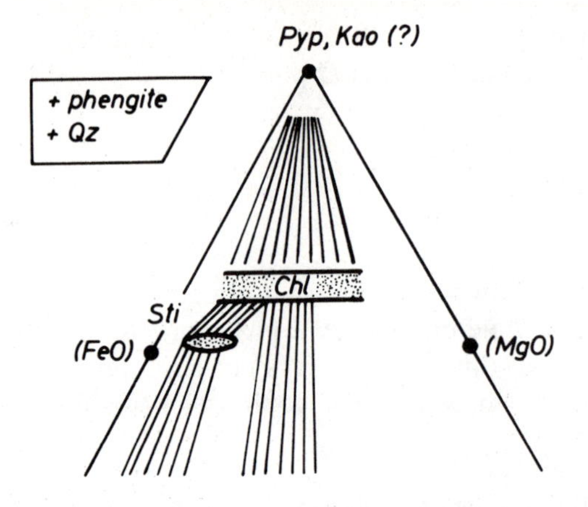

Fig. 14-2a Parageneses of very-low-grade. Conventional method of graphical representation.

geneses shown in Figure 14-2a are represented according to this method. Tie lines connecting two minerals are drawn as heavy lines. The composition of a single mineral may plot anywhere within its composition field, indicated by a stippled area.

In Figure 14-2b stilpnomelane is shown to coexist with chlorite and K feldspar. The two tie lines to these minerals must start from stilpnomelane of the same composition; they may be visualized as meeting at a point within the stilpnomelane composition field. The same consideration applies to other minerals, *e.g.*, biotite in Figure 14-3. It may be

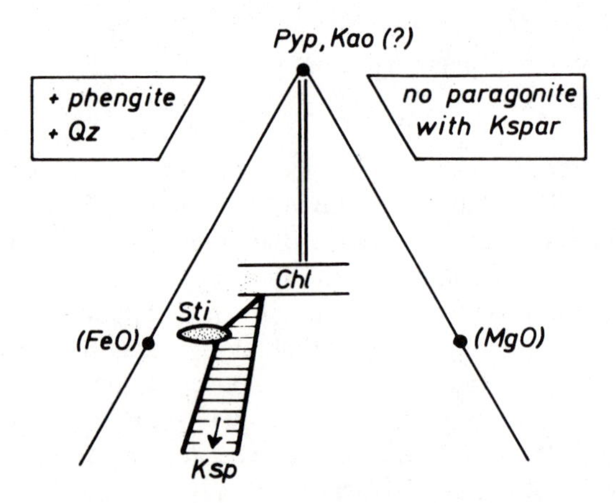

Fig. 14-2b Parageneses of very-low-grade. Graphical representation used in this book.

disturbing to see a mineral composition field projecting into the area of a three-mineral assemblage (*e.g.*, Figure 14-11a and c). However this feature does not interfere with a schematic representation of mineral assemblages and permits an indication of the common range of mineral compositions.

Figure 14-2 represents parageneses of very-low-grade metamorphic rocks, *e.g.*,

Chlorite + phengite + quartz
Pyrophyllite + chlorite + phengite + quartz
Stilpnomelane + chlorite + phengite + quartz
Stilpnomelane + chlorite + K feldspar + phengite + quartz.

These parageneses are not restricted to very-low-grade but are encountered as well in the lower temperature range of low-grade metamorphism. The boundary between very-low-grade and low-grade metamorphism cannot be determined on the basis of phase relations in pelitic metamorphic rocks; rocks of mafic composition or sediments containing debris of mafic rocks are necessary to give this information.

Figure 14-3 shows that the assemblage stilpnomelane + biotite may occur in very-low-grade metamorphism:

Stilpnomelane + biotite + chlorite + quartz
Stilpnomelane + biotite + K feldspar + quartz

This is based on petrographic observations by Frey (1973) and Brown (1971). The parageneses apparently are restricted to special bulk compositions. The parageneses of Figure 14-3, as well as those of Figure 14-2, may persist into the lower temperature range of low-grade metamorphism.

Chloritoid forms in low-grade metamorphism. At low-grade metamorphic conditions chloritoid is formed in pelites of special composition. Under the microscope, very small colorless chloritoid grains may easily be overlooked. However, x-ray diffraction permits easy identification. According to Frey (1978), the P-T conditions of the formation of chloritoid are approximately those of the boundary between very-low- and low-grade metamorphism. However, the temperatures may actually be somewhat lower, because Seidel *et al.* (1975) report chloritoid-bearing metapelites in association with very-low-grade lawsonite-glaucophane schists. The special chemical factors permitting the formation of chloritoid are a large Fe/Mg ratio and a relatively high Al content and simultaneously low contents of K, Na, and Ca (see Figure 14-7 on p. 222). This particular bulk chemistry of the rocks leads to the absence of chloritoid from assemblages including stilpnomelane, biotite, albite, and/or K feldspar. In other words, if chloritoid is present in a low-grade metapelite, these minerals do not coexist with chloritoid.

Figure 14-4 represents parageneses of the lowest temperature range of low-grade metamorphism, including chloritoid in rocks of appropriate composition. Chloritoid persists throughout the range of low-grade

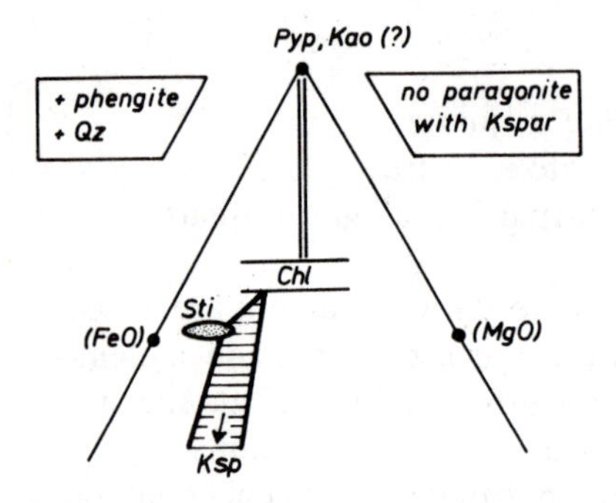

Fig. 14-2 Very-low-grade, persisting into low-grade up to Figure 14-4.

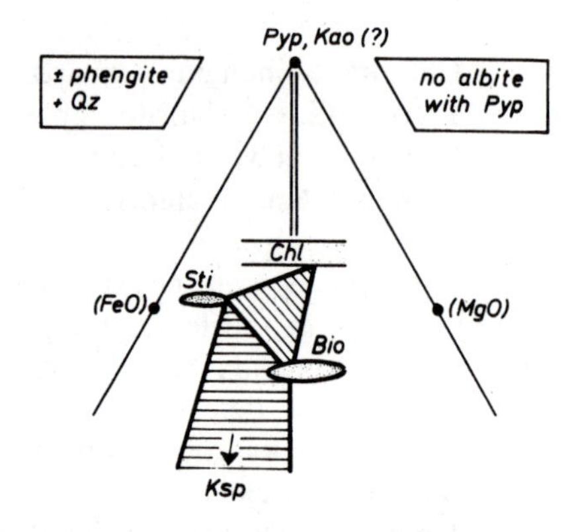

Fig. 14-3 Very-low-grade, special composition; persisting into low-grade up to Figure 14-5.

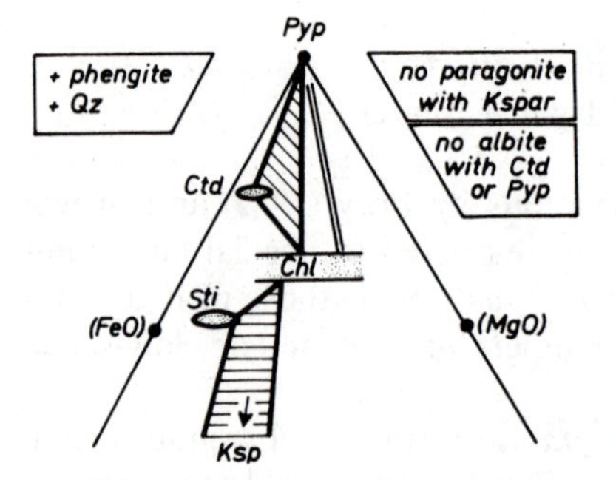

Fig. 14-4 Low-grade, below the reaction-isograd Sti + Mus-out/Bio + Mus-in.

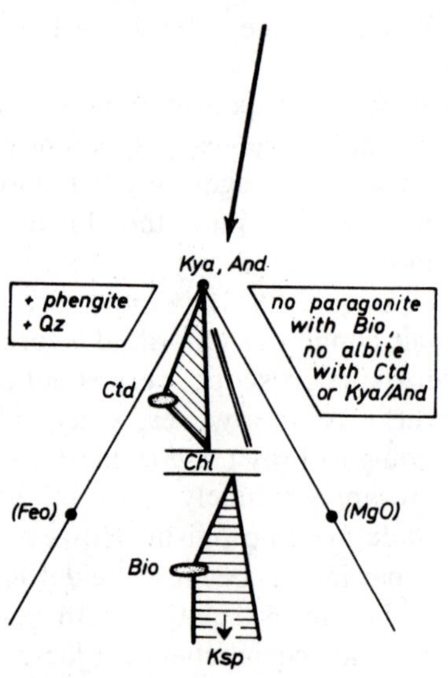

Fig. 14-5 Low-grade, above the reaction-isograd Sti + Mus-out/Bio + Mus-in, but below almandine stability.

metamorphism. The following two reactions leading to the formation of chloritoid have been suggested. The first one probably is responsible for the first appearance of chloritoid (Frey, 1972, 1978), whereas the second one may take place at slightly higher temperature, for which Figure 14-5 is valid.

$$\{\text{pyrophyllite} + \text{iron-rich chlorite}\} = \{\text{chloritoid} + \text{quartz} + H_2O\}$$

$$\{\text{hematite} + \text{iron-rich chlorite}\} = \{\text{chloritoid} + \text{magnetite} + \text{quartz} + H_2O\}$$

Thompson and Norton (1968), who suggested the second reaction, point out that "it is basically a dehydration reaction although easily mistaken for a deoxidation reaction on casual examination of the rocks. It is probably one of the key reactions involved in the disappearance of purple or red color from phyllites in the lower biotite zone."

With increasing temperature, paragenetic changes are illustrated by the following sequence of AFM diagrams:

a. Commonly: Figure 14-2 → Figure 14-4 → Figure 14-5
b. Rarer: Figure 14-2 ——————→ Figure 14-5

Figure 14-5 illustrates typical paragenesis of the so-called "biotite zone" above the reaction-isograd Sti + Mus-out/Bio + Mus-in and below the stability of almandine. When compared with the almandine-bearing zone shown in Figure 14-6, there is only one difference: the

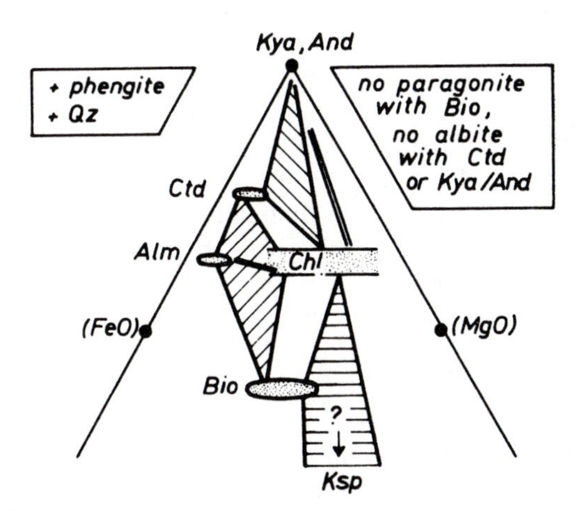

Fig. 14-6 [almandine]-low-grade, or almandine + chlorite + muscovite zone.

absence of almandine-rich garnet. The following parageneses without almandine occur in both Figures 14-5 and 14-6:

Bio + Chl + Mus + Qz
Bio + Chl + Ksp + Mus + Qz

Some parageneses of Figures 14-5 and 14-6 are already shown in Figure 14-4 and remain, in fact, stable throughout the range of low-grade metamorphism. Although they are typical of low-grade metamorphism, they are not diagnostic of specific zones. These parageneses are

Ctd + Chl + Mus + Qz
Ctd + Chl + Kya/And + Mus + Qz
Chl + Kya/And + Mus + Qz

In the lowest temperature range of low-grade metamorphism, pyrophyllite takes the place of kyanite and andalusite. Pyrophyllite together with kyanite has been observed occasionally in the higher temperature range (Albee, 1968); this would not be expected on the basis of curve 2 in Figure 14-1.

The formation of almandine depends significantly on the prevailing pressure:

a. At relatively low pressures the parageneses of Figure 14-5 persist to the beginning of medium-grade metamorphism.
b. At relatively high pressures almandine is formed in rocks of the appropriate composition giving rise to the parageneses shown in Figure 14-6. These parageneses are typical of the higher temperature range of low-grade metamorphism, commonly known as the "garnet zone" in metapelites. Figure 14-6 is valid to the beginning of medium-grade metamorphism. The progression from Figure 14-5 to 14-6 always takes place if pressures are sufficiently high to stabilize almandine (Albee, 1968). Low-grade metamorphism characterized by the presence of almandine is very common and may be designated conveniently by [almandine]-low-grade or *almandine + chlorite + muscovite zone*.

The characteristic assemblages of this metamorphic zone are

Almandine + chlorite + chloritoid + muscovite + quartz
Almandine + chlorite + biotite + muscovite + quartz

The assemblage almandine + chlorite + muscovite is diagnostic. The garnet + chlorite join precludes the coexistence of chloritoid + biotite (see Figure 14-6). Earlier reports of this mineral pair have been shown to be incorrect (Albee, 1968). The assemblage almandine + chlorite + muscovite remains stable to the beginning of medium-grade metamorphism. It then reacts to form biotite + staurolite. This is one of the diagnostic assemblages indicating the beginning of medium-grade. Before proceeding to a discussion of medium- and high-grade metamorphism, some comments will be made about the first formation of biotite and almandine in low-grade metamorphism.

Formation of biotite.[3] The various reactions leading to the first formation of biotite in pelitic rocks are not well understood. Certainly, two reactions mentioned earlier take place:

$$\{\text{stilpnomelane + phengite}\} = \{\text{biotite + chlorite + quartz} + H_2O\}$$

$$\{\text{stilpnomelane + phengite + actinolite}\} = \{\text{biotite + chlorite + epidote} + H_2O\}$$

However, there must be other biotite-producing reactions because the amount of stilpnomelane available in most metapelites in the stilpnomelane-low-grade zone is far too small to account for the amount of biotite in rocks of the biotite + muscovite zone.

Thompson and Norton (1968) have suggested the following reactions for the formation of biotite:

$$3\ \text{dolomite (or ankerite)} + \text{Ksp} + H_2O = \text{Bio} + 3\ \text{Cc} + 3\ CO_2$$

$$3\ \text{Chl} + 13\ \text{Ksp} = 7\ \text{Bio} + 6\ \text{Mus} + 12\ \text{Qz} + 5\ H_2O$$

But, as inferred by Frey (1973), Chl + Ksp should react at a lower temperature to produce biotite + stilpnomelane + quartz + H_2O.

The above reactions are inadequate to account for the bulk of biotite present in pelitic rocks. There should be common reactions which produce biotite from the dominant minerals in low-grade metapelites (devoid of biotite), *i.e.*, from phengite and chlorite. The following reactions have been suggested by Turner (1948) and Mather (1970), respectively:

$$\{\text{phengite + chlorite}\} = \{\text{biotite + Al-richer chlorite + quartz}\}$$

[3]At times a phyllocilicate mineral, looking like biotite, may be observed. X-ray and chemical studies demonstrate that this is an oxidized iron-rich chlorite (Chatterjee, 1966). Whenever dealing with metapelites in the low temperature range of low-grade metamorphism, a precise identification of biotite may be significant.

{phengite + chlorite + microcline} =
{biotite + quartz + phengite with less Mg,Fe}

The latter reaction may be significant in metagraywackes as well as in metapelites.

Formation of almandine-rich garnet. Almandine-rich garnet is a complex solid solution of predominantly almandine component with a minor (about 10%) amount of pyrope, grossularite, or/and spessartine components. Spessartine-rich garnet should be regarded as a separate species because it can be formed at very low temperature and pressure. Therefore, in petrology a clear distinction is necessary between spessartine-rich garnet and almandine-rich garnet. The first appearance of garnet in metapelitic rocks in the so-called "garnet zone" refers to almandine-rich garnet. The *P-T* conditions of the first appearance of a complex garnet must depend on its composition. Therefore, the first appearance of almandine-rich garnet cannot furnish a sharp boundary in terms of *T* and *P*. Nevertheless, almandine-rich garnet in pelitic rocks first appears in the higher temperature part of low-grade metamorphism, *i.e.*, distinctly above the boundary Sti + Mus-out/Bio + Mus-in. Apparently, the differences in the temperature of formation of almandine-rich garnet are very small, probably only 20° to 30°C at any given pressure.

The following reactions leading to the first appearance of almandine have been inferred from petrographic observations by Chakraborty and Sen (1967) and Brown (1969), respectively:

{chlorite + biotite$_{(1)}$ + quartz} =
{almandine-rich garnet + biotite$_{(2)}$ + H_2O}

{chlorite + muscovite + epidote} =
{almandine-rich garnet + biotite + H_2O}

Thompson and Norton (1968) suggest the following reactions in aluminous rocks and in more typical pelites, respectively:

{chloritoid + chlorite + quartz} = {almandine + H_2O}
{chlorite + muscovite + quartz} = {almandine + biotite + H_2O}

Hirschberg and Winkler (1968), using an iron-rich chlorite with $Fe^{2+}/Fe^{2+} + Mg = 8/10$ and only 0.1% MnO, produced almandine-rich garnet according to

{iron-rich chlorite + muscovite + quartz} =
{almandine-rich garnet + biotite + Al_2SiO_5 + H_2O}

Due to sluggish reaction rates, it was not possible to determine the low temperature boundary of this reaction, but from the data obtained at temperatures above 600°C it may be inferred that the following pressures must be exceeded to produce garnet: 4 kb at 500°C and 5 kb at about 600°C. The values will be lower by an estimated 2 kb if the garnet contains an appreciable amount of spessartine component; on the other hand, pressure will be higher if the Fe/Mg ratio is smaller.

Yet another reaction has been inferred from petrographic observations by Hoffer (1978, personal communication).

$$\{\text{chlorite}_{(1)} + \text{anorthite component of oligoclase}\} = \{\text{almandine-rich garnet} + \text{chlorite}_{(2)} + \text{quartz} + H_2O\}$$

$\text{Chlorite}_{(i)}$ has relatively larger Fe/Mg and Mn/Fe ratios than $\text{chlorite}_{(2)}$. In addition, $\text{chlorite}_{(2)}$ is expected to be somewhat richer in Al. The anorthite component is the only source available in the studied rocks to supply Ca and Al for the grossularite component present in the garnets. When garnet has formed, the composition of coexisting garnet and chlorite tends to equilibrate with rising temperature; thus, $\text{chlorite}_{(2)}$ in the above reaction is not meant to designate a fixed composition.

It should be noted that garnets commonly are zoned, the core being richer in Mn and Fe than the rim. Chloritoid, chlorite, and biotite, when coexisting with almandine-rich garnet, invariably have a higher Mg/Fe and a lower Mn/Fe ratio than garnet.

Metamorphism of Pelitic Rocks at Medium- and High-Grade

The transition from low-grade metamorphism is best recognized in metapelites by the diagnostic first appearnce of staurolite or cordierite. A band of various reaction-isograds briefly designated as the

"staurolite-in" and "cordierite-in"

isograd is used to define the beginning of medium-grade metamorphism. The *P-T* data are shown in Figure 14-1, and the arguments for this definition are discussed on p. 76ff. The reader is referred to that section because many facts important in the present discussion are given there and will not be repeated here. Attention is also drawn to the paragraph "Practical determination of the boundary" at the end of that section. There, it is pointed out that the assemblage non Mg-rich chlorite + muscovite disappears in medium-grade rocks, whereas chlorite not in contact with muscovite may persist to higher temperatures. Since AFM diagrams are projections through muscovite (and quartz), chlorite is absent in AFM diagrams showing mineral assemblages of medium-grade rocks.

The possible formation of staurolite is governed by the bulk composition of the rocks. Staurolite, as well as chloritoid at lower grade, is confined to rocks of a restricted composition. These restrictions have been summarized by Hoschek (1967). He collected a large number of rock analyses of metapelites and metapsammites both with and without either chloritoid or staurolite. The weight percentages of each analysis were converted to molecular proportions by dividing the weight percentage of an oxide by its molecular weight and the following molecular ratios were calculated (note that all Fe_2O_3 has been added to FeO):

$$Al_2O_3\text{: }(K_2O + Na_2O)\text{: }(FeO + Fe_2O_3 + MgO)$$
$$Al_2O_3\text{: }CaO\text{: }(FeO + Fe_2O_3 + MgO)$$
$$Al_2O_3\text{: }(FeO + Fe_2O_3)\text{: }MgO$$

The sum of each of the three values was converted to 100 and thus the ratios can be plotted as percentages in an equilateral triangle. The appropriate parts of these triangles are shown in Figure 14-7. The more restricted heavily dotted fields represent chloritoid-bearing rocks, and the same fields, somewhat enlarged by the lightly dotted areas, represent

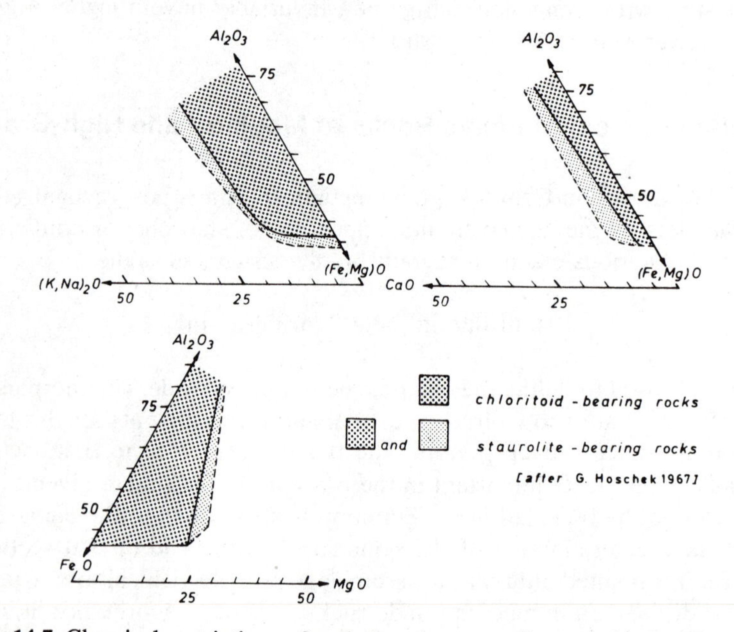

Fig. 14-7 Chemical restrictions of rocks bearing staurolite and chloritoid.

staurolite-bearing rocks. The assignment of staurolite and chloritoid to their respective fields is true on a statistical basis only; there are individual exceptions.

A knowledge of the chemical restrictions is useful in looking for rocks which potentially could contain staurolite. This may be of great help in locating the "staurolite-in" isograd and, at higher temperature, the "staurolite-out" isograd. In some areas of progressive metamorphism, the absence of staurolite has been erroneously attributed to its instability rather than to the lack of rocks of appropriate composition. Indeed, the chemical composition of about two-thirds of the pelitic and psammitic sediments does not allow the formation of staurolite in medium-grade metamorphism and more than three-quarters are unsuitable for chloritoid. The rare occurrences of chloritoid-bearing low-grade metamorphic rocks reflect these chemical restrictions.

Reactions producing staurolite. Many reactions have been suggested to account for the first appearance of staurolite (*e.g.*, Thompson and Norton, 1968; Hoschek, 1967). If chloritoid is present in low-grade rocks, it is the principal reactant in staurolite-producing reactions at the beginning of medium-grade metamorphism; see p. 78f.[4] Data for the oxidation reaction at the oxygen fugacities of the hematite-magnetite buffer

$$\text{chloritoid} + O_2 = \text{staurolite} + \text{magnetite} + \text{quartz} + H_2O$$

are given by Ganguly and Newton (1969) as 575°C at 10 kb and about 545°C at 5 kb. These data are very close to those indicated by curve 5 in Figure 14-1. However, staurolite is more common in medium-grade rocks than is chloritoid in low-grade rocks. Therefore, additional reactions not involving chloritoid must take place. These include chlorite and muscovite as reactants:

$$\{\text{chlorite} + \text{muscovite}\} = \{\text{staurolite} + \text{biotite} + \text{quartz} + H_2O\}$$

$$\{\text{chlorite} + \text{muscovite} + \text{almandine}\} = \{\text{staurolite} + \text{biotite} + \text{quartz} + H_2O\}$$

Experimental data for the first reaction have been supplied by Hoschek (1969) (see p. 79). They are shown as curve 5 in Figure 14-1, close to the curve representing the first appearance of cordierite. The second reac-

[4]Albee (1972) has developed the possible reaction relations between chloritoid and staurolite in the presence of quartz and muscovite.

tion has been suggested by Froese and Gasparrini (1975) from petrographic observations.

Reactions producing cordierite. It has been pointed out that the formation of staurolite is possible only in rocks of a certain composition. The Mg/(Fe+Mg) ratio of chlorite is of particular importance. If this ratio is larger than about 0.25 (see Figure 14-7), cordierite will form instead of staurolite. A common reaction, already mentioned on p. 77, is

$$\{\text{chlorite} + \text{muscovite} + \text{quartz}\} = \{\text{cordierite} + \text{biotite} + Al_2SiO_5 + H_2O\}$$

Therefore, staurolite and cordierite generally do not coexist in medium-grade rocks. On the other hand, almandine-rich garnet, which also forms in rocks of low Mg/(Fe+Mg) ratio, may coexist with staurolite in medium-grade metamorphism if the pressure is high enough to stabilize the garnet. Only at special conditions can cordierite, garnet, and staurolite coexist (see p. 230).

Staurolite is not limited to medium and high pressure but may form at low pressure as well. Therefore, staurolite occurs both in [almandine]-medium-grade and in [cordierite]-medium-grade rocks of appropriate composition. The different parageneses in metapelites are shown in Figures 14-9* and 14-9 for [almandine]-medium-grade and in Figure 14-8 for that part of [cordierite]-medium-grade in which staurolite is stable. At pressures lower than about 5 kb, staurolite breaks down before the upper boundary of medium-grade is reached. In this case, Figures 14-10a or 14-10b are valid instead of Figure 14-8.

The sequence of zones, as illustrated in AFM diagrams, from low-grade into [cordierite]-medium-grade is

either (at low pressures): Figures 14-5 → 14-8 → 14-10a → 14-11a (high-grade);
or (at somewhat higher pressures): Figures 14-5 → 14-8 → 14-10b → 14-11b (high-grade).

At still higher pressures, the sequence from low-grade into [almandine]-medium-grade is represented by:

Figures 14-5 → 14-6 → 14-9* → 14-9 → 14-11c (high-grade)
or
Figures 14-5 → 14-6 → 14-9* → 14-9 → 14-10b → 14-11b (high-grade).

The figures are arranged in these sequences on pages 225 and 226.

Some comments about certain parageneses shown in the AFM dia-

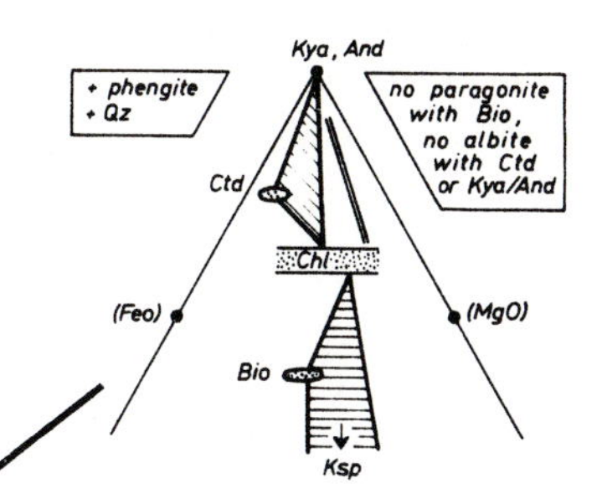

Fig. 14-5 Low-grade, above Sti + Mus-out/Bio + Mus-in, but below almandine stability.

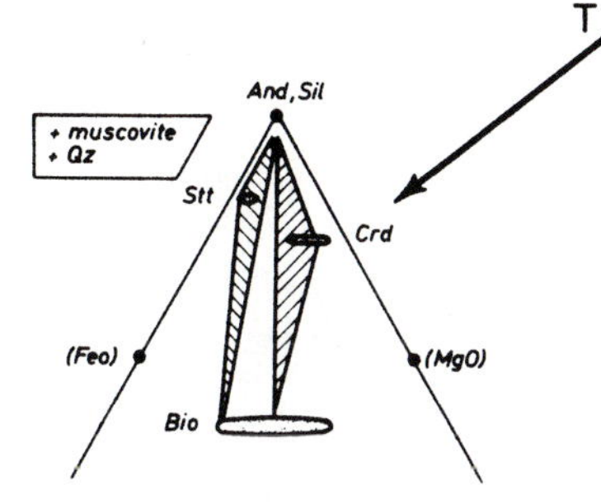

Fig. 14-8 [Cordierite]-medium-grade, with staurolite.

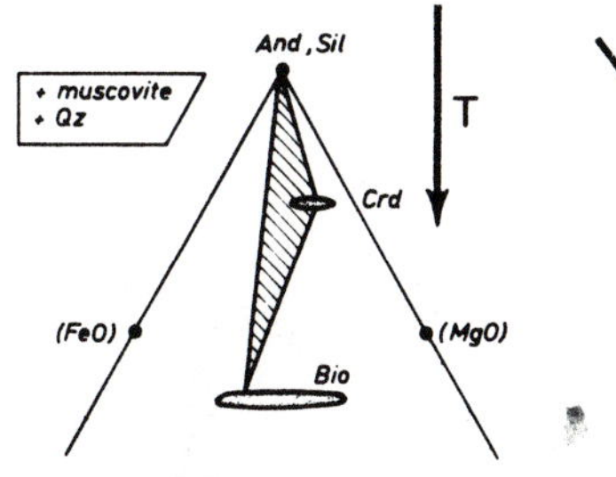

Fig. 14-10a [Cordierite]-medium-grade, without staurolite.

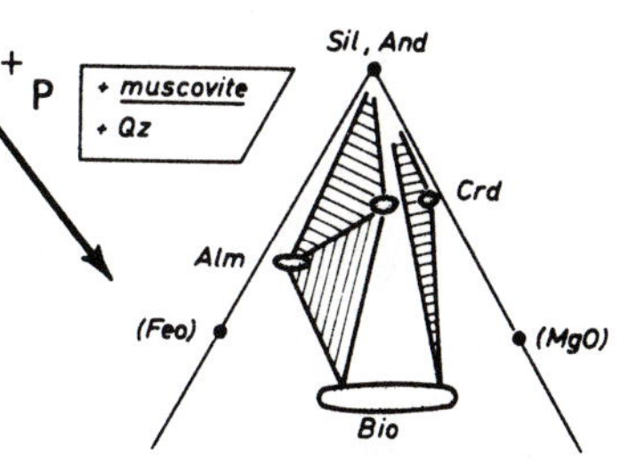

Fig. 14-10b [Cordierite-almandine]-medium-grade. (Staurolite may also be present.)

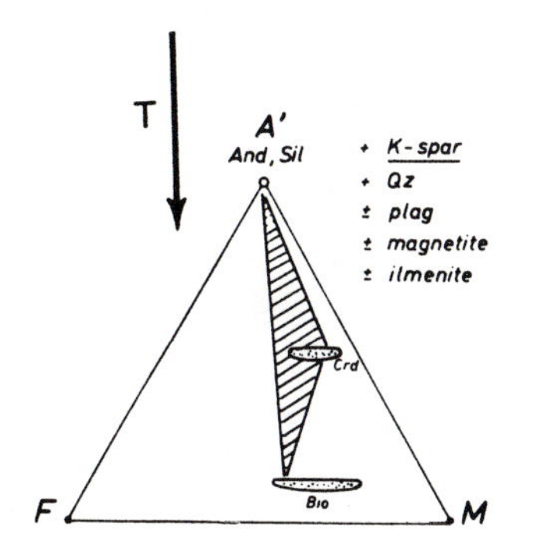

Fig. 14-11a [Cordierite]-high-grade.

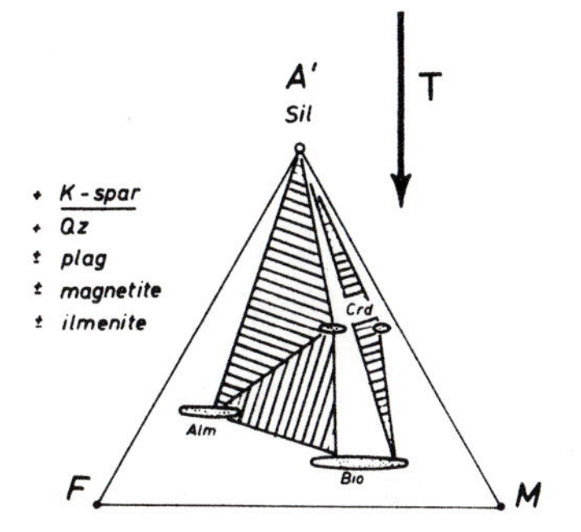

Fig. 14-11b [Cordierite-almandine]-high-grade.

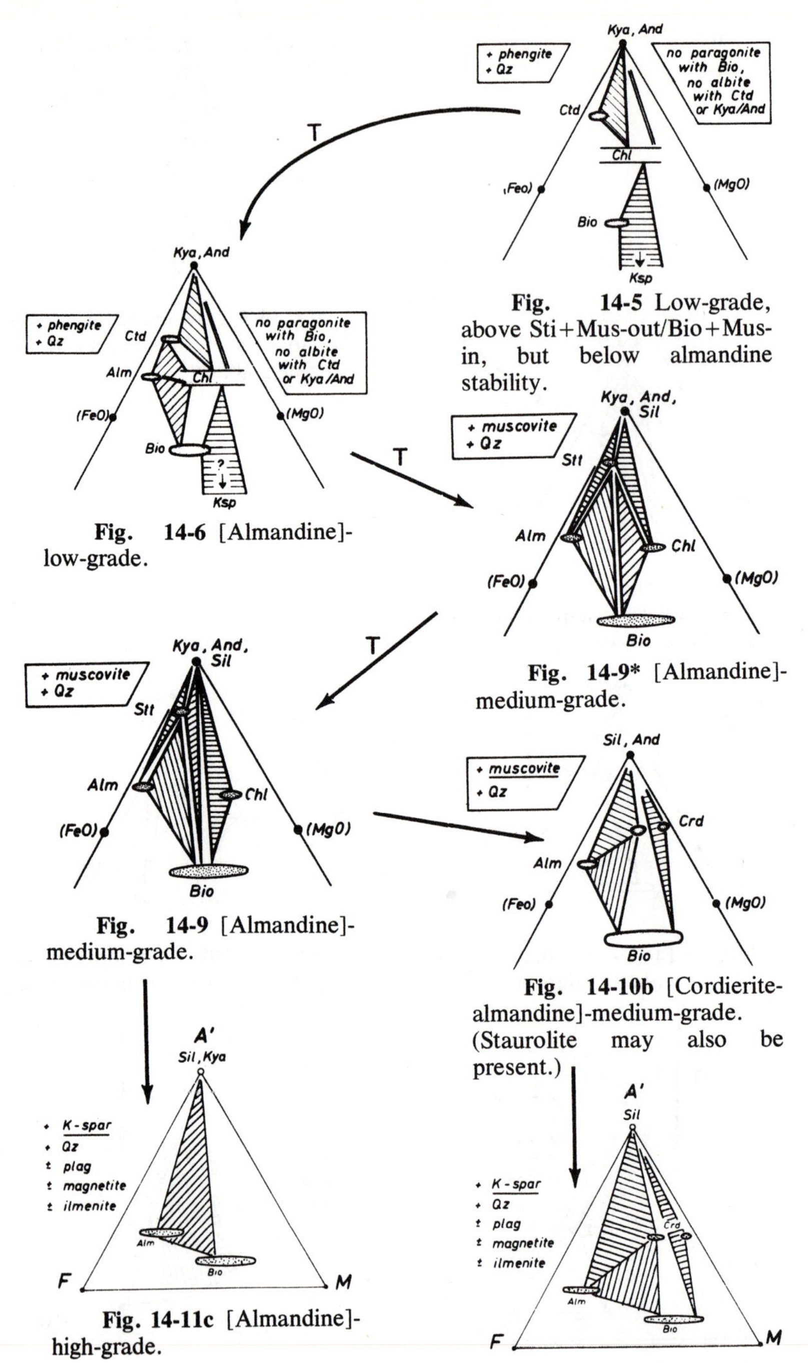

Fig. 14-5 Low-grade, above Sti + Mus-out/Bio + Mus-in, but below almandine stability.

Fig. 14-6 [Almandine]-low-grade.

Fig. 14-9* [Almandine]-medium-grade.

Fig. 14-9 [Almandine]-medium-grade.

Fig. 14-10b [Cordierite-almandine]-medium-grade. (Staurolite may also be present.)

Fig. 14-11c [Almandine]-high-grade.

Fig. 14-11b [Cordierite-almandine]-high-grade.

grams are necessary. It should be repeated here that the *beginning of medium-grade metamorphism* defined by the first appearance of staurolite or cordierite does not necessitate the disappearance of chlorite and/ or chloritoid in the presence of muscovite (see p. 81). In fact, this cannot be expected because the minerals involved in all reactions are solid solutions. Therefore, chloritoid and staurolite, chlorite + muscovite and staurolite, and chlorite + muscovite and cordierite may coexist within a narrow temperature range. Such cases are indicative of proximity to the boundary between medium- and low-grade metamorphism. This should be taken into account when a sequence from Figures 14-5 to 14-8 or from Figures 14-6 to 14-9* is encountered in the field.

From petrographic work, Carmicheal (1970), Guidotti (1974), and Froese and Gasparrini (1975) have ascertained that in relatively Mg-rich metapelites (with MgO/MgO + FeO being close to 0.5) a Mg-rich chlorite can be stable together with staurolite + muscovite + quartz ± biotite well within the medium-grade (up to approximately 50–60°C above the boundary of the beginning of medium-grade). This case is shown in Figurc 14-9*. The coexistence of Mg-rich chlorite and staurolite is terminated when the following reaction has taken place (Thompson and Norton, 1968):

$$\{\text{Mg-rich chlorite + staurolite + muscovite + quartz}\} = \{\text{biotite} + Al_2SiO_5 + H_2O\}$$

This reaction permits the coexistence of biotite + Al_2SiO_5, as is shown in Figure 14-9. In addition, Hoffer (1978) has pointed out that even in the absence of staurolite (but still within the range of medium-grade metamorphism) biotite + Al_2SiO_5 are produced from Mg-rich chlorite. The relevant reaction, first suggested by Carmicheal (1970), is

$$\{\text{Mg-rich chlorite + muscovite + quartz}\} = \{Al_2SiO_5 + \text{biotite} + H_2O\}$$

Experimental calibrations of the two reactions are not yet available. However, they will take place over a considerable temperature and pressure range because not only sillimanite but, in rare cases, kyanite or andalusite instead of sillimanite has been recorded.

Parageneses of [*almandine*]-*medium-grade*. Whenever relatively Mg-rich metapelites are present, a low temperature zone within the [almandine]-medium-grade can be distinguished: the staurolite-chlorite-biotite zone with the parageneses shown in Figure 14-9*. This paragenesis is diagnostic. At somewhat higher temperature the parageneses of Figure 14-9 are valid. Figure 14-9 shows, among others, the parageneses Stt + Alm + Al_2SiO_5, Stt + Alm + Bio, and Stt + Al_2SiO_5 +

Bio. However, the four phases Stt + Alm + Al_2SiO_5 + Bio in addition to Mus + Qz have commonly been observed. As a rule, garnet is attributed to significant amounts of additional components such as MnO or CaO. This, however, may not always be the reason. For example, Albee (1968) suggested "that the garnet is relatively nonreactive, as implied by its pronounced zoning[5] and persists as a relic even after changing conditions or compositions have placed the rock out of the stability field of garnet." Apparently, there are various reasons for the existence of the assemblage Stt + Alm + Al_2SiO_5 + Bio + Mus + Qz in addition to the assemblages shown in Figure 14-9. It should also be noted that in Figure 14-9 Mg-rich chlorite may stably coexist with muscovite, quartz, biotite, and/or Al_2SiO_5.

Disappearance of staurolite. Increasing temperature leads to the decomposition of staurolite in pelitic rocks. In many areas this takes place within the stability field of sillimanite. The following reactions probably are significant:

$$\{\text{staurolite} + \text{muscovite} + \text{quartz}\} = \{Al_2SiO_5 + \text{biotite} + H_2O\}$$

and at some higher pressure

$$\{\text{staurolite} + \text{muscovite} + \text{quartz}\} = \{Al_2SiO_5 + \text{almandine} + \text{biotite} + H_2O\}$$

The latter reaction has been suggested by Thompson and Norton (1968) and by Froese and Gasparrini (1975). However, a more complex and probably more realistic reaction has been inferred from petrographic studies by Guidotti (1970):

$$\{\text{staurolite} + \text{sodic muscovite} + \text{quartz}\} = \{Al_2SiO_5 + \text{biotite} + \text{K-richer muscovite} + \text{albite} + \text{almandine} + H_2O\}$$

Clearly, these reactions take place in medium-grade metamorphism where muscovite + quartz are stable; these minerals are necessary reactants. If muscovite is absent, as for instance in certain metagraywackes, these reactions cannot proceed. Therefore, staurolite may persist into high-grade rocks and eventually decompose according to the following reactions:

[5]Zoning may not always be present.

$$\{\text{staurolite} + \text{quartz}\} = \{Al_2SiO_5 + \text{almandine} + \text{cordierite} + H_2O\}$$

$$\{\text{staurolite} + \text{quartz}\} = \{Al_2SiO_5 + \text{almandine} + H_2O\}$$

The latter reaction has been investigated by Richardson (1968) using a pure Fe staurolite; it takes place at temperatures a little below 700°C and is only slightly pressure dependent. Common staurolites, however, contain about 10 to 30 atom % Mg substituting for Fe. Therefore, the temperature up to which staurolite (without muscovite) may persist will be somewhat higher, judging from the work of Schreyer (1968) on the Mg-staurolite end member.

Reactions involving muscovite and quartz occur at lower temperatures and the temperatures decrease markedly with decreasing pressure. This follows from the preliminary data for the reaction:

$$\{\text{staurolite} + \text{muscovite} + \text{quartz}\} = \{Al_2SiO_5 + \text{biotite} + H_2O\}$$

which has been investigated by Hoschek (1969). He showed that staurolite of composition $MgO/(MgO+FeO) = 0.4$ and 0.2 both break down at practically the same conditions:

2 kb H_2O pressure and 575° ± 15°C
5.5 kb H_2O pressure and 675° ± 15°C

The stated temperatures are maximum values and may be somewhat too high because reversals have not yet been achieved. Therefore, 10°C was subtracted before plotting the reaction curve 6a in Figure 14-1. This reaction is not valid at water pressures greater than 5 kb if anatexis in gneisses takes place and muscovite in the presence of quartz and plagioclase breaks down to form a granitic melt.

If muscovite and quartz do not occur together with staurolite, the following reaction may take place at a temperature somewhat higher than that of the previous reaction (Hoffer, 1975; personal communication):

$$\text{staurolite} + \text{biotite} + \text{quartz} = \text{cordierite} + \text{garnet} + \text{muscovite} + H_2O$$

If neither biotite nor muscovite is present, the reaction

$$\{\text{staurolite} + \text{quartz}\} = \{Al_2SiO_5 + \text{almandine} + H_2O\}$$

shown as curve 6b in Figure 14-1 will take place only when the temperature has reached about 700°C; or, as seems more likely and has been suggested by Froese (1973, personal communication), the following reaction may take place:

$$\{\text{staurolite} + \text{K feldspar component in an anatectic melt}\} = \{Al_2SiO_5 + \text{almandine} + \text{biotite}\}$$

This needs experimental verification.

It follows from this discussion that three different cases have to be distinguished:

1. *"Staurolite-out in the presence of Mus + Qz"* in medium-grade metamorphic rocks, implying H_2O pressure smaller than about 5 kb;
2. *"Staurolite-out in the presence of Bio + Qz";*
3. *"Staurolite-out in migmatitic areas."*

It is clear from this discussion that staurolite need not be restricted to medium-grade metamorphism but may persist into high-grade. Ashworth (1975) has indeed observed staurolite together with sillimanite in high-grade migmatized semipelites.

Coexistence of almandine-rich garnet and cordierite. Wynne-Edwards and Hay (1963) were probably the first to point out the worldwide occurrence of cordierite + almandine assemblages in high-grade terrains. It is now evident that this coexistence is possible only at high temperatures and within a restricted range of intermediate pressures.

The coexistence of cordierite and almandine is also known in some medium-grade rocks, as shown in Figure 14-10b; these belong to the zone of [cordierite-almandine]-medium-grade. This zone occurs invariably in some higher temperature part of medium-grade metamorphism. The following parageneses are diagnostic of this special higher temperature zone of medium-grade, designated as [cordierite-almandine]-medium-grade:

Crd + Alm + Bio + Mus + Qz ± plagioclase
Crd + Alm + Sil + Mus + Qz ± plagioclase
Crd + Alm + Sil + Bio + Mus + Qz ± plagioclase

It should be noted that generally sillimanite is the stable Al_2SiO_5 species but it may be accompanied in medium-grade rocks by metastably persisting andalusite. However, in an area described by Osberg (1968), andalusite seems to be the only stable modification.

The last paragenesis is not shown in Figure 14-10b but it has been observed in nature. In some instances it is attributed to a further component such as MnO or CaO stabilizing almandine-rich garnet. Such parageneses have been recorded from the Abukuma region in Japan (Miyashiro, 1958; Shido, 1958; Shido and Miyashiro, 1959), where the almandine-rich garnet contains less than 20 mole percent spessartine component. In the same metamorphic zone but from a different region, Okrusch (1971) has observed almandine-rich garnets with a spessartine component between 25 and 10 mole percent and an almandine component between 66 and 79. Both regions were metamorphosed at an estimated pressure of about 3 kb. It is obvious that cordierite + almandine assemblages can be formed only in rocks of appropriate composition. If the Mg/Fe ratio is high and the Mn content is low, the almandine-free assemblage Crd + Sil + Bio + Mus + Qz will form; this is also shown in Figure 14-10b. Increasing temperature brings about the transition from [cordierite-almandine]-medium-grade to [cordierite-almandine]-high-grade metamorphism.

High-grade metamorphism is characterized by the disappearance of primary muscovite in the presence of quartz (and plagioclase, at higher pressures) which makes possible the compatibility of K feldspar with Al_2SiO_5, almandine, and/or cordierite. Pertinent reactions have been discussed on p. 83ff. The breakdown reaction of muscovite in the presence of quartz is plotted in Figure 14-1 as curve 7; also shown is the boundary of anatexis in gneisses which, at pressures above about 4 kb, designates the beginning of high-grade metamorphism.

Parageneses diagnostic of high-grade pelitic gneisses are

Ksp + Sil/Kya + Alm + Bio ± plagioclase + Qz
for [almandine]-high-grade; Figure 14-11c
Ksp + Sil + Crd + Bio ± plagioclase + Qz
for [cordierite]-high-grade; Figure 14-11a
Ksp + Sil + Crd + Alm ± plagioclase + Qz
Ksp + Bio + Crd + Alm ± plagioclase + Qz
Ksp + Crd + Alm + Qz
Ksp + Bio + Sil + Crd + Alm ± plagioclase + Qz
for [cordierite-almandine]-high-grade; Figure 14-11b

Figures 14-11a-c are AFM diagrams as used by Reinhardt (1968); cf. p. 53.

The last paragenesis has even been recorded in a contact aureole by Okrusch (1971) where it apparently developed from the paragenesis Mus + Sil + Crd + Alm + Bio + Qz. Garnet from the high-grade rocks contains less spessartine component (only 5.5 to 8%) and more almandine component (79 to 85%).

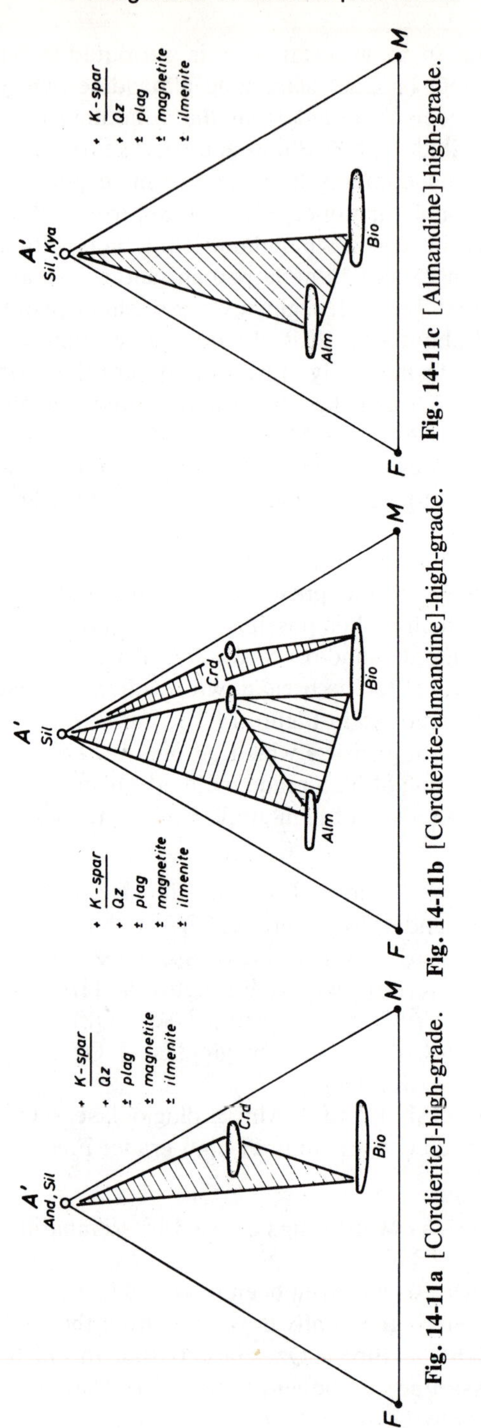

Fig. 14-11a [Cordierite]-high-grade.

Fig. 14-11b [Cordierite-almandine]-high-grade.

Fig. 14-11c [Almandine]-high-grade.

Diagrams are A'FM diagrams. A' = Al_2O_3 - (K_2O+Na_2O+CaO); F = FeO - (Fe_2O_3+TiO_2); M = MgO.

On the other hand, Reinhardt (1968) does not regard the paragenesis

$$\text{Ksp} + \text{Bio} + \text{Sil} + \text{Crd} + \text{Alm}$$

as stable. He studied phase relations in cordierite-bearing pelitic gneisses and concluded that only the parageneses Crd + Alm + Sil and Crd + Alm + Bio are stable, as shown in Figure 14-11b. This discrepancy in interpretation may be significant. Therefore, it is advisable to consider the problem of determining stable mineral assemblages. For this purpose, we quote a paragraph from Reinhardt's publication.

> As Zen [1963, p. 936] pointed out, the enumeration of phases in the physicochemical sense differs from the straight-forward identification and tabulation of all minerals present within a given rock sample. Most natural assemblages are masked by minor prograde and retrograde crystallization so that one must carefully assess which minerals represent the main metamorphic equilibration. This is often a difficult task because a certain reliance must be placed on textural criteria. The following account deals with the more important petrographic observations used to separate stable from metastable and retrograde phase associations.
>
> Sillimanite occurs in several ways in gneisses containing cordierite and garnet. Many cordierite-garnet-biotite gneisses contain small sillimanite needles as inclusions in the cordierite and garnet. The mode of occurrence does support the contention that sillimanite is a relict phase in almost all cordierite-garnet-biotite gneisses. This can be further explained if sillimanite and its host (either garnet or cordierite) are regarded as a two-phase subsystem that is stable in itself but unstable for the overall system made up of the minerals cordierite, garnet, and biotite. This two-phase subsystem compares with the *mosaic equilibrium* envisioned by Korzhinskii (1959, p. 19). A small number of cordierite-garnet-biotite gneisses were observed to contain large independent grains of sillimanite, but these rocks are also virtually devoid of plagioclase. Where recognized, the plagioclase was fine grained and altered. This indicates that plagioclase and sillimanite are incompatible phases in the presence of cordierite, garnet, biotite, quartz, alkali feldspar, and opaque oxides.
>
> Sillimanite appears to be a compatible associate of cordierite and garnet in plagioclase-bearing gneisses that lack stable biotite. Where biotite is present in these gneisses, it occurs as felted intergrowths armouring much larger grains of cordierite. The biotite therefore can be interpreted as a retrograde mineral, and the stable assemblage in terms of indicator minerals is cordierite-garnet-sillimanite.

It follows from this description that in the presence of plagioclase, cordierite + almandine + sillimanite + biotite is not considered to be a stable paragenesis, but if plagioclase is absent it apparently is stable. The necessity for careful observation of textural criteria is obvious. It is also expedient to recall the discussion in Chapter 4 concerning the determination of stable parageneses. It must be ascertained which minerals in a thin section are in contact as only those in contact may be regarded as an assemblage of coexisting minerals, *i.e.*, a paragenesis. Alteration products, of course, are exceptions.

The coexistence of cordierite and almandine-rich garnet is a significant phenomenon. Except for rare occurrences in certain rocks from the high temperature zones of medium-grade metamorphism, the pair cordierite + almandine is restricted to a specific *P-T* range in high-grade metamorphism and in the regional hypersthene zone (granulite facies). Information about these restricted *P-T* conditions is available from a number of recent experimental investigations.

Hensen and Green (1970, 1971, 1972) pointed out that a sharp distinction must be made between compositions with a ratio (MgO + FeO)/Al_2O_3 smaller than 1 and greater than 1 in the K_2O-free system MgO-FeO-Al_2O_3-SiO_2. In the first case, the assemblage Crd + Alm + Sil + Qz is formed, whereas in the second case, the assemblage Hyp + Crd + Alm + Qz is formed. The second assemblage is formed at pressures about 3 kb lower than those required to stabilize the first assemblage. The hypersthene-bearing assemblage is diagnostic of the regional hypersthene zone (granulite facies) treated in Chapter 16. Here we are predominantly concerned with the paragenesis

Crd + Alm + Sil + Qz,
commonly accompanied by biotite, K feldspar ± plagioclase.

The paragenesis Crd + Alm + Sil + Qz is attributed by Hensen and Green to the following reaction:

$$\text{Crd} = \{\text{Alm} + \text{Sil} + \text{Qz}\}$$

If biotite is present as well, the reaction

$$\{\text{Crd} + \text{Bio}\} = \{\text{Alm} + \text{Ksp} + H_2O\}$$

will take place, together with the former as a coupled reaction, at the same conditions (Currie, 1971). These are "sliding reactions," *i.e.*, reactants and products coexist over a limited pressure-temperature range (see Figure 14-13). Therefore, there is a field of coexisting

$$\text{Crd} + \text{Alm} + \text{Sil} + \text{Qz} \pm \text{Bio} \pm \text{Ksp}$$

It is of special importance that the reaction Crd = {Alm + Sil + Qz}, unlike most other solid-solid reactions, has a negative slope, *i.e.*, temperature decreases with increasing pressure.

As well documented by Hensen and Green, and Currie,

> the position and width of the divariant field (in terms of pressure and temperature) in which cordierite and garnet coexist, is a function of the MgO/(MgO+FeO) ratio.

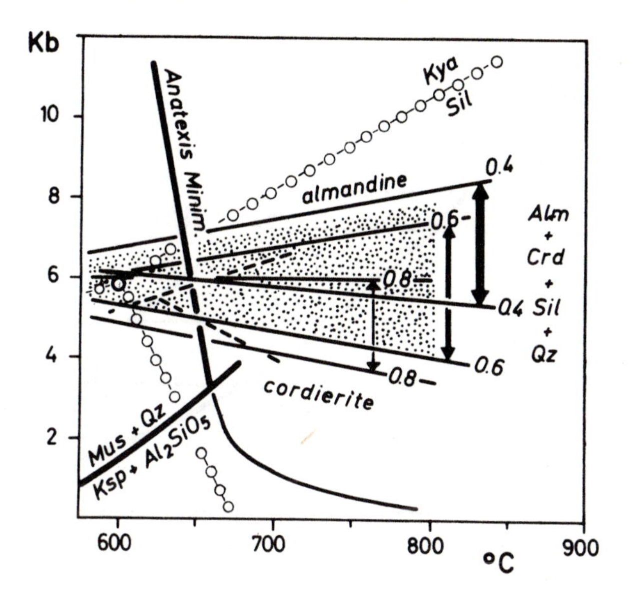

Fig. 14-12 *P-T* diagram of coexisting Crd + Alm + Sil + Qz for various FeO/(MgO+FeO) ratios of the bulk composition. Shown also are the reactions defining the boundary between medium-grade and high-grade metamorphism and the approximate phase boundaries kyanite/sillimanite and andalusite/sillimanite. Coexistence ranges are valid only if the Mn and Ca contents of garnet are negligible.

If this ratio is increased then the stability field of garnet is reduced and that of cordierite extends towards higher pressure. (Hensen and Green, 1971.)

This is demonstrated in Figures 14-12 and 14-13. Hensen and Green carried out experiments from 800°C upward, whereas Currie worked in the more relevant temperature range of 600° to 900°C. In addition, he used an internally heated pressure vessel and methane as a pressure medium which assures high accuracy of *P* and *T* measurements. This may be the reason why Currie's reported pressures are about 2 kb lower than those of Hensen and Green. The data given by Currie permit the *P-T* fields of the coexistence of Crd + Alm + Sil + Qz to be outlined for various FeO/(MgO+FeO) ratios of the bulk composition. In Figure 14-12, 0.8, 0.6, and 0.4 refer to this ratio. Each pair of lines designated by the same number outlines the area of stability of the assemblage at a given Fe/Mg ratio. It is a wedge-shaped area spreading out with increasing temperature.[6] This explains why cordierite-almandine-sillimanite assemblages

[6]The reality of the wedge-shaped area (as opposed to a band bounded by two lines both with positive slope) has recently been questioned by Hutcheon *et al.* (1974).

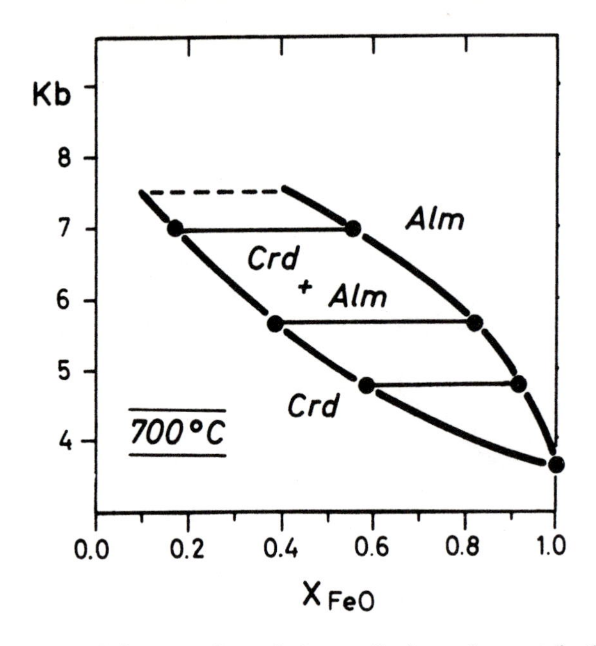

Fig. 14-13 *P-X* diagram of coexisting cordierite and garnet (in the presence of sillimanite and quartz) at 700°C (from Currie, 1971).

are much more common in high-grade than in medium-grade rocks. An even more pronounced wedge-shaped area has been found by Hirschberg and Winkler (1968). This area, for a ratio of FeO/(MgO+FeO) = 0.8, is outlined by broken lines in Figure 14-12.

The assemblage cordierite, almandine, sillimanite, biotite, quartz ± K feldspar + small amounts of plagioclase is particularly common in rocks with FeO/(MgO+FeO) ratios of 0.5 to 0.6. Therefore, this composition is shown stippled in Figure 14-12. This figure is valid for Mn-free of Mn-poor bulk compositions; the presence of more MnO probably will shift the area toward lower pressure.

Although the FeO/(MgO+FeO) ratio of the bulk composition determines the nature of the mineral assemblage, the compositions of coexisting garnet and cordierite, in the presence of quartz and sillimanite, are functions only of temperature and pressure. This was experimentally demonstrated by Hensen and Green (1971) and by Currie (1971). From the work of the latter author, a *P-X* diagram at 700°C is reproduced here as Figure 14-13. X = FeO/(MgO+FeO). Below the lower curve only cordierite is stable; above the upper curve only garnet is stable. In between the two curves cordierite and garnet coexist, together with sillimanite and quartz. The tie lines join compositions of coexisting cordierite and garnet at different pressures and constant temperature.

From additional experimental determinations at different temperatures, Currie obtained sufficient data to calibrate a geological thermometer and barometer. He pointed out:

> In equilibrated metamorphic rocks containing garnet, cordierite, quartz, and sillimanite, the exchange of iron and magnesium between cordierite and garnet offers a highly favourable geological thermometer and barometer, because this exchange reaction is insensitive to pressure.

The investigations presented so far by Currie suggest that the presence of biotite is inconsequential. On the other hand, Hensen (1971) suggests that biotite may have a disturbing influence. Also, the effects of grossularite and/or spessartine component in garnet are not known yet. Nevertheless, where these factors can be neglected, the cordierite-garnet geothermometer is potentially very useful in metamorphic petrology. For details, the reader is referred to Currie's paper and also to the more recent papers by Currie (1974), Hutcheon *et al.* (1974), Thompson (1976), and by Holdaway and Lee (1977).

References

Albee, A. L. 1968. *In* E-An Zen *et al.*, eds. Chap. 25. *Studies of Appalachian Geology*. Interscience Publisher (John Wiley & Sons), New York.
——— 1972. *Geol. Soc. Am. Bull.* **83:** 3249–3268.
Ashworth, J. R. 1975. *Contr. Mineral. Petrol.* **53:** 281–291.
Brown, E. H. 1969. *Am. Mineral.* **54:** 1662–1677.
——— 1971. *Contr. Mineral. Petrol.* **31:** 275–299.
——— and Fyfe, W. S. 1971. *Contr. Mineral. Petrol.* **33:** 227–231.
Carmicheal, D. M. 1970. *J. Petrol.* **11:** 147–181.
Chakraborty, K. R. and Sen, S. K. 1967. *Contr. Mineral. Petrol.* **16:** 210–232.
Chatterjee, N. D. 1966. *Contr. Mineral. Petrol.* **12:** 325–339.
——— 1971. *Neues Jahrb. Mineral. Abhand.* **114:** 181–245.
——— 1972. *Contr. Mineral. Petrol.* **34:** 288–303.
——— 1973. *Contr. Mineral. Petrol.* **42:** 259–271.
Currie, K. L. 1971. *Contr. Mineral. Petrol.* **33:** 215–226.
——— 1974. *Contr. Mineral. Petrol.* **44:** 35–44.
Dunoyer de Segonza, G. 1970. *Sedimentology* **15:** 281–346.
Frey, M. 1969. *Contr. Mineral. Petrol.* **24:** 63–65.
——— 1970. *Sedimentology* **15:** 261–279.
——— 1972. *Geol. Soc. Am. Ann. Meeting* (abs.).
——— 1973. *Contr. Mineral. Petrol.* **39:** 185–218.
——— 1978. *J. Petrol.* **19:** 95–135.
Froese, E. and Gasparrini, E. 1975. *Canad. Mineral.* **13:** 162–167.
Ganguly, J. and Newton, R. C. 1969. *J. Petrol.* **9:** 444–466.

Guidotti, C. V. 1968. *Am. Mineral.* **53:** 963–974.
——— 1970. *J. Petrol.* **11:** 277–336.
——— 1974. *Geol. Soc. Amer. Bull.* **85:** 475–490.
Hemley, J. J. 1967. *Am. Geophys. Union Trans.* **48:** 224 (abs.).
——— and Jones, W. R. 1964. *Econ. Geol.* **59:** 538–569.
Hensen, B. J. 1971. *Contr. Mineral. Petrol.* **33:** 191–214.
——— and Green. 1970. *Phys. Earth Planet. Interiors* **3:** 431–440.
——— 1971. *Contr. Mineral. Petrol.* **33:** 309–330.
——— 1972. *Contr. Mineral. Petrol.* **35:** 331–354.
Hirschberg, A. and Winkler, H. G. F. 1968. *Contr. Mineral. Petrol.* **18:** 17–42.
Hoffer, E. 1978. *Contr. Mineral. Petrol.* **67:** 209–219.
Holdaway, M. J. and Lee, S. M. 1977. *Contr. Mineral. Petrol.* **63:** 175–198.
Hoschek, G. 1967. *Contr. Mineral. Petrol.* **14:** 123–162.
——— 1969. *Contr. Mineral. Petrol.* **22:** 208–232.
Hutcheon, I., Froese, E., and Gordon, T. M. 1974. *Contr. Mineral. Petrol.* **44:** 29–34.
Kerrick, D. M. 1968. *Am. J. Sci.* **266:** 204–214.
Korzhinskii, D. S. 1959. *Physicochemical Basis of the Analysis of the Paragenesis of Minerals.* New York, Consultants Bureau.
Mather, J. D. 1970. *J. Petrol.* **11:** 253–275.
Miyashiro, A. 1958. *J. Fac. Sci. Univ. Tokyo* **11:** 219–271.
Nitsch, K.-H. 1970. *Fortschr. Mineral.* **47:** 48–49.
Okrusch, M. 1971. *Contr. Mineral. Petrol.* **32:** 1–23.
Osberg, P. H. 1968. *Maine Geol. Surv. Bull.* **20.**
Reinhardt, E. W. 1968. *Can. J. Earth Sci.* **5:** 455–482.
Richardson, S. W. 1968. *J. Petrol.* **9:** 467–488.
Schreyer, W. 1968. *Carnegie Inst. Yearbook* **66:** 381–384.
Seidel, E., Okrusch, M., and Schubert, W. 1975. *Contr. Mineral. Petrol.* **49:** 105–115.
Seki, Y., *et al.* 1969. *J. Japan Assoc. Mineral. Petrol. Econ. Geol.* **61:** 1–75.
Shido, F. 1958. *J. Fac. Sci. Univ. Tokyo* **11:** 131–217.
——— and Miyashiro, A. 1959. *J. Fac. Sci. Univ. Tokyo* **12:** 85–102.
Thompson, A. B. 1970. *Am. J. Sci.* **268:** 454–458.
Thompson, A. B. 1976. *Am. J. Sci.* **276:** 425–454.
Thompson, J. B. and Norton, S. A. 1968. Paleozoic regional metamorphism in New England and adjacent areas. *In* E-An Zen *et al.* eds. *Studies of Appalachian Geology.* Interscience Publisher (John Wiley & Sons), New York.
Turner, F. J. 1948. *Evolution of Metamorphic Rocks.* Geol. Soc. Amer., Memoir No. 30.
Weber, K. 1972. *Neues Jahrb. Geol. Paläon. Abhand.* **141:** 333–363.
Wenk, C. and Keller, F. 1969. *Schweiz. Mineral. Petrog. Mitt.* **49:** 157–198.
Winkler, H. G. F. 1967. *Petrogenesis of Metamorphic Rocks.* 2nd edit. Springer-Verlag, New York-Berlin.
Wynne-Edwards, H. R. and Hay, P. W. 1963. *Can. Mineral.* **7:** 453–478.
Zen, E-An. 1960. *Am. Mineral.* **45:** 129–175.
——— 1963. *Am. J. Sci.* **261:** 929–942.

Chapter 15

A Key to Determine Metamorphic Grades and Major Reaction-Isograds or Isograds in Common Rocks

This chapter summarizes some highlights given in previous chapters in order to provide a key for the determination of metamorphic grades and reaction-isograds. High-grade granulitic rocks diagnostic of the regional hypersthene zone are not considered here because they are treated later. Sequences of isograds are indicated by selected *mineral reactions* in common rocks. A more precise determination of metamorphic grade may be achieved by referring to the details presented in each relevant chapter. Earlier, in Chapter 7, arguments were given for setting up major divisions of metamorphic grade within the large *P-T* field of metamorphic conditions. These are very-low-grade, low-grade, medium-grade, and high-grade. With respect to pressure the metamorphic grades have been further subdivided:

[Laumontite]- or [Wairakite]-very-low-grade	low-grade	[Cordierite]-medium-grade	[Cordierite]-high-grade
[Lawsonite]-very-low-grade		[Cordierite-almandine]-medium-grade	[Cordierite-almandine]-high-grade
	[Almandine]-low-grade		
[Glaucophane-lawsonite]-very-low-grade	[Glaucophane-clinozoisite]-low-grade	[Almandine]-medium-grade	[Almandine]-high-grade
[Jadeite-quartz]-very-low-grade			

With the exception of the laumontite/lawsonite boundary, no quantitative data are available for the other pressure boundaries. Nevertheless, these divisions of metamorphic grade are very useful as a first approximation. They are easily determined in the field, but it must be realized that the divisions of very-low-grade metamorphism can only be identified in mafic rocks and graywackes containing mafic debris and that the grades indicated by cordierite and/or almandine can be identified only in pelitic rocks of appropriate composition.

Very-Low-Grade Metamorphism

In addition to laumontite, wairakite, and lawsonite, the minerals prehnite and pumpellyite are diagnostic of very-low-grade metamorphism (see p. 174 ff and Figure 12-11). The details given earlier allow the distinction of a number of reaction-isograds or isograds and consequently of metamorphic zones within very-low grade. In the present survey only a gross division of the *P-T* field of very-low-grade metamorphism is shown in Figure 15-1. The stability fields of laumontite, wairakite, lawsonite, and aragonite are marked. The boundary above which the conspicuous assemblage glaucophane + lawsonite is formed in mafic rocks may be estimated from field relations (see p. 89f) to be situated at approximately the *P-T* values indicated in Figure 15-1. Similarly, the boundary of the formation of jadeitic pyroxene + quartz in nonmafic psammitic albite-bearing rocks also is only an approximation; it is known that the equilibrium pressure at any given temperature must be somewhat lower than that required for the formation of pure jadeite (see Chapter 13). Therefore, the curve in Figure 15-1 has arbitrarily been drawn at a pressure 0.5 kb lower than the reaction boundary for pure jadeite + quartz.

Note the reaction taking place only in SiO_2 richer ultramafic rocks:

$$\text{serpentine} + 2\ \text{quartz} = \text{talc} + H_2O \qquad (1)$$

This reaction enables a distinction to be made between a field of coexisting serpentine + quartz, on the one hand, and a field with the assemblages talc + quartz or talc + serpentine, on the other. The reaction curve, designated as 1 in Figure 15-1, practically coincides at pressures above 4 kb with that of reaction 2.

Reaction (2)

$$\{\text{pumpellyite} + \text{chlorite} + \text{quartz}\} = \{\text{clinozoisite} + \text{actinolite}\} \qquad (2)$$

is one of the reactions defining the boundary between very-low- and low-grade metamorphism. This reaction is limited to rocks of mafic compo-

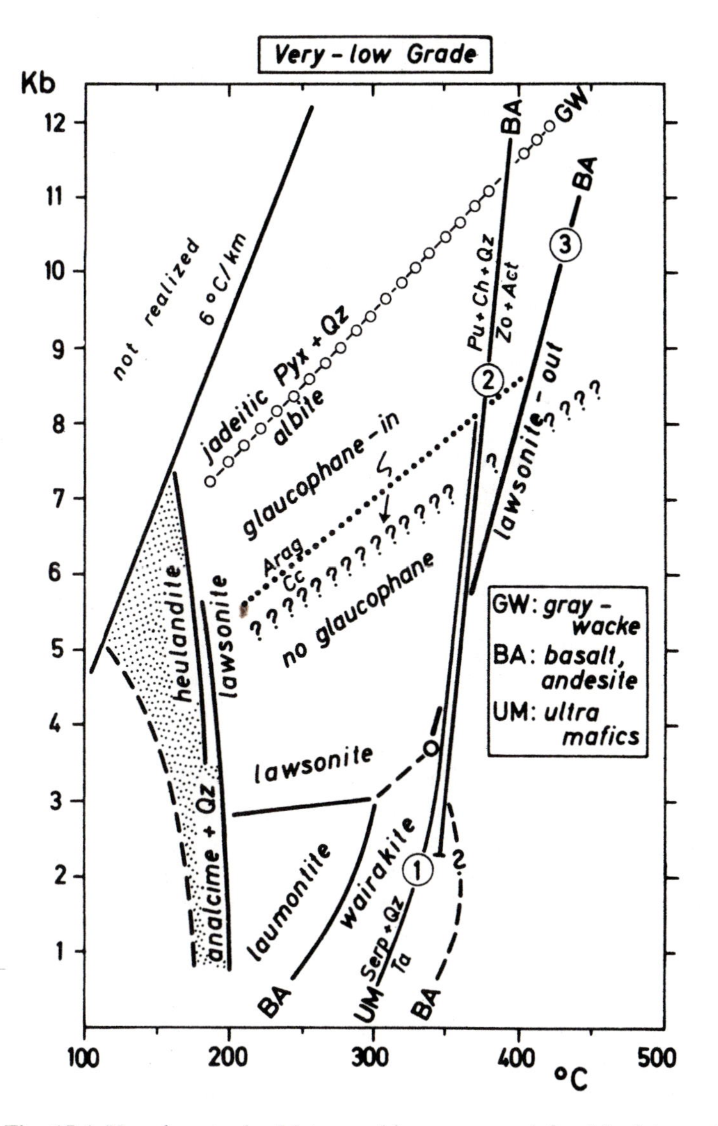

Fig. 15-1 Very-low-grade. Metamorphic zones are defined by laumontite, wairakite, lawsonite + albite, lawsonite + glaucophane ± albite or chlorite, lawsonite + jadeitic pyroxene + quartz (± albite), serpentine + quartz, and talc + serpentine or talc + quartz. In addition, the following parageneses are diagnostic: prehnite-bearing assemblages, prehnite + pumpellyite, and pumpellyite + chlorite + quartz; although not shown here, they are given in Figure 12-11.

sitions. If these are not available in a given terrain, the boundary cannot be determined accurately. However, in ultramafic rocks, reaction (1) indicates conditions which are very close to the boundary between very-low-grade and low-grade.

It is important to realize that mafic rocks are the diagnostically significant rocks in very-low-grade metamorphism. Of course, clastic sediments containing such rock debris serve the same purpose. On the other hand, although pelitic rocks may give certain indications of very-low-grade metamorphism (see Chapter 14), they do not by themselves allow a clear-cut determination.

In Figure 15-1 and in subsequent figures the rock type in which a reaction takes place has been designated by the following symbols:

UM = ultramafic rocks
BA = basalt and andesite (mafic rocks)
GW = graywacke
PG = pelitic rocks and graywacke
DOL = dolomitic siliceous limestone

Low-Grade

The disappearance of lawsonite, prehnite, and pumpellyite and the formation of clinozoisite and zoisite (as contrasted to iron-rich epidote) mark the transition from very-low-grade to low-grade conditions. The range of low-grade *P-T* conditions, as shown in Figure 15-2, is limited by reactions (1), (2), and (3) at low temperatures and reaction (6) at high temperatures. Within this temperature range of about 200°C, reactions in pelitic rocks give rise to the mineral assemblages of the classic Barrovian "chlorite, biotite, and garnet zones." The absence or presence of the assemblage biotite + muscovite determines the boundary between the chlorite and biotite zones.

As discussed on p. 219, the reactions leading to the first *formation of biotite* are not well understood. However, the following reaction certainly is relevant:

$$\{\text{stilpnomelane} + \text{phengite}\} = \{\text{biotite} + \text{chlorite} + \text{quartz} + H_2O\} \quad (4)$$

This reaction provides a reaction-isograd, which is designated as

(stilpnomelane + muscovite)-out/(biotite + muscovite)-in

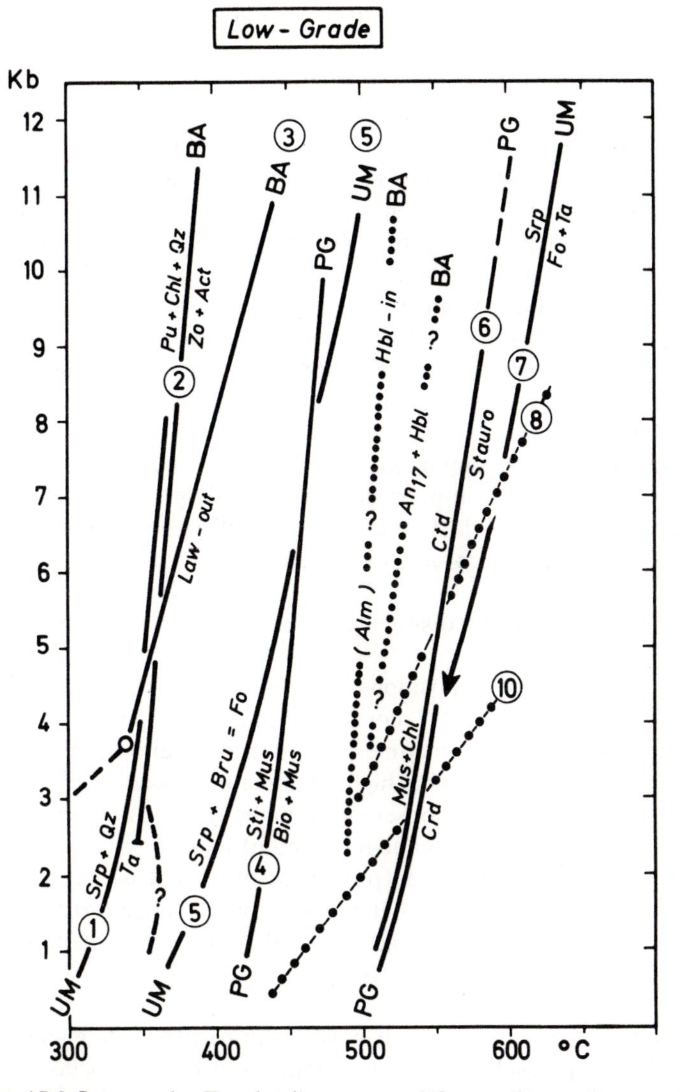

Fig. 15-2 Low-grade. For details see text. PG = pelites and graywackes. UM = ultramafic rocks. BA = basalts and andesites (mafic rocks).

The *P-T* conditions of this reaction are shown as curve 4 of Figure 15-2. This reaction-isograd corresponds closely to the boundary between the classic "chlorite and biotite zones."

Somewhat below (and at very high pressure, somewhat above) the beginning of the biotite zone, reaction (5) takes place, forming forsterite in metaserpentinites, at the lowest possible temperatures:

$$\text{antigorite} + \text{brucite} = \text{forsterite} + 3\ H_2O \qquad (5)$$

At higher temperatures and relatively high pressures, almandine-rich garnet is formed in rocks of the appropriate composition. From the work of Hirschberg and Winkler (1968), referred to on p. 220, it follows that a pressure of 4 kb at 500°C must be exceeded to stabilize almandine-rich garnet with a very low content of spessartine component. The temperature at which garnet first forms is not yet known; it will vary depending on the composition of the garnet. In spite of the variable composition of garnet, its first appearance, at sufficiently high pressures, invariably takes place in the higher temperature range of low-grade metamorphism. From the compilation by Turner (1968), it is evident that this coincides very nearly with the first appearance of hornblende (instead of actinolite) in mafic rocks.

Neither are the conditions of the first formation of hornblende exactly known. If an iron-poor, colorless or pale-green actinolite is present in a mafic greenschist, the formation of green hornblende would probably take place according to the reaction

$$\{\text{actinolite} + \text{clinozoisite} + \text{chlorite} + \text{quartz}\} = \text{hornblende}$$

Due to the complex compositions of the minerals taking part in the reaction, a certain range of *P-T* conditions must be expected. This would explain why actinolite sometimes persists, together with newly formed hornblende, until a somewhat higher temperature is attained. Therefore, the reaction-isograd band observed in mafic rocks has been designated in Figure 15-2 as "hornblende-in" (hbl-in) and not as "actinolite-out/hornblende-in." As stated on p. 174, at present it is only possible to estimate, on the basis of other metamorphic changes, that the hornblende probably forms at temperatures of about 500°C, the temperature rising only slightly with increasing pressure; this is shown in Figure 15-2.

Initially, hornblende coexists with clinozoisite-epidote and a plagioclase. In low pressure contact metamorphism the plagioclase is an oligoclase but at medium and high pressures the plagioclase is albite. If generally valid, this observation indicates that the boundaries for "hornblende-in" and for the abrupt change in plagioclase composition cross over at low pressure and diverge with increasing pressures, as schematically shown in Figure 15-2 by the two dotted lines. The composition of plagioclase changes abruptly from albite with up to 5 molecular percent anorthite to oligoclase with 17 to 20 percent anorthite. Albite may coexist with newly formed oligoclase over a narrow range of temperature

(Wenk and Keller, 1969). Therefore, it is correct to designate the isograd as "(An_{17-20} + hornblende)-in."

The discussed petrographic observations

in pelitic rocks:	almandine-in
in mafic rocks:	hornblende-in, and further
	(An_{17} + hornblende)-in

are very common but exact *P-T* data are not available. However, it is known that these transformations take place in the higher temperature range of low-grade metamorphism, as indicated in Figure 15-2.

Attention is again drawn to the reaction

$$1 \text{ antigorite} + 1 \text{ brucite} = 2 \text{ forsterite} + 3 \, H_2O \qquad (5)$$

This reaction curve, shown as 5 in Figure 15-2, is well suited to subdivide the lower temperature range of low-grade metamorphism. This is especially valuable since the only other means of distinguishing that range precisely is based on the presence of stilpnomelane + phengite; this, however, is not a common assemblage.

A glance at Figure 15-2 shows that the whole *P-T* range of low-grade metamorphism is subdivided by a number of reactions which are approximately parallel and are only slightly influenced by pressure. Therefore, we can determine rather well the temperatures of metamorphism in a low-grade terrain, especially if a pressure estimate from adjacent very-low- or medium-grade areas is available. However, it is not possible from the reactions taking place in low-grade metamorphism to make pressure determinations. The necessary crossing reaction curves are not available, except close to the beginning of medium-grade metamorphism.

Medium- and High-Grade

The first formation of cordierite and/or staurolite, in pelitic rocks of appropriate composition, defines the beginning of medium-grade metamorphism. With the exception of a small range of coexistence, *chloritoid and the assemblage muscovite + chlorite (when not Mg-rich) are absent in medium-grade rocks*. (Note that chlorite without muscovite, e.g., in rocks of mafic composition, may persist into medium-grade.)

The following petrographic features in pelitic rocks determine the beginning of medium-grade:

(muscovite + non Mg-rich chlorite)-out
chloritoid-out
(cordierite and/or staurolite)-in

This situation is represented in Figures 15-2 and 15-3 by curve 6. Details are given on p. 77ff.

At higher temperature, the boundary between medium and high grade is defined by

a. The breakdown of muscovite in the presence of quartz
b. At pressures greater than 3 to 4 kb, by the breakdown of muscovite in the presence of quartz, plagioclase, and water, *i.e.*, by the formation of migmatites in gneiss terrains.

This has been explained on p. 83ff.

Most reactions indicative of medium-grade metamorphism are present at high-grade as well. A distinction between the medium- and high-grade part of a specific reaction-isograd or zone can easily be made with the help of Figure 15-3. This figure summarizes the major reactions in common rocks which are useful indicators of metamorphic zones in medium- and high-grade terrains. It is to be expected that detailed petrographic work combined with relevant experimental studies will suggest additional reactions and establish their *P-T* conditions.

A glance at Figure 15-3 shows that a number of reaction curves or bands intersect each other. Two implications of this feature are the basis for using mineral assemblages as geothermometers and geobarometers:

a. The sequence and nature of metamorphic zones developed in response to rising temperature are different for various pressures.
b. On the assumption that $P_{total} = P_{fluid}$, the intersection point of two curves determines both pressure and temperature.

In the following, the various reactions and significant intersections are discussed and examples are given of different sequences of reaction-isograds or isograd bands at different pressures.

The relevant reactions. Reaction (8), taking place in marls, is

$$\{\text{margarite + quartz}\} = \{\text{anorthite + And/Kya} + H_2O\} \quad (8)$$

The anorthite produced increases the content of anorthite in coexisting plagioclase. Andalusite or kyanite may have been present before this

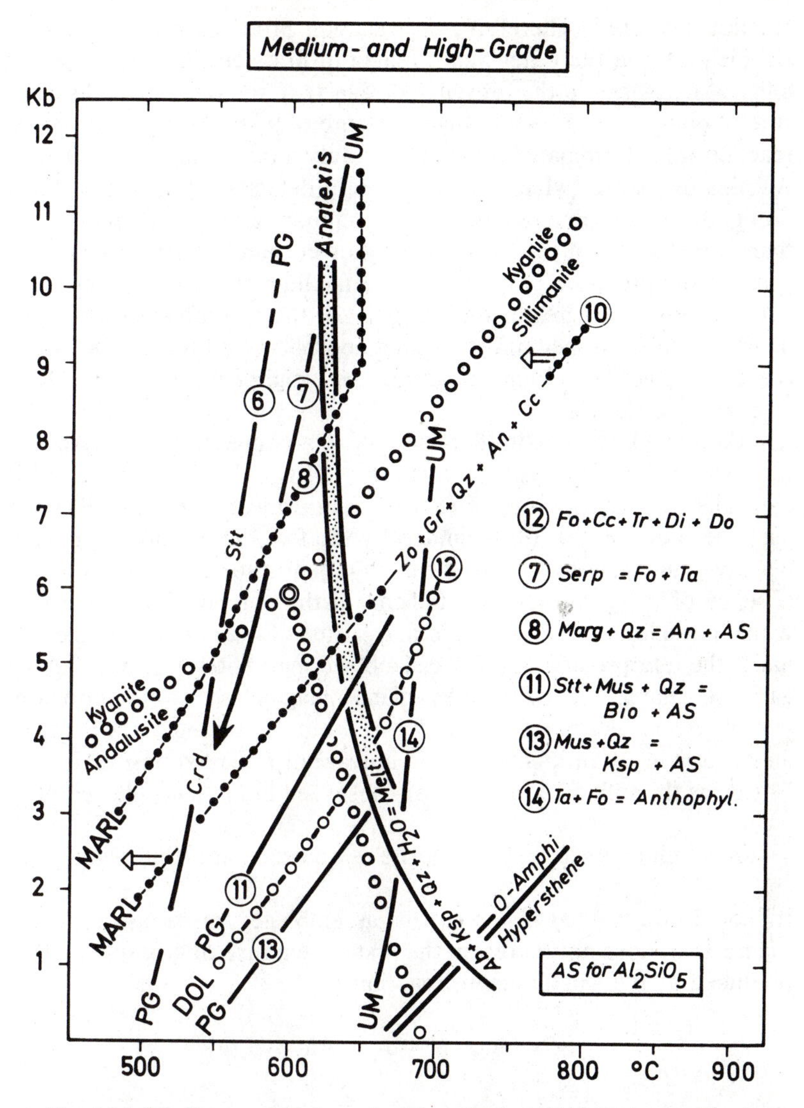

Fig. 15-3 Medium-and-high grade. The stippled band designated by "anatexis" (*i.e.*, beginning of anatexis in gneiss) and reaction (13) indicate the boundary between medium-grade and high-grade metamorphism. Common rock types are MARL, PG = pelite and graywacke, DOL = siliceous dolomitic limestone, UM = ultramafic compositions. The approximate phase boundaries kyanite/sillimanite and kyanite/andalusite shown in this figure are values intermediate between those given by Althaus (1967, 1969) and by Richardson *et al.* (1968, 1969). Curve (9) is lacking here because the data on the univariant equilibrium paragenesis Tr + Ta + Cc + Do + Qz have yet to be agreed upon.

reaction occurred. Therefore, the reaction products are of no special significance. On the other hand, reaction (8) determines the upper stability of margarite in the presence of quartz in the range from low pressure of possibly 2 to 3 kb to high pressure of 9 kb. At higher pressures, reaction (8) is terminated by another reaction (see Chapter 10). The Ca, Al-mica margarite, which is conveniently detected by its x-ray diffraction pattern, is probably much more common in low-grade marly rocks than previously thought. At pressures lower than 5 kb, margarite disappears in the presence of quartz before medium-grade is reached. However, at pressures greater than 5 to 6 kb, the assemblage margarite + quartz is stable in medium-grade metamorphism. Although such occurrences have not been found in nature, they might be discovered in future petrographic studies.

Above 9 kb, reaction (8) is not stable and margarite + quartz react to form zoisite/clinozoisite + kyanite + H_2O. The assemblage zoisite/clinozoisite + kyanite is an *indicator of pressures greater than 9 kb*. The usefulness of reaction (8) is limited by the fact that it supplies only the negative information "(margarite + quartz)-out." Therefore, the absence of margarite is not significant if the composition of the rocks would not allow the formation of margarite at lower grade. On the other hand, the interpretation of the assemblage margarite + quartz and margarite + quartz + andalusite-kyanite + plagioclase is straightforward.

Paragenesis Zo + Gr + Qz + An + Cc. As outlined in Chapter 10, an isobaric invariant point, *i.e.*, an univariant *P-T* relationship, is established for the following paragenesis observed in metamorphosed marls:

zoisite-clinozoisite + grossularite + quartz + anorthite + calcite

It should be repeated here that this assemblage, shown as curve 10 in Figure 15-3, does not represent the first appearance of grossularite but is produced by two simultaneous reactions:

$$\{Zo + Qz\} = \{Gr + An + H_2O\}$$

and

$$\{Zo + CO_2\} = \{An + Cc + H_2O\}$$

For details consult Figure 10-2, p. 143.

The paragenesis Zo + Gr + Qz + An + Cc has been observed in medium-grade rocks in the Damara Belt of southwest Africa. In most rocks plagioclase, although anorthite-rich, is not pure anorthite. In such cases the reduction of the anorthite activity by the albite component will create an additional degree of freedom and displace curve 10 to lower temperature, which is schematically indicated by the two arrows in Figure 15-3. This effect probably will be small if plagioclase with high anorthite content is present. With this restriction in mind, curve 10 is significant in the *P-T* grid of Figure 15-3. It has a relatively gentle slope and therefore crosses several other reaction-isograds or isograd bands.

Paragenesis Fo + Di + Cc + Do. The metamorphic conditions for the formation of the paragenesis

forsterite + diopside + calcite + dolomite

are restricted to large mole fractions of CO_2 in the fluid phase. This makes the paragenesis particularly interesting. It represents the equilibrium of the following reaction

$$1\ \mathrm{Di} + 3\ \mathrm{Do} = 2\ \mathrm{Fo} + 4\ \mathrm{Cc} + 2\ \mathrm{CO_2}$$

This is reaction (14) in Chapter 9. As shown there in Figure 9-4, the equilibrium paragenesis of reaction (14) is bivariant and occurs in a narrow temperature band at a fluid pressure not exceeding about 4 kb; at a greater pressure, the temperature range increases considerably. Paragenesis (14) seems to be of limited value. However, it is most useful to find paragenesis (14) in the field because it facilitates the search for the univariant paragenesis

Fo + Cc + Tr + Di + Do

This is the paragenesis of the isobaric invariant point IV of Chapter 9; its P_f-T data are shown as curve (12) in Figure 15-3.

Staurolite-out. Line 11 in Figure 15-3 represents the breakdown of staurolite in the presence of quartz and muscovite according to reactions discussed on p. 228ff. Since muscovite, in addition to quartz, is a necessary reactant, the reactions are commonly restricted to medium-grade metamorphism because, in the presence of plagioclase, muscovite plus quartz form an anatectic melt. Thus, all the muscovite is consumed at the beginning of high-grade metamorphism. Only if plagioclase is absent, as in some muscovite-bearing quartzites, does staurolite persist together with muscovite and quartz in high-grade rocks at high pressures. This case indicates pressures above about 4.5 kb, as deduced from Figure 15-3. As noted earlier in Chapter 14, staurolite without muscovite will persist well into the range of high-grade metamorphism.

The breakdown products biotite, Al_2SiO_5, and ± almandine could have been present in the metamorphic rock before the breakdown of staurolite; therefore, it is not easy to prove that staurolite was originally present. The composition of some rocks will not allow its formation in the first place. In order to be sure that curve 11 has been surpassed, only rocks that have a composition suitable for the formation of staurolite should be considered.

Muscovite-out. The band of reaction-isograds represented by 13 in Figure 15-3 has been discussed on p. 84ff. This constitutes the boundary between medium-grade and high-grade metamorphism. In high-grade metamorphism primary muscovite + quartz are not stable. This statement is valid only for pressures lower than 3 to 4 kb. At pressures exceeding this value, muscovite disappears at the beginning of high-grade metamorphism by the following reaction only if, as is commonly the case, plagioclase is also present:

$$\text{muscovite} + \text{quartz} + \text{plagioclase} + \text{water} = \text{melt}$$

A granitic melt is formed producing migmatitic gneisses. The stippled band with the designation "Anatexis" in Figure 15-3 indicates the beginning of melting; for this and other possible reactions involving muscovite and quartz, see Figure 7-3. The positive criterion of muscovite + quartz-out is the coexistence of K feldspar and Al_2SiO_5 (mostly sillimanite), commonly with cordierite and/or almandine-rich garnet.

Reaction (14).

$$9 \text{ talc} + 4 \text{ forsterite} = 5 \text{ Mg-anthophyllite} + 4\, H_2O \qquad (14)$$

takes place in ultramafic rocks. Either bracketed by the assemblages Ta + Fo on one side and Antho + Fo or Antho + Ta on the other or determined directly by the equilibrium paragenesis.

$$\text{Antho} + \text{Ta} + \text{Fo}$$

this reaction-isograd subdivides the *P-T* field of high-grade metamorphism. As seen in Figure 15-3, curve 14 intersects curves 12 and 10. However, the *P-T* data of reaction (14) shown in Figure 15-3 are valid only for small values of X_{CO_2} as discussed in Chapter 11. This situation has indeed prevailed during metamorphism of serpentinites (see Figure 11-7).

Geothermometers and Geobarometers

From the compilation of reaction curves or bands in Figure 15-3, the points of intersection can easily be read off; they constitute geothermometers and geobarometers. But each pair of intersecting curves can supply only one specific set of P and T values. Beyond this, the sequence of reaction curves in a given terrain only indicates whether P and T has been higher or lower than that of a given intersection point.

Unfortunately, the intersections with the phase boundaries andalusite-kyanite and andalusite-sillimanite cannot be used with confidence. For reasons given on p. 92ff, special care is required in petrogenetic deductions based on the presence of Al_2SiO_5 species. But with adequate precaution, valuable information may be obtained. However, intersections with the Al_2SiO_5 species phase boundaries have not been listed in the following compilation.

The points of intersection deserve special attention in any metamorphic area; the relevant reaction-isograds should be determined in the field and their point or points of intersection should be located. This enables a determination of the physical conditions at the time of metamorphism with considerable precision. As seen in Figure 15-3, the intersections of different reaction-isograds are relevant at different metamorphic grades and load pressures.

The following compilation gives a survey of P-T indicators. The numbers of intersecting reaction curves, given in Table 15-1, correspond to those in Figure 15-3. In each case, the common rock type that allows the formation of the appropriate mineral assemblage is given.

Sequences of Reaction-Isograds or Isograds

Although the sequence of isograds in pelitic rocks is commonly recorded, such sequences in other rocks, *e.g.*, serpentinites, siliceous dolomitic marbles, mafic rocks, and marls, have not been investigated in

Table 15-1

Case	Reaction	Rock	Reaction	Rock	T, °C	P, kb
			At Boundary Low/Medium Grade			
a	(8)	Marl	(6)	PG	550–570	5–6
b	(10)	Marl	(6)	PG	530–540	3–ca. 4
			At Boundary Medium/High Grade			
c	(8)	Marl	Anatexis	PG	620–630	8–8.5
d	(10)	Marl	Anatexis	PG	630–625	5.5–ca. 6.5
e	(11)	PG	Anatexis	PG	640–650	4.5–5
f	(12)	Dol	Anatexis	PG	645–655	ca. 4
g	(13)	PG	Anatexis	PG	650–665	ca. 3
			In High Grade			
h	(14)	UM	(10)	Marl	695–700	6.5–8
i	(14)	UM	(12)	Dol	680	5
j	(14)	UM	Anatexis	PG	675–680	2.5–3.5

comparable detail. As progress is made in this direction, it is hoped that this chapter will be a helpful guide.

In the following, some examples will be given of sequences of reaction-isograds (or isograds) in the range from low-grade to high-grade metamorphism, which are suggested by reactions shown in Figures 15-2 and 15-3. It is certain that future petrographic and laboratory work will improve our knowledge of metamorphic reactions and provide more precise sequences of reaction-isograds. Some deductions are possible even now on the basis of information given in previous chapters, and a few examples are given here.

Shallow Contact Metamorphism

According to Turner (1968), the conventional division of metamorphic conditions at shallow depth, *i.e.*, in low pressure contact metamorphism, is

Albite-epidote-hornfels facies	Belonging to low-grade
Hornblende-hornfels facies	Belonging to medium-grade
Pyroxene-hornfels facies	Belonging to high-grade

Winkler (1967) proposed the more informative name "K feldspar-cordierite-hornfels facies" to replace the term "pyroxene-hornfels facies." Furthermore, he distinguished two subfacies or zones: (a) K-spar + cordierite zone with orthoamphibole and (b) K-spar + cordierite zone with orthopyroxene, where hornblende may still be present but orthoamphibole is absent. The appearance of orthoamphibole and the appearance of orthopyroxene in low pressure, high-grade rocks is indicated by a double line in Figure 15-3. The data are based on experiments by Akella and Winkler (1966) and Choudhuri and Winkler (1967).

In place of the three facies and the two subfacies, it is suggested that low pressure contact metamorphism (as well as very low pressure regional metamorphism) be characterized by the following sequence of reaction-isograds or isograds. The following sequence from low to high grade, Table 15-2, is *valid for pressures up to about 2.5 kb,* corresponding to a depth of 9 km or less. In addition, wollastonite will be formed in high-grade rocks and possibly other minerals which have been discussed on pp. 133–137. These pages should be consulted for a discussion of relevant reactions at very low pressures and high temperatures.

Sequence at Medium Pressures

At *pressures between about 3 and 5.5 kb,* the sequence in Table 15-3 will develop. In the upper part of this pressure range, metamorphic

Table 15-2. Low pressures.

Paragenesis	Curve	Rock	Grade
Srp + Qz + Ta	(1)	UM	Very-low
Srp + Bru + Fo	(5)	UM	Low
(Bio + Mus)-in	(4)	PG	Low
Zo + Gr + An + Cc + Qz	(10)	Marl	Low
Hornbl-in	—	BA	Close to
(An_{17} + Hbl)-in	—	BA	boundary
Crd-in; Stt-in (Chl + Mus)-out	(6)	PG	Medium
Srp + Fo + Ta	(7)	UM	Medium
(Stt + Mus + Qz)-out	(11)	PG	Medium
Fo + Tr + Cc + Do + Di	(12)	Dol	Medium
(Mus + Qz)-out (K-spar + Crd)-in	(13)	PG	High
Ta + Fo + Anthophyl	(14)	UM	High
Orthoamphib-out/Hypersthene-in	—	BA	High

conditions will correspond to [almandine]-low-grade, [almandine]-medium-grade, and [cordierite-almandine]-high-grade. In appropriate pelitic rocks, andalusite is the common species that will be replaced or accompanied by sillimanite near the beginning of high-grade metamorphism.

Table 15-3. Medium pressures.

Paragenesis	Curve	Rock	Grade
Srp + Qz + Ta	(1)	UM	Very-low/low
Srp + Bru + Fo	(5)	UM	Low
(Bio + Mus)-in	(4)	PG	Low
Hornbl-in;	—	BA	Low
possibly Alm-in	—	PG	
(An_{17} + Hbl)-in	—	BA	Low
Marg + Qz + AS + Plag	(8)	Marl	Low/medium
Stt-in; possibly Crd-in, (Chl + Mus)-out	(6)	PG	Medium
Srp + Ta + Fo	(7)	UM	Medium
Zo + Gr + An + Cc + Qz	(10)	Marl	Medium
(Stt + Mus)-out	(11)	PG	Medium
Begin anatexis; (K-spar + Sil)-in	—	PG	High
Fo + Tr + Cc + Do[a]	(12)	Dol	High
Ta + Fo + Anthophyl	(14)	UM	High

[a]May be interchanged with previous reaction.

Sequence at a Pressure of 7 kb

As an example, the sequence of reaction-isograds at 7 kb is given in Table 15-4. Appropriate compositions of pelitic rocks will contain kyanite. Approximately at the medium—high-grade boundary sillimanite crystallizes either as an additional constituent or as a transformation product of kyanite. In this high pressure series the metamorphic grades are low grade,[almandine]-low-grade, [almandine]-medium-grade, and [almandine]-high-grade or [cordierite-almandine]-high-grade, if the pressure is about 7 kb or somewhat lower.

At *pressures higher than 7 kb,* the sequence will be very similar, only the following difference has to be taken into account:

Instead of reaction (8), the assemblage zoisite/clinozoisite + kyanite occurs at pressures greater than 9 kb; it appears close to the medium—high-grade boundary.

Table 15-4. Pressure about 7 kb

Paragenesis	Curve	Rock	Grade
Srp + Qz + Ta	(1)	UM	Very-low/low
(Bio + Mus)-in	(4)	PG	Low
Srp + Bru + Fo	(5)	UM	Low
Hornbl-in; }	—	BA	Low
Alm-in }	—	PG	
(An_{17} + Hbl)-in	—	BA	Low
Stt-in; (Chl + Mus)-out	(6)	PG	Medium
Srp + Fo + Ta }	(7)	UM	Medium
Marg + Qz + AS + Plag }	(8)	Marl	Medium
Begin anatexis; (K-spar + Sil)-in	—	PG	High
Zo + Gr + An + Cc + Qz }	(10)	Marl	High
Ta + Fo + Anthophyl }	(14)	UM	High
Fo + Tr + Cc + Do	(12)	Dol	High

Indicators of Very High Pressure

Apart from the high pressure indicators summarized in the section on very-low-grade metamorphism, the paragonite + zoisite + quartz paragenesis (p. 151), and the zoisite/clinozoisite + kyanite assemblage [indicating a pressure greater than 9 kb (p. 148)], the pair kyanite + talc has been shown to be a reliable indicator for H_2O pressures exceeding 10 kb (Schreyer and Seifert, 1969). Recently Schreyer (1974) pointed out that this rather uncommon association has been found in a few localities. It may well be that in the future more localities will become known where the kyanite + talc association has been formed in metasedimentary schists that are low in alkalies and calcium but rich in MgO and Al_2O_3.

General Comment

Metamorphic rocks now exposed at the surface may have been brought to their present position by different tectonic mechanisms. They may have been uplifted as separate crustal blocks of various sizes and from different depths. In this case, no large scale metamorphic zonation or clear-cut isograd sequence will be detectable. On the other hand, a large metamorhpic terrain may have been uplifted as a coherent unit without having been disrupted into smaller blocks. In these cases, we have to distinguish between an even uplift, a tilted uplift, and a flexural uplift.

If an even uplift has taken place the rocks now exposed at the earth's surface have been metamorphosed at the same depth. If large tilting has taken place, the rocks presently exposed have been metamorphosed at progressively greater depth depending on the amount of tilting. Postmetamorphic tilting accompanied by flexural uplift of a large crustal segment has been deduced from a study of metamorphic conditions in the Damara orogen in southwestern Africa by Hoffer and Puhan (1978, personal communication). Indeed, uneven postmetamorphic uplift seems to be common. Therefore, we must not assume *a priori* that a metamorphic terrain exposes rocks that have been formed at approximately the same pressure.

References

Akella, J. and Winkler, H. G. F. 1966. *Contr. Mineral. Petrol.* **12:** 1–12.

Althaus, E. 1967. *Contr. Mineral. Petrol.* **16:** 29–44.

Choudhuri, A. and Winkler, H. G. F. 1967. *Contr. Mineral. Petrol.* **14:** 239–315.

Hirschberg, A. and Winkler, H. G. F. 1968. *Contr. Mineral. Petrol.* **18:** 17–42.

Richardson, S. W., Bell, P. M., and Gilbert, M. C. 1968. *Am. J. Sci.* **266:** 513–541.

Richardson, S. W, Gilbert, M. C. and Bell, P. M. 1969. *Am. J. Sci.* **267:** 259–272.

Schreyer, W. 1974. *Geol. Rundschau* **63:** 597–609.

——— and Seifert, F. 1969. *Am. J. Sci.* **267:** 371–388.

Turner, F. J. 1968. *Metamorphic Petrology,* McGraw-Hill Book Company, New York.

Wenk, C. and Keller, F. 1969. *Schweiz. Mineral. Petrog. Mitt.* **48:** 455–457.

Winkler, H. G. F. 1967. *Petrogenesis of Metamorphic Rocks.* 2nd edit. Springer-Verlag, New York-Berlin.

Chapter 16

Regional Hypersthene Zone (Granolite High-Grade)

As pointed out in Chapter 7 (p. 88f), a distinction must be made in high-grade metamorphism between two situations: (1) water pressure being approximately equal to load pressure, or (2) considerably less than load pressure. In the latter case and at medium to high load pressure, rocks called granulites are formed (the granulite facies is developed). Since hypersthene is the most diagnostic mineral in both mafic and acidic granulites, it is suggested that granulite high-grade metamorphism be designated as the "regional hypersthene zone." This is in contrast to the low pressure, *contact metamorphic* high-grade zone, where orthopyroxene is also present in rocks of appropriate composition.

Before dealing with the mineral parageneses which are formed under the conditions of the regional hypersthene zone, it is desirable first to consider the problem of nomenclature of granulites and other high-grade rocks occurring with granulites.

Nomenclature and Mineralogical Features of "Granulites"

There is considerable confusion in the use of the term "granulite." For instance, the loose definition consistent with that successfully employed in Canadian reconnaissance geology is: "Granulites may be regarded as high-grade metamorphic rocks that characteristically contain pyroxene and that exhibit distinctive olive-green or brown colors on weathering" (Reinhardt and Skippen, 1970). A very different meaning of the term granulite has been proposed by Macdonald (1969): "Any rock which is of the granulite facies grade, as defined by Eskola, or which retains evidence of having reached this grade." These different definitions reflect only some of the many aspects of these metamorphic rocks.

In an attempt to devise a definition acceptable to petrologists in many countries, in 1968 an ad hoc international group proposed a definition and characterization of granulites; this was sent to many geologists for discussion. The definition and comments were published by Behr *et al.* (1971). The discussion continued in print and by correspondence. Mehnert (1972) undertook the task of analyzing the accumulated information. The following revised definition of the term granulite, as presented by Mehnert, was the result:

> Granulite is a fine- to medium-grained metamorphic rock composed essentially of feldspar with or without quartz. Ferromagnesian minerals are predominantly anhydrous. The texture is mainly granoblastic (granuloblastic), the structure is gneissose to massive. Some granulites contain lenticular grains or lenticular aggregates of quartz ("disc-like quartz"). Granulite is the type rock of the granulite facies. The composition of the minerals correspond to granulite facies conditions. The following rock types should *not* be included in the definition of "granulite": Medium- to coarse-grained rocks (> 3 mm) of corresponding composition and origin should be termed *granofels*. Granulites being rich in ferromagnesian minerals (> 30 vol%) should be termed *pyriclasites, pyribolites, or pyrigarnites,* depending on their composition.

However, Winkler and Sen (1973) felt that this new proposal is unlikely to improve matters; therefore, they advanced another system of nomenclature based on the following premises:

1. A rock should be named on the basis of its own properties and not according to its association with other rocks.
2. The mineral assemblage is the most important criterion in the nomenclature of rocks and should, in the case of metamorphic rocks, be used together with a supplementary textural and, if necessary, structural term. In the resulting *compound rock names,* one part refers to the mineral assemblage and the other part to texture and some additional information. (Modal composition may be the basis for a more detailed classification.)
3. Mehnert's statement "granulite is composed essentially of feldspar" is modified to "feldspar(s) are major constituents (about 20% or more)."
4. We agree with Mehnert: "Granulite is the type rock of the granulite facies"; and "granulite should refer only to rocks occurring within granulite facies terrain" (see Oliver and Wynne-Edwards in Behr *et al.,* 1971) but, of course, not all rocks in the granulite facies will be granulites. Thus, the term granulite bears a distinct genetic connotation. This may be unfortunate but is unavoidable. Therefore, taking into account point 1, the following restriction is essential to ensure that all rocks termed granulites belong

only to the "granulite facies": Only metamorphic rocks with mineral assemblages diagnostic of that facies, of the regional hypersthene zone, should be called granulites.

Point 4 is fundamental to the new proposal. In order to minimize possible confusion, it is preferable to coin and define a new term rather than redefine an old one. By simply changing the spelling from "granulite" to "granolite" a new term is suggested which closely shows its immediate connection to the subject discussed here. *Granolite* is not a misprint, it is meant to have an *o!* By making this change, the mental association of the old term granulite (being derived from granulum = small grain) with fine-grained rocks no longer exists. The new term granolite does not put a restriction on grain size; this is an important aspect of the new definition.

Since its introduction in 1973, the term granolite has occasionally been used in the literature; however, it has also been criticized as being "awkward". I do not agree with this because "grano" (not "granu") is a common prefix in petrographic terms such as granodiorite, granophyre, and granoblastic.

5. All high-grade regional metamorphic rocks similar to granolites but lacking a diagnostic mineral assemblage must, consequently, not be called granolite. Because of the granoblastic textures which they have in common with the granolites, Winkler and Sen (1973) suggested calling high-grade rocks with mineral assemblages *not diagnostic* of the regional hypersthene zone (granulite facies) by the new name *Granoblastite*.
6. Granoblastite and Granolite are group names of high-grade regional metamorphic rocks. Any specific rock will have a compound name, *e.g.*,

   X or Y granoblastite
   A or B granolite

   where X and Y stand for a mineral assemblage *not* diagnostic of the regional hypersthene zone, and A and B stand for a paragenesis diagnostic of that zone.
7. Grain size, distinction between felsic and mafic composition, and amount of mafic minerals are not regarded as significant enough to preclude the term as a *group name*. Compositional differences are indicated by stating the mineral assemblage or the modal composition (see Figure 16-2).

Table 16-1

GRANULITE ROCKS REDEFINED

1. Felsic to mafic *high-grade regional metamorphic* rocks of any grain size. Excluded are ultramafic rocks, calcsilicate rocks and marbles, amphibolites, and biotite micashists.
2a. Feldspar(s) are major constituents (about 20% or more)
2b. No primary muscovite associated with quartz and feldspar.
3. Mafic minerals are often predominantly anhydrous.
4. Crystalloblastic, mainly granoblastic textures and their varieties. Structural distinctions 4a and 4b are made.

	(4a) Massive, no gneissose structure	(4b) With gneissose structure
Mineral assemblages *diagnostic* of the regional hypersthene zone.	A, B, or C Granolite[a]	Gneissose A, B, or C Granolite[a]
Mineral assemblages *not diagnostic* of the regional hypersthene zone.	X, Y, or Z Granoblastite	Gneissose X, Y, or Z Granoblastite

[a]See Figure 16-2 for hypersthene-bearing granolites.

A, B, and C stand for a definite mineral assemblage diagnostic of the regional hypersthene zone; X, Y, or Z, for a mineral assemblage which is not diagnostic.

The X, Y, or Z granoblastites commonly form in *high-grade* metamorphism in granolite terrains but occasionally also in [almadine]-high-grade metamorphism.

Table 16-1 summarizes the new proposal for a nomenclature of granulite rocks. All four points listed are important.

In order to apply the proposed terminology it is necessary to recognize the diagnostic mineral assemblages of the regional hypersthene zone. Also, it is expedient to draw attention to some typical mineralogical features that are common in granolites and in most granoblastites. This has been compiled in Table 16-2.

Textures of granolites and granoblastites. The description of the textures of granolites and related rocks has been presented in many different ways. In an effort to standardize the terminology, Moore (1970) proposed a scheme which apparently is acceptable to many petrologists and, therefore, is quoted here:

> The texture of the rock here refers to the mutual arrangement of the constituent minerals in the rock and to their size and shape.
>
> The textures are divided into three main groups (Figure 16-1), which are described below.

1. *Granoblastic:* Here the rock consists of a mosaic of xenoblastic mineral grains. Depending on the relative sizes of these grains the group is further divided into three subgroups:
   (a) *Equigranular:* where the majority of constituent grains are approximately the same size.
   (b) *Inequigranular:* where the frequency distribution of grain sizes is distinctly bimodal such that large grains of approximately the same size occur in a finer grained equigranular matrix. A special case would be *platy granoblastic* (which is characteristic of many quartzofeldspathic granulites) where long, platelike crystals, generally of quartz or feldspar, occur in a finer grained, equigranular matrix. These lenticles may be single crystals or, more often, groups of several crystals with lattice orientations differing only slightly (*i.e.*, separated by low-angle grain boundaries).
   (c) *Seriate:* where the grain size frequency distribution is such that a complete gradation from the finest to the coarsest is represented.
2. *Flaser:* This applies to rocks in which lenses and ribbonlike crystals, usually of quartz, often showing undulose extinction are separated by bands of finely crystalline, usually strain-free, material. Ovoid crystals, commonly feldspar or mafic minerals, may occur within the matrix as augenlike megacrysts around which the foliation bends. Although this may be regarded as an extreme version of the platy

Table 16-2

DIAGNOSTIC MINERAL ASSEMBLAGES OF THE REGIONAL HYPERSTHENE ZONE

(a) All parageneses with hypersthene
(b) The paragenesis clinopyroxene + almandine-rich garnet + quartz.

MINERALOGICAL CHARACTERISTICS[a]	
Alkali feldspar	Is typically perithitic; the Na content is commonly high so that the albite component may amount to about 50%.
Plagioclase	May be antiperthitic.
Orthopyroxene	Is mainly hypersthene and contains up to 10% Al_2O_3; it is pleochroic in greenish, pinkish, and yellowish colors.
Clinopyroxene	Is light-green diopside-hedenbergite low in ferric iron. Some clinopyroxenes may be Al-rich.
Garnet	Is usually almandine-rich but may contain appreciable amounts of pyrope and grossularite component, *i.e.*, up to 55% pyrope and up to 20% grossularite. Commonly, the amount of spessartine component is very low. The grossularite content is highest in garnet coexisting with clinopyroxene.
Hornblende	Is olive-green to brown due to a relatively high content of Ti.
Biotite	Usually is rich in Mg and Ti.
Al_2SiO_5	Occurs as sillimanite or kyanite.

[a]For more details, see Mehnert (1972, p. 148).

granoblastic texture, it is separated because rocks having flaser textures usually have a very strong lamination and lineation, and the length-to-width ratios of the "ribbons" are large, giving the rock a distinctive appearance.

3. *Mylonitic:* This term has the strongest genetic implications of the three and is applied to rocks which, in appearance, conform as nearly as possible to mylonites as defined by Lapworth (1885) and Christie (1960). The rocks are very fine-grained and have a marked lamination ("fluxion structure"). Small megacrysts may occur within the rock and the lamination bends around these.

For all three texture types described above in terms of grain size and over-all shape, the mutual relationship of the grains requires description. Following Berthelsen (1960) and Katz (1968), the grain boundary relationships are described as polygonal, interlobate, and amoeboid. These are shown diagrammatically in Figure 16-1.

Examples of the use of this proposed terminology may be given here. A rock showing *granuloblastic* texture (Binns, 1964) would, using this proposed scheme,

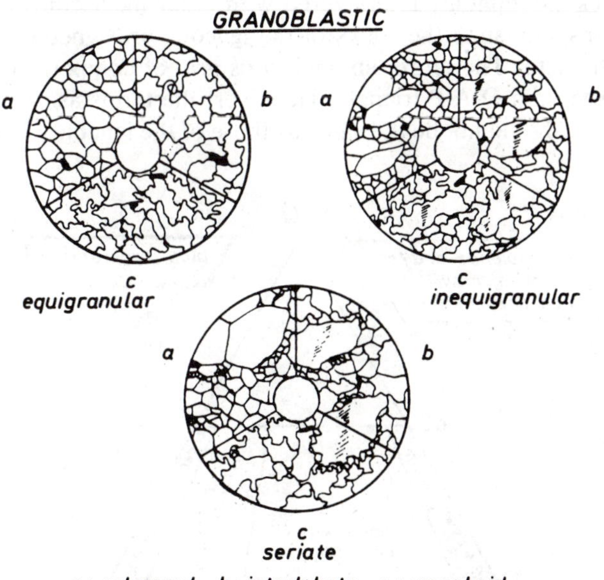

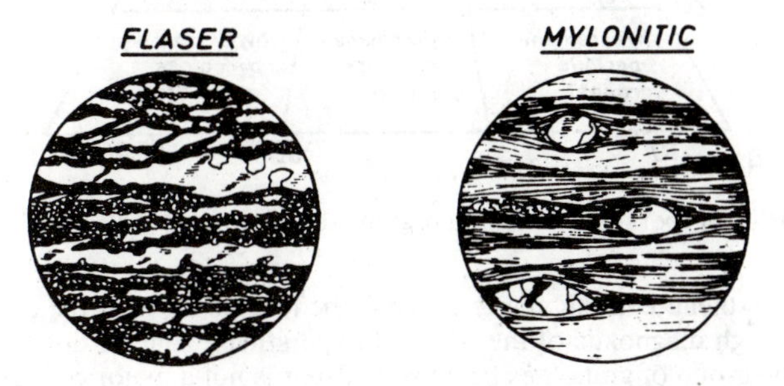

Fig. 16-1 Textures of granolites and granoblastites (after Moore, 1970).

be described as having a granoblastic equigranular polygonal texture. The mortar gneiss and augen gneiss illustrated by Katz (1968) would both be described as having granoblastic inequigranular interlobate textures. The only difference between these rocks is in the proportion of finer grained matrix.

Nomenclature of hypersthene-bearing granolites. Massive, gneiss-ose (foliated), flaser, and mylonitic varieties of granolites with hypersthene, plagioclase, and/or perthitic alkalifeldspar, with or without quartz, may be classified according to a scheme of nomenclature which includes the composition range of common felsic to mafic granolites. The scheme is presented in Figure 16-2. The Streckeisen classification triangle of quartz (Q), alkali feldspar (A) and plagioclase (P) has proved its usefulness in classifying the igneous granite-gabbro and charnockite-norite suites of rocks. It is well known that *metamorphic* hypersthene-bearing rocks, commonly interlayered with other metasediments, have compositions similar to the rocks of the igneous or "igneous looking" charnockite suite. Therefore, an analogous grid of compositional areas can be drawn in a Q-A-P triangle. However, the terms assigned to the various fields should not be the same as those of the igneous charnockite

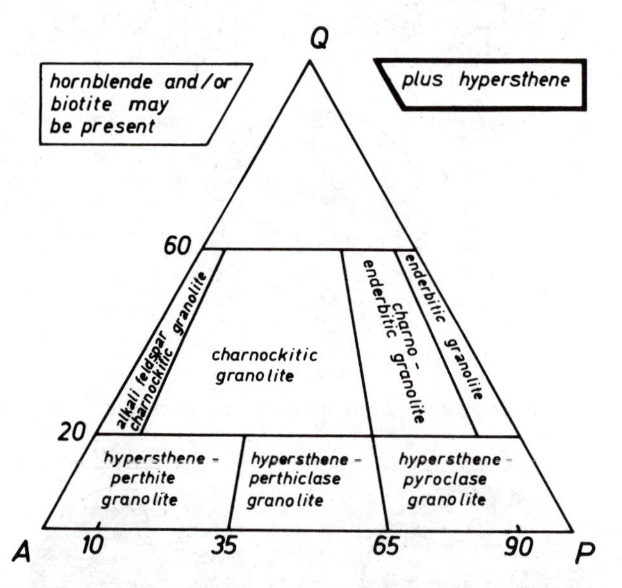

Fig. 16-2 Hypersthene-bearing regional metamorphic rocks of the granolite group.

Note: Hypersthene-bearing quartzite would be located close to the corner at Q. Although diagnostic of the regional hypersthene zone, it does not belong to the group of granolites because feldspar is not a major constituent.

suite of rocks in order to indicate that these are metamorphic rocks of the granolite group. All rocks of Figure 16-2 are either massive granolites or gneissose, flaser, and mylonitic granolites.

Apart from hypersthene, the fields are named according to the predominant mineral assemblage. Hypersthene is present in all assemblages. Thus, when perthite (alkali feldspar) is predominant and plagioclase and quartz are present in subordinate amounts, the rocks will be classified as hypersthene-perthite granolite. A rock with hypersthene + clinopyroxene + plagioclase ± quartz falls into the field of hypersthene-pyroclase granolite; this term has been abbreviated by a contraction of clino*pyro*xene + plagio*clase* = pyroclase. The field of hypersthene-pyroclase granolites includes the rock called "pyriclasite" by Berthelsen (1960). If clinopyroxene is absent no abbreviation is necessary and the rock should be called "hypersthene-plagioclase granolite."

The field intermediate between hypersthene-perthite granolite and hypersthene-pyroclase granolite is named hypersthene-perthiclase granolite; here perthite and plagioclase are present in about equal amounts, and hypersthene commonly is the only pyroxene. The contraction perthiclase stands for the assemblage *perthi*te + plagio*clase*.

In the composition fields with appreciable amounts of quartz ($>$ 20%) the terms charnockitic and enderbitic designate mineral assemblages and modal compositions. Thus charnockitic granolite is a *metamorphic rock* with the same mineral assemblage as igneous charnockite.[1] The "granolite" part of the rock name makes clear the metamorphic origin and the "charnockitic" part indicates only the mineral assemblage and modal composition. If the rock is gneissose, it may be called "gneissose charnockitic granolite"; this is a long term but it is easy to comprehend.

In Figure 16-2, from left to right, the following hypersthene- and quartz-bearing granolites are distinguished:

Alkalifeldspar charnockitic granolite
Charnockitic granolite
Charno-enderbitic granolite
Enderbitic granolite
and their gneissose, mylonitic, or flaser varieties.

If hornblende and/or biotite is additionally present in appreciable amounts, this should be stated as "hornblende-bearing," etc.

[1]Some petrographers state that a distinction can be made between metamorphic and igneous rocks on the basis of texture. Other workers regard this to be often impossible and rely on observations, like interbedding with other metasediments, or intrusive contacts.

Figure 16-2 is applicable to metamorphic rocks only. Igneous and "igneous-looking" rocks of the charnockite suite have their own (different) nomenclature. The most recent nomenclature recommendation has been published by Streckeisen (1974).

Metamorphism of Granolites and Related Granoblastites

Calcareous rocks occurring in granolite terrains have developed the same parageneses as in high-grade terrains where they are associated with biotite gneisses, migmatites, high-grade amphibolites, etc. Parageneses like calcite + dolomite + forsterite + diopside, diopside + scapolite + calcite, diopside + grossularite-andradite are not affected by the specific condition of granolite metamorphism, *i.e.*, water pressure is considerably less than load pressure. This is easy to understand because the above assemblages can be formed, whether in a granolite or other high-grade terrain, only in the presence of a very CO_2-rich fluid phase.

The specific condition of $P_{H_2O} << P_{total}$ has a large effect only on reactions including the transformation of hydroxyl-bearing minerals into anhydrous ones. Reactions in which hornblende and/or biotite are completely or partly converted into anhydrous minerals like pyroxenes and garnet would require very high temperatures at medium and high pressure if P_{total} would equal P_{H_2O}. As seen in Figure 15-3, the lowest possible reaction of orthoamphibole plus quartz to form orthopyroxene takes place at about 700°C and 1 kb $P_{H_2O} = P_{total}$, but at 3 and 5 kb the equilibrium temperature is 800° and 900°C, respectively; these temperatures are not reached in regional metamorphism. On the other hand, lowering P_{H_2O} relative to P_{total} decreases the reaction temperature at constant P_{total}. This allows, for instance, the partial or complete breakdown of hornblende in amphibolites to ortho- and clinopyroxene. Thus, a hypersthene-bearing mafic granolite is formed, *i.e.*, a hypersthene-pyroclase granolite, consisting mainly of hypersthene + clinopyroxene + plagioclase.

Hornblende and biotite are the hydroxyl-bearing minerals involved in reactions leading to formation of granolites and granoblastites; other (OH)-bearing minerals, such as chlorite and paragonite, become unstable even before high-grade conditions are reached. The breakdown of muscovite + quartz defines the medium-grade—high-grade boundary and muscovite is therefore absent from granolite terrains.

A number of reactions have been suggested by De Waard (1965; see also 1967a, 1971) to account for the diagnostic appearance of orthopyroxene (mostly hypersthene, rarely bronzite), the formation of additional almandine-rich garnet, and the decrease or disappearance of hornblende and biotite.

In mafic rocks these are:

$$\{\text{hornblende} + \text{quartz}\} = \{\text{hypersthene} + \text{clinopyroxene} + \text{plagioclase} + H_2O\}$$

$$\{\text{hornblende} + \text{biotite} + \text{quartz}\} = \{\text{hypersthene} + \text{K feldspar} + \text{plagioclase} + H_2O\}$$

$$\{\text{hornblende} + \text{almandine} + \text{quartz}\} = \{\text{hypersthene} + \text{plagioclase} + H_2O\}$$

In pelitic gneisses the following reactions occur:

$$\{\text{biotite} + \text{sillimanite} + \text{quartz}\} = \{\text{almandine} + \text{K feldspar} + H_2O\}$$

$$\{\text{biotite} + \text{quartz}\} = \{\text{hypersthene} + \text{almandine} + \text{K feldspar} + H_2O\}$$

Hypersthene is produced by several reactions, and it may occur in many of the common rock types but certainly not in all rocks of granolite grade. For example, the felsic type "granulite" from Saxony contains garnet + kyanite but not hypersthene and is not diagnostic[2] of the "granulite facies" or the regional hypersthene zone. However, in regional metamorphosed terrains, hypersthene is the diagnostic mineral of the regional hypersthene zone. Therefore, in agreement with De Waard (1965), the best boundary between common high-grade metamorphism (high-grade almandine-amphibolite facies) and granolite metamorphism is the hypersthene isograd, based upon the first appearance of hypersthene (or bronzite), regardless of rock type and the nature of the hypersthene-producing reaction. If a specific reaction can be recognized, a specific reaction-isograd can be mapped.

All reactions involving hornblende and/or biotite are "continuous" reactions or, in other words, with a "sliding" equilibrium, *i.e.*, at a given total pressure and water pressure, the reaction starts at a certain temperature but all equilibrium phases will be present over a temperature range. The Fe/Mg ratios of the ferromagnesian phases will vary with temperature. Thus, although orthopyroxene will be produced from hornblende, the reaction need not proceed to completion, and decreasing amounts of hornblende will coexist with the newly formed hypersthene until, with rising grade, all hornblende has disappeared. Therefore, De

[2]Therefore, this nondiagnostic "granulite" is called here garnet-kyanite-quartz *granoblastite*.

Waard (1965) made the distinction between (a) hornblende-orthopyroxene-plagioclase subfacies of the granulite facies, and (b) orthopyroxene-plagioclase subfacies (free from hornblende and biotite).

The hornblende-free assemblage is an end stage which is stable when a certain temperature at some given water pressure has been reached. Therefore, the boundaries (which also depend on bulk composition) will be very difficult to assess. However, the hornblende-free assemblage may be interpreted as lying in a prograde direction.

Mineral assemblages in mafic rocks. Mineral assemblages formed in mafic rocks (but not only in these rocks) are diagnostic of granolite-forming conditions of the regional hypersthene zone. The diagnostic mineral assemblages developed in the regional hypersthene zone may be represented in an ACF diagram. Figure 16-3 shows the assemblages of the *hypersthene-plagioclase-granolite subzone*. The possible presence of hornblende, biotite, and rarely zoisite/clinozoisite need not be graphically recorded, because these minerals are not diagnostic; however, they should be stated in the petrographic description.

According to petrographic observations of De Waard (1965), the assemblage orthopyroxene + plagioclase becomes unstable if load pressure exceeds a particular value at constant temperature. The two minerals react to produce clinopyroxene + almandine-pyrope-grossularite garnet + quartz:

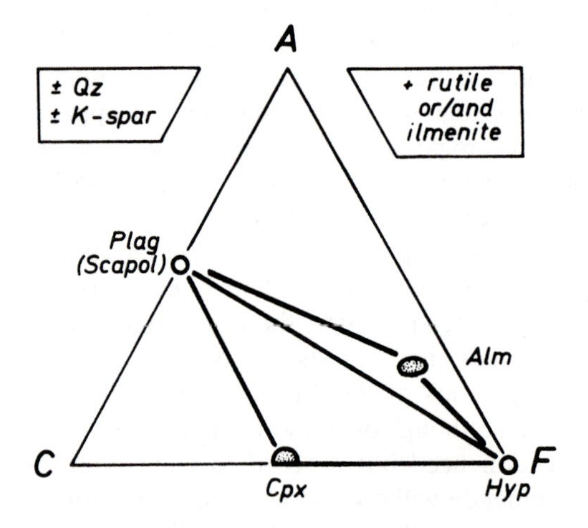

Fig. 16-3 Hypersthene-plagioclase-granolite subzone of the regional hypersthene zone of high-grade metamorphism. Only parageneses of mafic rocks are represented. Hornblende and/or biotite may also be present.

{hypersthene + plagioclase} =
{clinopyroxene + garnet + quartz}

$$3(Fe,Mg)_2Si_2O_6 + 2CaAl_2Si_2O_8 =$$
$$Ca(Fe,Mg)Si_2O_6 + 2Ca_{0.5}(Fe,Mg)_{2.5}Al_2(SiO_4)_3 + SiO_2$$

Again, this reaction is continuous over a certain range of physical conditions and, during a transitional stage, reactants and products coexist. Eventually, the new mineral assemblages

clinopyroxene + garnet + quartz + either plagioclase or hypersthene,

shown in Figure 16-4, become stable. They represent the higher pressure *clinopyroxene-almandine-quartz granolite subzone* of the regional hypersthene zone. In rocks of the transitional stage, hornblende and biotite may also be present. The paragenesis hypersthene + garnet + diopside + plagioclase ± hornblende ± biotite + quartz has been recorded from several different areas (Turner, 1968).

It is worth noting that in mafic rocks the grossularite content of almandine-rich garnets is consistently high, amounting up to about 20 molecular percent. On the other hand, almandine-rich garnets in pelitic metamorphic rocks contain very little grossularite component.

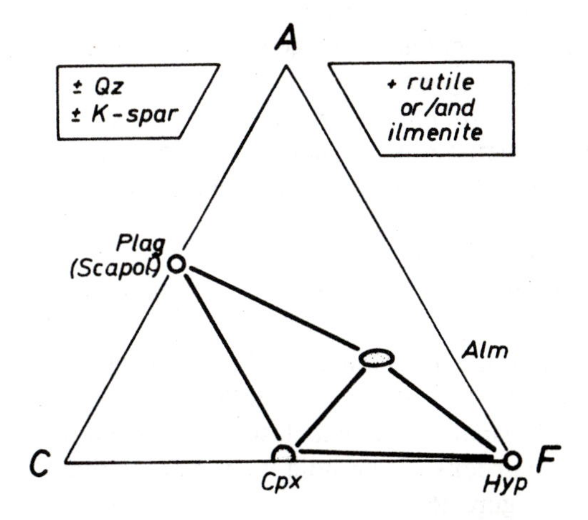

Fig. 16-4 Clinopyroxene-almandine-quartz-granolite subzone of the regional hypersthene zone. Only parageneses in mafic rocks are represented. Hornblende and/or biotite may also be present. Because the join clinopyroxene-garnet-quartz is produced by a continuous reaction, the assemblage including reactants and products is common, *i.e.*, clinopyroxene + garnet + hypersthene + plagioclase + quartz.

The reaction

{orthopyroxene + plagioclase} =
{clinopyroxene + garnet + quartz}

has been investigated experimentally (*e.g.*, Green and Ringwood, 1967). At 700°C, a pressure of 8 to 10 kb will stabilize the five-phase assemblage clinopyroxene + garnet + quartz + hypersthene + plagioclase. The association clinopyroxene + garnet is known in eclogites as well but in eclogites plagioclase is absent, indicating pressures even greater than 8 to 10 kb.

When interpreting a given granolite terrain it should be taken into account that the association clinopyroxene + almandine-rich garnet, relative to the orthopyroxene + plagioclase assemblage, does not necessarily indicate a higher pressure at constant temperature, but could alternatively represent a lower temperature at constant pressure. Thus, the "retrograde" formation of the assemblage Ga + Cpx + Qz, postulated by Martignole and Schrijver (1971), can be understood. Further, Manna and Sen (1974) made the observation that mafic granolites with orthopyroxene + clinopyroxene + plagioclase are intimately associated with granolites containing orthopyroxene + clinopyroxene + garnet + plagioclase; however, the latter rocks have a higher Fe/Mg ratio. This suggests that, at a given temperature, the pressure necessary to initiate the clinopyroxene + garnet assemblage is somewhat lower when the Fe/Mg ratio of the rock is higher.

Mineral assemblages in pelitic rocks and graywackes. In the regional hypersthene zone, hypersthene may also form in pelitic and semipelitic rocks due to the partial decomposition of biotite according to the reaction

{biotite + quartz} = {hypersthene + almandine + K feldspar}

or possibly according to a more complicated reaction. Rocks with hypersthene, even if the amount is very small, are classified as granolites. If quartz, alkali feldspar, and plagioclase in variable amounts are the major constituents, the rocks are granolites of charnockitic to enderbitic compositions (see Figure 16-2).

However, many semipelitic rocks in a granolite terrain do not contain either hypersthene or the association clinopyroxene + garnet + quartz, which is diagnostic as well. A large variety of mineral assemblages are formed at granolite metamorphic conditions which are not diagnostic of these conditions. These rocks, according to our terminology, are called granoblastites.

Reinhardt and Skippen (1970) have presented a tentative interpretation of phase relations in granolites and granoblastites, all containing quartz, alkali feldspar (abbreviated K feldspar), and plagioclase as major constituents, from the Westport area in the Grenville Province (Figure 16-5). The observed mineral assemblages are summarized in the form of an extended AFM diagram previously used by Reinhardt. Various mineral assemblages are represented by a single mineral, two minerals connected by a tie line, or three minerals, shown in Figure 16-5, together with those minerals common to all assemblages and indicated at the upper right of the figure.

Three-mineral assemblages of granolites proper are

$$\text{Hyp} + \text{Alm} + \text{Bio} \quad (3)$$
$$\text{Hyp} + \text{Hbl} + \text{Bio} \quad (4)$$
$$\text{Hyp} + \text{Cpx} + \text{Alm} \quad (5)$$
$$\text{Hyp} + \text{Cpx} + \text{Hbl} \quad (6)$$

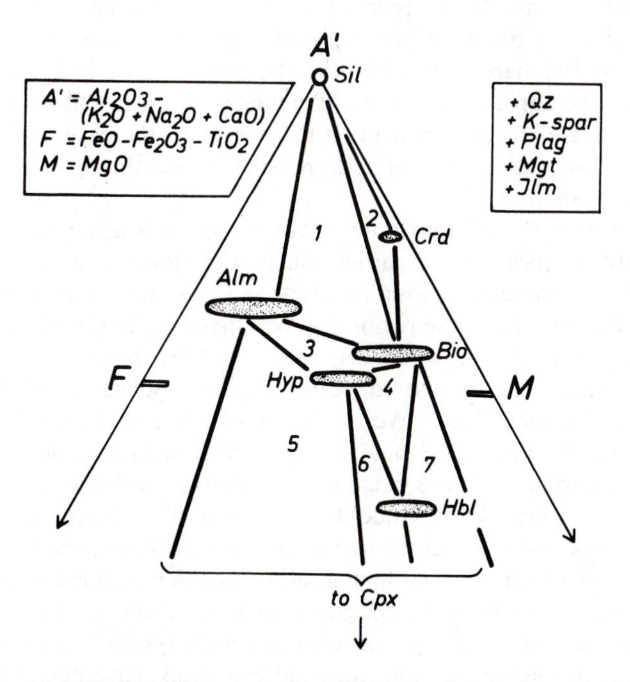

Fig. 16-5 Tentative interpretation of phase relations in gneissic granolites and granoblastites from the Westport area, Grenville Province; valid for constant P_{H_2O}, P_{total}, and T. (After Reinhardt and Skippen, 1970). Cpx = clinopyroxene. Plag stands for plagioclase of nearly constant composition.

Three-mineral assemblages of granoblastites are

$$\text{Sil} + \text{Alm} + \text{Bio} \quad (1)$$
$$\text{Sil} + \text{Crd} + \text{Bio} \quad (2)$$
$$\text{Cpx} + \text{Hbl} + \text{Bio} \quad (7)$$

Kyanite, instead of sillimanite, may be present in other terrains.

The exact reactions producing the various assemblages are not known. However, Figure 16-5 might give some indication of possible reactions among phases. This provides valuable guidance in petrographic studies.

The coexistence of almandine and cordierite in pelitic rocks has been recorded in several granolite terrains. However, the occurrence of cordierite without almandine is rare (Turner, 1968) being restricted to rocks having a very low FeO/MgO ratio (Figure 16-5).

There is another point which deserves attention: It is commonly observed that biotite decreases in amount and obviously contributes to the formation of anhydrous ferromagnesian minerals. It might be expected that eventually the reaction would go to completion leading to the elimination of biotite in the regional hypersthene zone. However, according to Reinhardt and Skippen, the assemblages in Figure 16-5 emphasize "the apparent necessity of having biotite as a stable associate in granulite facies metamorphism. If this were not the case, we should expect the association sillimanite-hypersthene and this appears to be extremely uncommon."

The coexistence of sillimanite and hypersthene is also precluded by the assemblage cordierite + garnet which is common in granoblastites and in some granolites. About 15 years ago, petrographers found the presence of cordierite in the regional hypersthene zone (granulite facies) puzzling because cordierite was regarded as a "low pressure mineral." This has since been shown to be wrong; see Schreyer (1965) and Schreyer and Yoder (1964). We know from further experimental work that almandine + cordierite may coexist at high metamorphic temperatures over a certain pressure range which increases with increasing temperature (see Figure 14-12). Undoubtedly, cordierite and almandine-rich garnet may coexist in rocks formed at conditions of the regional hypersthene zone as well as in other high-grade and even medium-grade rocks, *i.e.*, [cordierite-almandine]-high-grade and -medium grade. Various parageneses in high-grade pelitic gneisses with typically granoblastic texture, *i.e.*, gneissose granoblastites and granolites, have been reported by Reinhardt (1968). He distinguished two sets of parageneses which are related by the following "discontinuous" reaction:

{almandine-rich garnet + biotite} =
{cordierite + hypersthene}

(This reaction does not cause the first appearance of hypersthene.)

Figures 16-6 and 16-7 have been schematically drawn from Reinhardt's paper. The parageneses are:

In Figure 16-6:

Sil + Alm + Crd	(1)
Alm + Crd + Bio	(2)
Alm + Bio + Hyp	(3)

In Figure 16-7:

Sil + Alm + Crd	(1)
Alm + Crd + Hyp	(4)
Crd + Hyp + Bio	(5)

Petrogenetic Considerations

As emphasized in the first edition of this book (1965) the essential condition of granolite formation is $P_{H_2O} << P_{total}$. This has been widely accepted and supported by further work by De Waard (1967a), Althaus (1968), Weisbrod (1970), Newton (1972), and Chatterjee and Schreyer

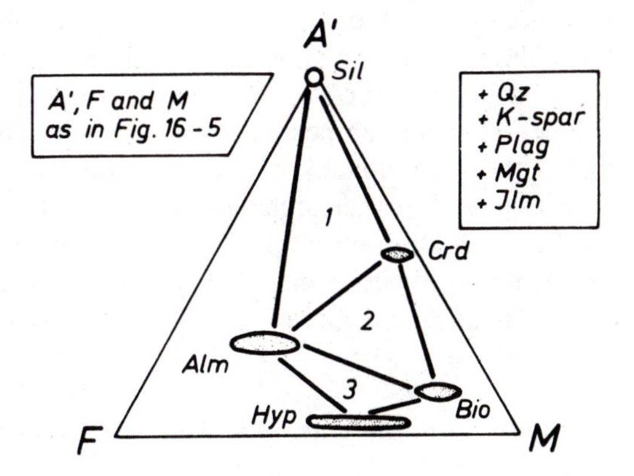

Fig. 16-6 Cordierite-almandine-bearing granoblastites and granolites and their associated rocks (after Reinhardt, 1968).

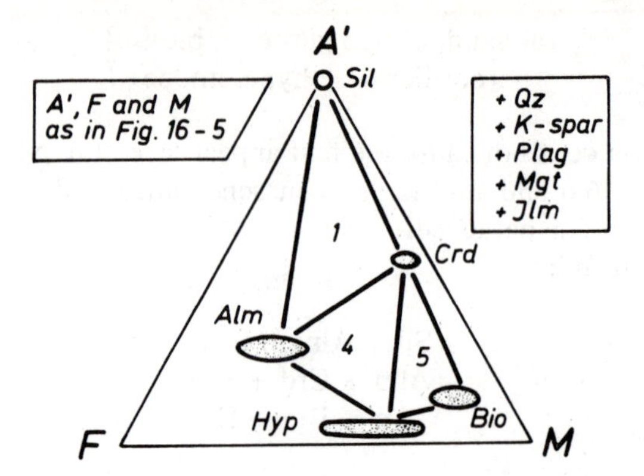

Fig. 16-7 Cordierite-almandine-bearing granoblastites and granolites and their associated rocks at a prograde stage compared to Figure 16-6 (after Reinhardt, 1968).

(1972). The latter authors concluded that the assemblage sapphirine[3] + quartz found in some high pressure granolite terrains indicates that water pressure was very much lower than total pressure. Newton (1972) attached the same significance to the paragenesis hypersthene + sillimanite + quartz.

The presence of either sillimanite or kyanite indicates medium to high pressures. The association sapphirine + quartz indicates pressures greater than about 8 kb. The assemblage hypersthene + sillimanite + quartz is the high pressure equivalent of cordierite and is stabilized at pressures of about 10 kb (Newton, 1972).

The temperature in granolite terrains is estimated to range from 650° to 800° or 850°C. This estimate is substantiated by the work of Bohlen and Essene (1977). On the basis of two-feldspar thermometry and oxide thermometry, they deduced a temperature of formation of 650° to 800°C for granolites from the Adirondacks; this temperature determination is independent of pressure. In other areas and in a few cases only, even higher temperatures (at high pressures) are indicated by the rare occurrence of the parageneses hypersthene + sillimanite + quartz and sapphirine + quartz (Hensen and Green, 1973). These assemblages are related by the reaction

$$\text{hypersthene} + \text{sillimanite} \rightarrow \text{sapphirine} + \text{quartz}$$

[3]Sapphirine is $(Mg,Fe)_2Al_4O_6(SiO_4)$ with Fe substituting for less than 20% of Mg. Under the microscope, this mineral can be mistaken for corundum.

Although very high temperatures may have been reached occasionally, it is most significant that granolites do not require exceptionally high temperatures and pressures for their formation; the ordinary temperature and pressure range of high-grade regional metamorphism is sufficient, provided that H_2O is much lower than P_{total}.

Conditions suitable for the formation of granolites will, in general, be fulfilled when magmatic rocks are metamorphosed without the introduction of additional water. The petrographic observations of Buddington (1963) on highly metamorphosed orthogneisses provide convincing evidence in favor of the existence of such metamorphic conditions. For his area in the northwest Adirondacks, Buddington concludes: "The rocks have, in general, been impermeable either to access or to loss of H_2O during plastic flow and any subsequent recrystallization." Thus, the premetamorphic H_2O-free pyroxene-syenites have been isochemically recrystallized to orthopyroxene-clinopyroxene garnet-bearing granolites. However, under the very same set of conditions, hornblende-quartz syenite, etc., have been recrystallized to hornblende-bearing granolites because the H_2O set free by the partial breakdown of hornblende was unable to leave the system. A certain H_2O pressure was hereby set up and most of the hornblende was preserved.

Thus, the association of granolites with or without hornblende is attributed not to differences in temperature at the same pressure but rather to the metamorphism of rocks having different original water contents in closed systems. On the other hand, if some water is available from the environment, the rocks behave as a partially open system. As pointed out by Buddington, on the basis of his field observations, this situation is also of great importance in the genesis of granolites:

> If we envision a series of geosynclinal sediments with intruded sheets of gabbro downfolded to a zone of high temperature and pressure, then the outer borders of the gabbro sheets may be subjected to recrystallization under conditions of T, P_f, and P_{H_2O} adequate to produce the mafic mineral assemblages hornblende + hypersthene + (clinopyroxene) or the same plus garnet, while the inner portions were being recrystallized to a mineral assemblage characteristic of the pyroxene-granulite subfacies.

In the inner portions, water did not have access and hydroxyl-bearing minerals could not form. In the outer portions, a little water had access, moving along grain boundaries; some hornblende (and biotite) formed where water was available. Therefore, orthopyroxene may coexist with hornblende and biotite. A sufficient amount of water would convert the mineral assemblages of the regional hypersthene zone to those of [almandine]-high-grade metamorphism.

Some granolites are products of the metamorphism of pelitic and psammitic rocks. It is difficult to understand how a single period of metamorphism could give rise to rocks of such low water content. It would be necessary to assume that most water produced by progressive metamorphism was able to leave the rocks during the process of metamorphism. This could occur if the period of high-grade metamorphism lasts for an exceptionally long period of time. However, it is more likely that metasedimentary granolites represent originally high-grade rocks which were subjected to a second period of metamorphism without access of water. On the basis of his extensive geological and petrographical studies of the classical "granulites" of Saxony, Scheumann (1961) came to this conclusion:

> The original geosynclinal sediments consisted of clays with graphitic and quartzitic intercalations, in which ophiolitic (gabbroic) intrusions were also incorporated. This whole complex, caught in an orogenesis, was transformed to crystalline schists with migmatitic gneisses in the core. In a subsequent period, this presumably anatectically granitized migmatitic gneiss was subjected to high temperatures and strong shear movements as a result of which it was altered to granulite. . . . The melanocratic rocks remained either amphibolitic or were thereby changed to garnetiferous amphibolites at their border zones. In the central part of the newly set up *P-T* field, they were uniformly reconstituted to granulitic mineral facies and were squeezed between the leucocratic granulites in the form of lenses or bands.

Therefore, the conditions leading to the formation of granolites are high temperatures, very low P_{H_2O}, and high or very high P_s. The condition of $P_{H_2O} << P_s$ can be realized only in the deeper (though not necessarily deepest) parts of the crust where very little water has had access to igneous rocks and to metamorphosed sediments which had lost most of their water either during an exceptionally long period or during a previous period of high-grade metamorphism.

The condition of $P_{H_2O} << P_s$ could also be produced by a dilution of the fluid phase with CO_2. This process is suggested by the nature of fluid inclusions studied by Touret (1971). He found that fluid inclusions in quartz from [almandine]-high-grade rocks consisted essentially of water, whereas in rocks from [granolite]-high-grade (*i.e.,* from the hypersthene zone) the predominant gas species was CO_2. This explains why P_{H_2O} was very much smaller than P_{total} during granolite formation. At the same time, this unexpected discovery raises a new problem: How did the CO_2 fluid phase originate in a variety of rocks that have been metamorphosed to granolites or related granoblastites in the absence of carbonate rocks?

References

Althaus, E. 1968. *Neues Jahrb. Mineral. Monatsh.* **1968:** 289–306.
——— 1969. *Neues Jahrb. Mineral. Abhand.* **111:** 74–161.
Behr, H. J., *et al.* 1971. *Neues Jahrb. Mineral. Monatsh.* **1971:** 97–123.
Berthelsen, A. 1960. *Medd. Groenland* **123:** 1–226.
Binns, R. 1964. *J. Geol. Soc. Australia.* **11:** 283–330.
Bohlen, S. R., and Essene, E. J. 1977. *Contr. Mineral. Petrol.* **62:** 153–169.
Buddington, A. F. 1963. *Geol. Soc. Am. Bull.* **74:** 1155–1182.
Chatterjee, N. D. and Schreyer, W. 1972. *Contr. Mineral. Petrol.* **36:** 49–62.
Christie, J. M. 1960. *Trans. Edin. Geol. Soc.* **18:** 79–93.
De Waard, D. 1965. *J. Petrol.* **6:** 165–191.
——— 1967a. *J. Petrol.* **8:** 210–232.
——— 1967b. *Koninkl. Ned. Akad. Wetenschap. Ser. B* **70:** 400–410.
——— 1971. *Freiberger Forschungsh.* **C268:** 33–39.
Green, D. H. and Ringwood, A. E. 1967. *Geochim. Cosmochim. Acta* **31:** 767–833.
Hensen, B. J. and Green, D. H. 1973. *Contr. Mineral. Petrol.* **38:** 151–166.
Katz, M. 1968. *Can. J. Earth Sci.* **5:** 801–812.
Lapworth, C. 1885. *Nature* **32:** 558–559.
MacDonald, R. 1969. *22nd Intern. Geol. Congr. India.* Pt. XIII, pp. 227–249.
Manna, S. S. and Sen, S. K. 1974. *Contr. Mineral. Petrol.* **44:** 195–218.
Martignole, J., and Schrijver, K. 1971. *Canadian J. Earth Sci.* **8:** 698–704.
Mehnert, K. R. 1972. *Neues Jahrb. Mineral. Monatsh.* **1972:** 139–150.
Moore, A. C. 1970. *Lithos* **3:** 123–127.
Newton, R. C. 1972. *J. Geol.* **80:** 398–420.
Reinhardt. E. W. 1968. *Can. J. Earth Sci.* **5:** 455–482.
——— and Skippen, G. B. 1970. *Geol. Surv. Can. Rept. Activities Paper 70-1.* Pt. B, pp. 48–54.
Scheumann, K. H. 1961. *Neues Jahrb. Mineral. Abhand.* **96:** 162–171.
Schreyer, W. 1965. *Beitr. Mineral. Petrol.* **11:** 197–322.
——— and Yoder, H. S. 1964. *Neues Jahrb. Mineral. Abhand.* **101:** 271–342.
Streckeisen, A. 1974. *Géologie des Domaines Cristallins.* Soc. Géol. Belgique, pp. 349–360.
Touret, J. 1971. *Lithos* **4:** 423–436.
Turner, F. J. 1968. *Metamorphic Petrology.* McGraw-Hill Book Company, New York.
Weisbrod, A. 1970. *Compt Rend. Acad. Sci. Paris* **270:** 581–583.
Winkler, H. G. F. and Sen, S. K. 1973. *Neues Jahrb. Mineral. Monatsh.* **1973:** 393–402.

Chapter 17

Eclogites

Eclogites are rocks of gabbroic-basaltic composition and consist primarily of two minerals: a grass-green clinopyroxene called omphacite and a red or red-brown garnet. In spite of the basaltic composition, plagioclase does not occur in eclogites. Petrogenetically, this is very significant. Eskola attributed the origin of eclogites to very high pressures, and this suggestion has been substantiated by modern experimental work. Eskola also established an "eclogite facies." However, eclogite zones cannot be mapped in metamorphic terrains. Eclogites are usually encountered as bands or lenses in metamorphic complexes of different grades: in the granolite zone, [almandine]-high-grade and [almandine]-medium-grade, low-grade, and [glaucophane-lawsonite]-very-low-grade. Eclogites also occur as inclusions in kimberlites and basalts and as bands and layers in dunites and peridotites. In some localities, they appear to have been tectonically squeezed into an alien environment. Eclogites therefore occur in a wide range of environments, as has been pointed out, among others, by Smulikowski (1964).

Three compositional groups of eclogites may be distinguished. Eclogites of group A consist almost entirely of omphacite and garnet. Also present in small amounts may be quartz and either hypersthene or kyanite (never sillimanite). Rutile is a very common accessory mineral. Eclogites of group B may contain primary hornblende and zoisite (Angel, 1957; Heritsch, 1963; Binns, 1967); the hornblende is a Na-Al-rich amphibole, called barroisite or "karinthin." Eclogites of group C may carry primary glaucophane and epidote.

Omphacite is a complex clinopyroxene containing significant amounts of jadeite, diopside, hedenbergite, Tschermak component, and acmite components. The chemical compositions of these components are $NaAlSi_2O_6$, $CaMgSi_2O_6$, $CaFeSi_2O_6$, $CaAl(SiAl)O_6$, and $NaFe^{+3}Si_2O_6$. The last component generally does not exceed 15 mole percent. Some typical metal ratios in omphacites reported by Clark and Papike (1968) are:

Na/(Na + Ca) between 0.2 and 0.8
Al/(Al + Fe^{+3}) greater than 0.5

The *garnet* of eclogites is primarily a solid solution of pyrope, almandine, and grossularite components; the content of chromium may also be appreciable. Details of garnet composition in relation to the conditions of eclogite formation will be given further on.

The chemical composition of basaltic rocks can be expressed by a combination of garent, omphacite, rutile, hypersthene or kyanite, and quartz, garnet and clinopyroxene being the major constituents. The transformation of gabbro to eclogite may be schematically represented (disregarding the FeO component) as

$$\underbrace{(NaAlSi_3O_8 + CaAl_2Si_2O_8)}_{\text{labradorite}} + \underbrace{CaMgSi_2O_6}_{\text{diopside}} + \underbrace{Mg_2SiO_4}_{\text{olivine}} =$$
$$\underbrace{(CaMgSi_2O_6 + NaAlSi_2O_6)}_{\text{omphacite}} + \underbrace{CaMg_2Al_2Si_3O_{12}}_{\text{garnet}} + \underbrace{SiO_2}_{\text{quartz}}$$

There is no doubt that eclogites have crystallized under very high pressures. Their high specific gravity (about 3.5 g/cm^3 as compared to about 3.0 g/cm^3 for isochemical gabbros), the presence of kyanite and jadeite-bearing pyroxene, the absence of plagioclase, and the occasional appearance of diamonds in group A eclogites all indicate a high pressure origin. Diamond-bearing eclogites (so-called griquaites) are associated with peridotite nodules in kimberlite pipes and are thought to originate within the mantle of the earth. Experimental work has shown that it is possible to transform basalt into eclogite at high pressure in the absence of water.

Using various basaltic compositions, the stability of eclogite has been determined by Ringwood and Green (1966), Green and Ringwood (1967, 1972), and Ito and Kennedy (1970). Figure 17-1 shows an extrapolation of experimental results obtained at high temperatures to lower temperatures and pressures of relevance in the metamorphic origin of eclogites. Figure 17-1 is based on a graph given by Green and Ringwood (1972).

The formation of garnet in rocks of basaltic composition produces the mineral assemblage of a granolite, *i.e.*, garnet, plagioclase, and pyroxenes. At any given temperature, the pressure required to stabilize garnet varies considerably, depending on slight differences in the composition of the rocks. A similar dependence of pressure on composition has been demonstrated by Ringwood and Green (1966) for the disappearance of plagioclase leading to the formation of eclogites. The pressure at which plagioclase disappears at 1100°C and garnet, clinopyroxene ± quartz make up the new assemblage is indicated in Figure 17-1 for (1) quartz-tholeiite, (2) alkali-olivine-basalt, and (3) olivine-tholeiite.

The lower *double line* indicates the lowest possible *P-T* conditions of the transition from gabbro to garnet-granolite observed in olivine-tholeiite. The *bold dotted line* gives an approximate indication of the lowest possible *P-T* conditions of the transition to eclogite (plagioclase-out reaction) observed in olivine-tholeiite. Also, this line shows approximately the uppermost transition boundary from gabbro to granolite observed in quartz-tholeiite.

In the *stippled area* eclogites of compositions (2) and (3) are stable as well as granolite of slightly different composition, if water is absent (Ringwood and Green). Above the line G + R (Green and Ringwood) only eclogite is stable. According to Ito and Kennedy (1971), this boundary is given by line I + K at a higher pressure. This line has been derived from the plagioclase-out reaction in rocks of olivine-tholeiite composition, which is close to that of oceanic tholeiite. This boundary implies a *P-T* field larger than the stippled area in Figure 17-1 where mafic garnet granolite (garnet-hypersthene-pyroclase granolite) and eclogite of slightly different composition are stable. The differences may be due, in part, to the extrapolation of experimental data.

The differences in the extrapolation of the experimental data of the plagioclase-out lines may be significant. Thus, Green and Ringwood (1972) believe eclogite to be stable "in dry basaltic rocks along normal

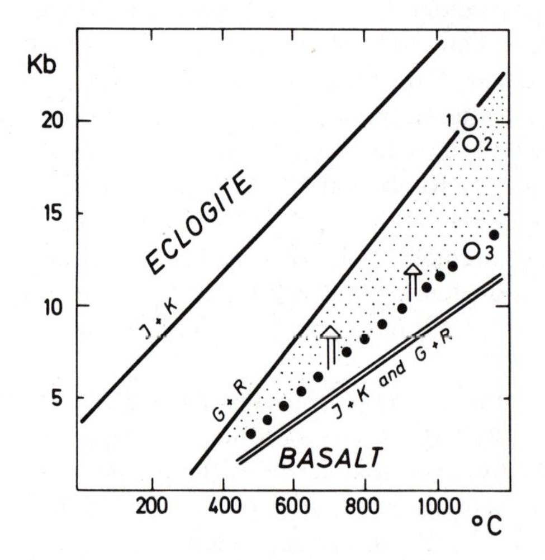

Fig. 17-1 Extrapolation to lower temperatures of experimentally determined boundaries between eclogite, garnet-granolite, and gabbro (basalt). Water pressure is zero.

geothermal gradients throughout the continental crust in stable or shield areas," whereas Kennedy and Ito (1972) "find feldspar-rich garnet-rich assemblages to be the stable mineralogy along normal geothermal gradients in the continental crust." However, this discrepancy is of no great concern in considering the metamorphic origin of eclogites, because so-called normal geothermal gradients are the exception rather than the rule in orogenic metamorphism.

Figure 17-1 illustrates that eclogites are stable over a very large temperature range if pressures are high and water is absent. If some water is present, zoisite and/or hornblende-bearing eclogite will form. Several occurrences are known (see for instance Heritsch, 1973) where such eclogites contain kyanite as well as zoisite. This assemblage indicates pressures exceeding 9 kb at temperatures higher than about 620–650°C (see Chapter 10, p. 148 and Figure 15-3).

Some eclogites originate under very high pressures and temperatures below the earth's crust and are incorporated as inclusions in kimberlites, basalts, and ultramafic rocks; others may originate within lower parts of the crust.

Eclogites are also formed at lower pressures and temperatures, as suggested by the *in situ* occurrence of eclogitic layers in [almandine]-medium-grade rocks. Bearth (1966) described eclogites as being genetically related to low-grade rocks, and Coleman *et al.* (1965) observed interlayered glaucophane schists and eclogites within the same block tectonically transported as a whole. Apparently the two rock types were formed under similar *P-T* conditions of very-low-grade metamorphism, but the eclogites reflect a local water-free environment.

The relative scarcity of eclogites compared to other metamorphic rocks of basaltic composition (*e.g.*, amphibolites and greenschists) indicates that the essential requirement for their formation, *i.e.*, a very low P_{H_2O} due to a very small water content of the rock, is commonly not realized in the environment of metamorphism. Thus, in the pennine zone of the Swiss Alps, Bearth (1966) found that eclogites are encountered only where eruptive rocks of very low water content predominate, whereas greenschists (prasinites) are found in other areas where abundant water-rich metasediments were able to supply H_2O during metamorphism.

The mineral assemblage of eclogites is stable only under unusual conditions. For this reason many eclogites are highly altered when they are exposed, at shallower depths, to lower pressures and the access of water.

In the field, complete gradation may be traced from unaltered eclogite through eclogite amphibolites containing relict garnet and omphacite together with newly gener-

> ated plagioclase and hornblende, to amphibolites of normal composition. In some cases myrmekite-like intergrowth of diopside and plagioclase first replace omphacite, and then in turn pass over into the amphibole-plagioclase association of the almandine-amphibolite facies. Elsewhere (*e.g.*, in the Franciscan of California), eclogites show every stage of retrogressive metamorphism to glaucophane schists . . . (Turner and Verhoogen, 1960).

In the inner part of the pennine zone of the Swiss Alps, the various occurrences of eclogites exhibit a transition to a glaucophanitic assemblage, *i.e.*, to lawsonite-free glaucophane-chloritoid-garnet-epidote-paragonite-muscovite schist. This, in turn, is transformed to the so-called prasinites having the albite-epidote-actinolite-chlorite assemblage typical of the lower temperature part of low-grade metamorphism (greenschist facies) (Bearth, 1966).

The composition of garnets. In considering the petrogenesis of eclogites, it is interesting to note that the chemical composition of eclogite garnets is related to the geological setting of the eclogites. This relationship was pointed out by Coleman *et al.* (1965) in a study based in part on Tröger's (1959) compilation of garnet analyses. Garnets from three genetic groups were distinguished. Group A eclogites associated with kimberlites, dunites, and peridotites are richer in pyrope component than garnets from group B eclogites, which are encountered in [almandine]-medium- and high-grade areas. Garnets from group C eclogites associated with glaucophane schists have the lowest pyrope content. All garnets have a very low spessartine content.

It is apparent that increasing *P-T* conditions of the formation of eclogites are reflected by greater pyrope contents. However, the correlation between garnet composition and environment of eclogites is *statistical* only. There is a considerable overlap in the composition of garnets from groups A and B and from groups B and C. This is well documented by a more recent compilation by Lovering and White (1969) shown in Figure 17-2. The composition is plotted in terms of the molecular proportions of pyrope, grossulatrie + andradite = Ca-bearing garnet, and almandine + very little spessartine. It is remarkable that garnets from those group A eclogites which are associated with kimberlite pipes may have a very high Ca garnet content. Garnets from other gorup A eclogites, as well as those from group B and group C eclogites, only contain about 25% Ca garnet component.

It is of interest to compare garnets from eclogites with those from granolites. Tröger reported a content of 24 ± 15 mole percent pyrope and 60 ± 11 mole percent almandine in garnets from granolites. This suggests a composition intermediate between garnets from eclogite groups B and C. However, more recent data (White, 1969) show that the

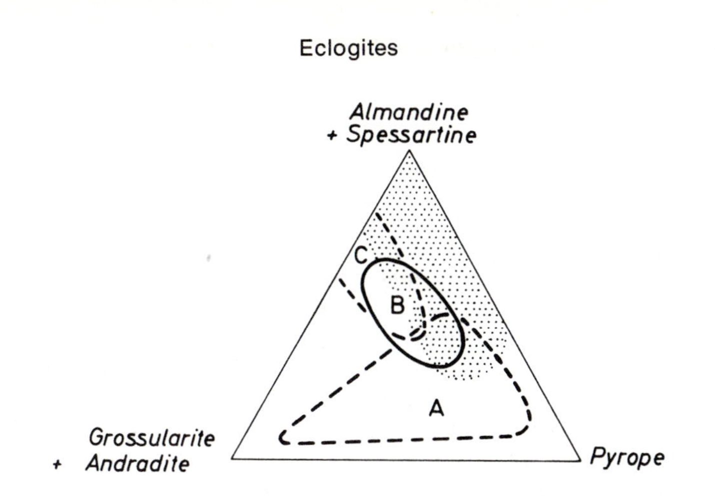

Fig. 17-2 Composition of garnets from eclogite groups A, B, and C. For comparison, the compositional field of garnets from granolites and granoblastites ("granulites") is also shown by the stippled area.

compositional field of granolite garnets is much larger and, as shown in Figure 17-2, overlaps the compositional field of all eclogite garnets.

According to Coleman *et al.* (1965), *omphacites* do not display any systematic variation in composition. The jadeite content of these pyroxenes ranges from a few percent to 40 mole percent. Only the omphacites of group C eclogites have a higher jadeite content of 28 to 40 mole percent. However, Smulikowski (1965) believes that the jadeite content of omphacites increases continuously from group A through group B to group C eclogites. Banno and Matsui (1965) have suggested methods of distinguishing the temperatures of formation of eclogites on the basis of the distribution of Mg, Fe, and Mn between omphacite and garnet; see also Chappell and White (1970).

References

Angel, F. 1957. *Neues Jahrb. Mineral. Abhand.* **91:** 151–192

Banno, S. and Matsuie, Y. 1965. *Proc. Jap. Acad.* **41:** 716–721.

Bearth, P. 1966. *Schweiz. Mineral. Petrog. Mitt.* **46:** 12–23.

Binns, R. A. 1967. *J. Petrol.* **8:** 349–371.

Chappell, B. W. and White, A. J. R. 1970. *Mineral Mag.* **37:** 555–560.

Clark, J. R. and Papike, J. J. 1968. *Am. Min.* **53:** 840–868.

Coleman, R. G., Beatty, L. B., and Brannock, W. W. 1965. *Geol. Soc. Am. Bull.* **76:** 483–508.

Eskola, P. 1939. *In* Barth, Correns, and Eskola, *Die Entstehung der Gesteine.* Springer-Verlag, Berlin.

Green, D. H. and Ringwood, A. E. 1967. *Geochim. Cosmochim. Acta* **31:** 767–833.

——— 1972. *J. Geol.* **80:** 277–288.

Heritsch, H. 1963. *Mitt. Naturw. Ver. Steiermark* **93:** 178–198.

——— 1973. Tschermaks Min. Petr. Mitt. **19:** 213–271.

Ito, K. and Kennedy, G. C. 1970. *Mineral Soc. Am. Special Paper* **3:** 77–83.

——— 1971. *In* J. G. Heacock, ed. *The Structure and Physical Properties of the Earth's Crust.* Geophys. Monograph Am. Geophysics Union, Washington.

Kennedy, G. C. and Ito, K. 1972. *J. Geol.* **80:** 289–292.

Lovering, J. F. and White, A. J. R. 1969. *Contr. Mineral. Petrol.* **21:** 9.

Ringwood, A. E. and Green, D. H. 1966. *Tectonophysics* **3:** 383–427.

Smulikowski, K. 1964. *Bull. Acad. Polon. Sci.* **12:** 17–34.

——— 1965. *Bull. Akad. Polon. Sci.* **13:** 11–18.

Tröger, E. 1959. *Neues Jahrb. Mineral. Abhand.* **93:** 1–44.

Turner, F. J. and Verhoogen, J. 1960. *Igneous and Metamorphic Petrology.* 2nd edit., McGraw-Hill Book Company, New York.

White, A. J. R. 1969. *Geol. Soc. Australia Special Publ.* **2,** 353.

Chapter 18

Anatexis, Formation of Migmatites, and Origin of Granitic Magmas

In metamorphic terrains, high-grade metamorphic rocks are spatially, and presumably genetically, related to migmatites. Read (1957) describes this general observation as follows: "One of the most firmly established facts in metamorphic geology is the close association in the field of highest-grade metamorphic rocks and migmatites."

Sederholm (1908) introduced the term, "migmatite," to designate certain gneissic rocks "which look like mixed rocks." These are coarse-grained, heterogeneous, and megascopically composite rocks, comprising portions of magmatic rocks (or looking like magmatic rocks) and portions of metamorphic rocks.

The following genetic characteristic features of migmatites have been proposed in a symposium on the nomenclature of migmatites[1]:

(a) Megascopically composite rocks that once consisted of geochemically mobile and immobile (or less mobile) parts (Dietrich and Mehnert). Mehnert designates as mobilization the increase of geochemical mobility of rocks or portions of rocks beyond the dimension of individual crystals, regardless of the state of aggregation of the rocks and of the mechanism of migration.
(b) Composite, heterogeneous rocks, consisting of pre-existing rock material and of granitic material, subsequently intruded or originated *in situ* (Polkanov).

A comment on point (b) is necessary: It is of major importance to describe and give a name to such migmatites which have attained their migmatitic appearance by the injection of granitic magmas in the form of

[1]Symposium on migmatite nomenclature (ed. by H. Sörensen). *Intern. Geol. Congress,* Copenhagen, 1960, Part 26, 54–78.

apophyses, smaller or larger dykes, etc. into a metamorphic country rock. It is suggested to call such migmatites *injection migmatites*. These should be clearly distinguished from *in-situ migmatites, i.e.,* from those migmatites whose light-colored granitic portions are overwhelmingly not of an intrusive origin from an outside source, but have formed within a high grade metamorphic gneiss or schist itself. The leucocratic portions in *in-situ* migmatites typically have the form of stringers and/or layers which are limited in lateral extent and are of such a scale that the composite appearance of the rock can frequently be ascertained in a hand specimen. On the other hand, injection migmatites often need outcrop size to display themselves, but they may also show very small scale injections which make a megascopic distinction from *in situ* migmatites impossible.

The most common type of migmatite is the *in-situ* migmatite, and it is only in *in-situ* migmatites that the formation of the light-colored granitic portions within the composite rock must have been governed by physicochemical laws.

Following Mehnert (1968), the following parts can generally be distinguished in *in-situ* migmatites (they are illustrated in Figure 18-1):

(1) The *paleosome, i.e.,* the unaltered or only slightly modified parent rock or country rock, a gneiss.
(2) The *neosome, i.e.,* the newly formed rock portion. Here again two rock types are distinguishable as a rule, *viz.,*
   (a) the *leucosome* containing more light minerals (quartz and/or feldspar) with respect to the paleosome.
   (b) the *melanosome* containing mainly dark minerals, such as biotite, hornblende, cordierite, garnet, sillimanite, and others. The melanosome seams surrounding a leucosome are commonly very narrow and may not even be discernible.

The minerals of the dark-colored layers, like those in gneisses and crystalline schists, generally have a preferred orientation, whereas the

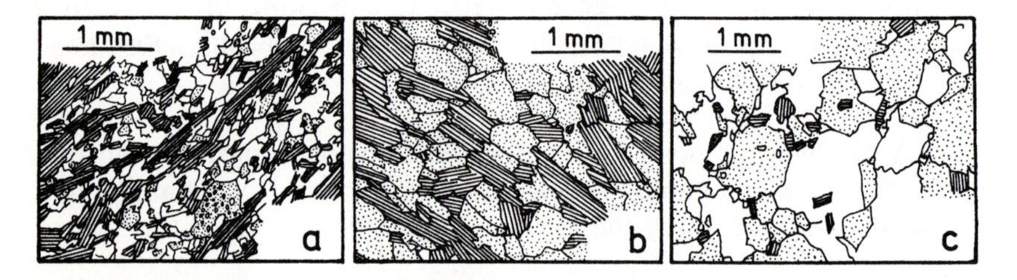

Fig. 18-1 Sketches of typical trondhjemitic migmatites, as seen in thin section (from Ashworth, 1976); (a) paleosome, (b) melanosome, (c) leucosome. Quartz shown blank, plagioclase stippled, and biotite ruled.

fabric of the leucosome is characteristic of a rock formed by magmatic crystallization. The structure of migmatites may vary considerably. The foliation characteristic of gneisses is modified so that individual light-and-dark-colored layers may have a thickness of a few centimeters or tens of centimeters or even meters. Some of the heterogeneous rocks show folds of various dimensions, others have the appearance of breccia; the latter type of migmatite is known as an agmatite. Nebulites are migmatites which have reached an advanced stage of homogenization of the heterogeneous portions of the rock, so that a parallelism of fabric elements of the metamorphic rocks is only vaguely preserved in the form of schlieren. For examples of migmatite structures, see Mehnert (1968), p. 10 ff.

Some time ago, Sederholm suggested that the "strongly contorted structure" generally observed in migmatites "originated when the rock was in a melting condition." The concept of a "melting condition" is indeed supported now by experiments. The leucosomes of granitic composition may be explained as products of partial melting of the original gneisses, a process known as anatexis. It is shown further on that partial melting of a gneiss gives rise to melts of granitic composition and to a crystalline residue composed mainly of mafic minerals like biotite, etc., which make up the melanosomes.

Other views concerning the origin of migmatites invoke an introduction of material by "emanations" or by a metasomatic exchange of material without the participation of a melt, *i.e.*, without an intervening magmatic stage. The mechanism of such an imagined process is not known, although some authors even see in it an explanation of the origin of granitic rocks in the basement complexes where characteristic features of magmatic intrusion are not evident (granitization). In this connection, Buddington (1963) remarks, "There are neither experimental physicochemical data nor dependable physicochemical theory upon which to base a valid hypothesis for the origin of granite (of normal average composition) by emanations." This is corroborated more recently by experiments by Luth and Tuttle (1969), see p. 332.

A discussion of the origin of granites and migmatites should take into account the well-established observation of geologists, which Read formulates as follows: "When we follow rocks into higher metamorphic grades, we finally end in a granitic core. This cannot be accidental; the association of metamorphites, migmatites, and granites must mean something." This worldwide observation suggests that the spatial association of these rock types is due to processes occurring at similar temperature and pressure conditions in the deeper part of the crust, giving rise to high-grade metamorphic rocks, as well as to migmatites and granites. The origin of granites and migmatites in deep-seated parts of orogenic belts must be considered as directly connected with high-grade metamorphism. Experiments attempting to elucidate the origin of migmatites and granites show that the process of anatexis is of major petrological importance.

Anatexis: General Considerations

Clays and shales are the commonest sediments, and graywackes are the most important clastic sediments. The metamorphism of these sediments was experimentally investigated by Winkler *et al.* (1957 ff). The highest-grade metamorphic rocks formed at P_{H_2O} = 2000 bars consist of quartz + plagioclase + alkali feldspar + biotite accompanied by cordierite, opaque minerals, and, in many cases, sillimanite. At higher pressures, garnet occurs instead of cordierite, and, at very high pressures, kyanite takes the place of sillimanite.

The amount of calcite originally present in the sediments determines the amounts of cordierite and biotite produced by metamorphism. Considering a shale to which various amounts of calcite are added, the experiments by Winkler and von Platen (1960) show that in highest-grade metamorphic rocks an increase of the amount of calcite in the original sediment decreases the amount of cordierite and increases the amount of biotite. The explanation is simple. The CaO of calcite leads to the formation of anorthite component. This, in turn, uses up much of the Al_2O_3 of the rock, decreasing the amount of Al_2O_3 available for the formation of cordierite. Consequently, more of the MgO and FeO of the rock is combined in biotite (causing a slight decrease in the amount of K feldspar). A shale having the average chemical composition of shales is transformed, when subjected to highest-grade metamorphism, into a paragenesis free of cordierite, consisting of quartz + plagioclase + K feldspar + biotite. This paragenesis, in which the amount of plagioclase greatly exceeds the amount of K feldspar, is a very common one in paragneisses.

The parageneses of these highest-grade paragneisses remain unchanged within a certain temperature interval (at constant pressure). However, when the temperature of metamorphism reaches a certain value, anatexis sets in, *i.e.*, the gneiss is partially melted in the presence of H_2O. The resulting melt is predominantly composed of quartz, plagioclase, and alkali feldspar. Only very small amounts of biotite, cordierite, and sillimanite can be dissolved in the melt; together with surplus plagioclase and/or some quartz, they form the crystalline residue. The temperatures required for anatexis are remarkably low; at P_{H_2O} = 2000 bars, the temperature is about 700°C, at P_{H_2O} = 4000 bars, it is about 680°C, and at higher water pressure, temperatures are even lower. These conditions are the same as those of high-grade metamorphism.

The composition and the amount of the melt formed by anatexis of gneisses depends on the chemical and, therefore, mineralogical composition of the gneisses, in addition to H_2O pressure and temperature. This is best realized by considering the equilibria between melt + crystals + gas in the system of components SiO_2, $NaAlSi_3O_8$, $CaAl_2Si_2O_8$, $KAlSi_3O_8$, and H_2O, *i.e.*, a system consisting of quartz, plagioclase, alkali feldspar, and H_2O. The system

$$SiO_2 - NaAlSi_3O_8 - CaAl_2Si_2O_8 - KAlSi_3O_8 - H_2O$$

may be designated as the granitic system, because these components constitute 90 to 95% of granitic rocks, comprising granite, adamellite, granodiorite, and tonalite or trondhjemite; the latter two are often grouped under the term quartz-diorite, but this should not be continued (see Streckeisen, 1973).

Approximate Granitic System

Until recently, the system SiO_2-$NaAlSi_3O_8$-$CaAl_2Si_2O_8$-$KAlSi_3O_8$-H_2O had not been investigated, but very important data for the system SiO_2-$NaAlSi_3O_8$-$KAlSi_3O_8$-H_2O, *i.e.*, the granitic system without the An component, have been available for some time (Tuttle and Bowen, 1958). Before proceeding to the more complex granitic system, their results in the approximate granitic system are reviewed. They studied the equilibria among crystals, melt, and gas in the system SiO_2-$NaAlSi_3O_8$-$KAlSi_3O_8$-H_2O at various pressures and in the presence of sufficient water to form an H_2O-rich gaseous phase in all experiments. The silicate melt, which under pressure dissolves some water, was in all cases saturated with water. In Figure 18-2 the system is shown in perspective view at a certain H_2O pressure. The concentration triangle, Q, Ab, and Or, forms the base of a trigonal prism (Q = quartz = SiO_2; Ab = albite = $NaSiAl_3O_8$; Or = orthoclase = $KAlSi_3O_8$). The temperature axis is parallel to the edges of the prism. H_2O is present "in excess," so that H_2O need not be represented graphically as a component. This system is bounded by the marginal systems SiO_2-$NaAlSi_3O_8$-H_2O, SiO_2-$KAlSi_3O_8$-H_2O, and $KAlSi_3O_8$-$NaAlSi_3O_8$-H_2O, represented by the three sides of the prism. Each of the first two marginal systems have a eutectic point; these are connected by a cotectic line. The cotectic line is curved and passes through a temperature minimum. The position of the cotectic line and the position of the temperature minimum of the cotectic line depend on pressure; an increase in pressure causes a shift of the temperature minimum toward the Ab corner (see Table 18-2 p. 295).

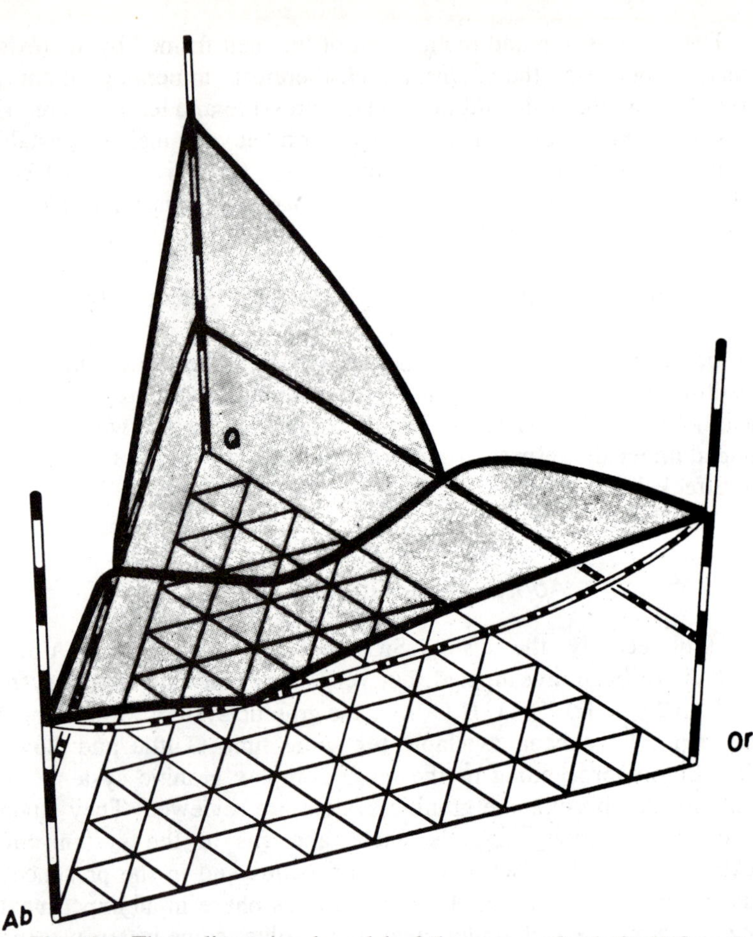

Fig. 18-2 Three-dimensional model of the system Q-Ab-Or-H_2O at P_{H_2O} = 1000 bars; H_2O is present in excess. Vertical direction = temperature axis (drawn on the basis of data from Tuttle and Bowen, 1958). Shown are the liquidus surfaces which intersect in the cotectic line. Along the cotectic line, quartz + alkali feldspar + melt + gas coexist in equilibrium. The three marginal systems are shown, including their solidus relations. The field of leucite, existing up to 2600 bars at and near the Or corner, is not shown.

Figure 18-3 shows a projection of the three-dimensional model (Figure 18-2) onto the base, Q-Ab-Or, at P_{H_2O} = 2000 bars. The isotherms of the liquidus surface, also shown in the projection, indicate the steep rise of the liquidus surface toward the Q and Or corners and the gentler rise toward the Ab corner. The gentle rise of the cotectic line from the temperature minimum M toward the eutectic points E_1 and E_2 should also be noted. At constant pressure, a cotectic line connects the eutectic points E_1 and E_2 in the system Q-Ab-Or-H_2O. This line divides the liquidus surface into two regions, the quartz field and the alkali feldspar field. If the composition of a melt is represented by a point lying within the quartz field, quartz begins to crystallize when the liquidus temperature is reached on cooling. Further cooling causes the crystallization of more quartz, and the point representing the melt composition migrates toward the cotectic line. When the cotectic line is reached, alkali feldspar crystallizes in addition to quartz, *i.e.*, alkali feldspar, quartz, melt, and gaseous phase are in equilibrium. Cotectic crystallization continues upon further cooling until all melt has crystallized. The point representing the melt composition migrates along the cotectic line toward the temperature minimum, which, however, is generally not reached. The last portion of melt has the composition corresponding to the cotectic minimum in the case of pronounced fractional crystallization, *i.e.*, if conditions are such that equilibrium between crystals and melt is not always established.

As a next step, the melting behavior at constant pressure of rocks

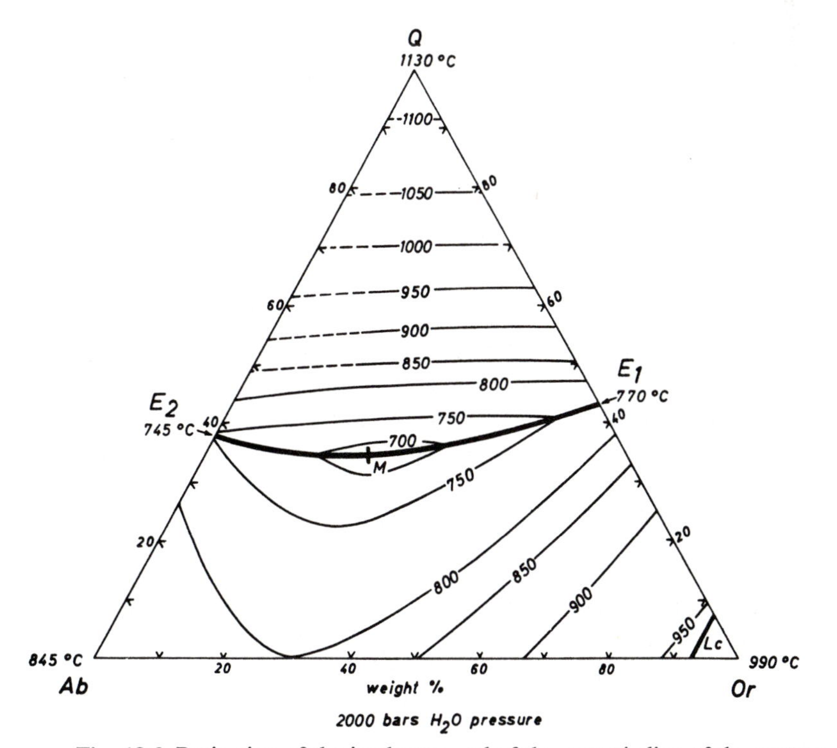

Fig. 18-3 Projection of the isotherms and of the cotectic line of the system SiO_2-$NaAlSi_3O_8$-$KAlSi_3O_8$-H_2O at P_{H_2O} = 2000 bars. E_1 and E_2 are eutectic points; M marks the composition of the temperature minimum on the cotectic line. Compositions are given in weight percent [drawn on the basis of data from Tuttle and Bowen (1958) and from Shaw (1963)]. Lc = leucite.

consisting of components Q, Ab, and Or will be considered. The formation of the first melt always takes place under equilibrium conditions. Therefore, the composition of the first melt, in general, is not that of the cotectic minimum. The composition does lie on the cotectic line, but the position on the line depends on the particular composition of the rock. With the aid of Figure 18-3, the melting of a rock with the composition Q:Ab:Or = 50:40:10 will be discussed as an example. The first melt is formed at a temperature slightly higher than 700°C, not at the temperature of the cotectic minimum. The composition of the first melt is given by a point on the cotectic line having an approximate component ratio of Q:Ab:Or = 35:50:15. Alkali feldspar and quartz have formed a melt. An increase in temperature of 5° to 10°C causes the melting of all alkali feldspar, together with the appropriate amount of quartz, to give a cotectic composition. When the last alkali feldspar disappears, the melt reaches its limiting composition on the cotectic line; its composition now is Q:Ab:Or = 36:52:12. Any additional melting of appreciable amounts of quartz, which is the only remaining crystalline phase, requires a considerable increase in temperature. In the case of the rock composition chosen as an example, 850°C is required to completely melt the mixture of quartz and alkali feldspar.

This consideration shows that in the process of partial melting, the cotectic minimum has no greater significance than any other cotectic composition in the system Q-Ab-Or-H_2O. However, at H_2O pressures greater than approximately 2.2 kb, two different alkali feldspars may exist together with quartz and the cotectic minimum is therefore replaced by an eutectic point in the isobaric system Q-Ab-Or-H_2O (Morse, 1970).

According to Seck (1971), the eutectic appears at approximately 3 kb rather than 2.2 kb. Even at water pressure greater than 3 kb, not all compositions within the system may attain the state of eutectic equilibrium; this is possible only for compositions in the central region of the composition triangle. The range of such compositions increases with pressure (see Luth *et al.*, 1964) because crystallization temperatures decrease and thus progressively restrict the amount of solid solution between Na and K feldspar components. Restriction of solid solution causes two alkali feldspar phases to coexist, an Ab-rich and an Or-rich phase; only in this case is an eutectic equilibrium possible.[2]

The presence of a eutectic phase relationship causes a fundamental difference in the melting behavior of the system: The eutectic point indi-

---

[2]Eutectic equilibrium is defined as a *univariant* heterogeneous equilibrium involving a liquid phase. In the four-component system Q-Ab-Or-H_2O, eutectic equilibrium demands the coexistence of the following five phases: quartz, Ab-rich alkali feldspar, Or-rich alkali feldspar, vapor, and liquid.

cates the temperature and composition (at a given water pressure) of the first melt formed upon heating of any composition in the system that consists of a mixture of the three crystalline phases, Or-rich alkali feldspar, Ab-rich alkali feldspar, and quartz. Eutectic temperatures and compositions at various pressures are given in Table 18-2, p. 295.

Formation of Melts in the Granitic System

Previously it was assumed that the phase relations in the system Q-Ab-Or-H_2O provide an adequate basis for an understanding of the crystallization of granitic melts and of the anatexis of gneisses. However, although phase relations in this system are a valuable demonstration of general principles, they do not supply quantitative data about anatectic melts. Concerning the anatexis of gneisses, it is absolutely necessary to consider the anorthite component present in the plagioclase of gneisses and granites in addition to other components, even though the anorthite component constitutes only 5 to 20% of gneisses and granites; its amount is always subordinate to that of the Ab component.

The phase relations in the five-component system Qz-Ab-An-Or-H_2O at a constant H_2O pressure of 5000 bars will be described with the aid of Figure 18-4. The tetrahedron is bounded by four triangles, each representing a four-component system; three components are indicated by the corners of the triangle and H_2O (not shown) is present as a fourth component. The four systems are:

1. *Qz-Ab-An-H_2O system.* A cotectic line connects the eutectic E_1 on the Qz-Ab edge of the tetrahedron and the eutectic E_2 on the Qz-An edge. E_2 has been determined by Stewart (1957, 1967). Yoder (1968) has investigated the system Qz-Ab-An-H_2O at 5 kb; he has shown that the melt compositions coexisting with quartz and plagioclase are located on a straight line extending from E_2 with decreasing temperature to E_1.
2. *Qz-Or-An-H_2O system.* There is a ternary eutectic E_5. From it three cotectic lines extend to binary eutectics: E_2 on the Qz-An edge, E_3 on the Qz-Or edge, and E_4 on the An-Or edge. The position of E_5 has been determined by Winkler and Lindemann (1972) and Winkler and Ghose (1974) at various pressures (see Table 18-1).
3. *Ab-An-Or-H_2O system.* A cotectic line passes from the eutectic E_4 on the Or-An edge (near the Or corner) toward the eutectic E_6 of the Or-Ab system. The Ab-An-Or system has been investigated at 5000 bars H_2O pressure by Yoder *et al.* (1957).

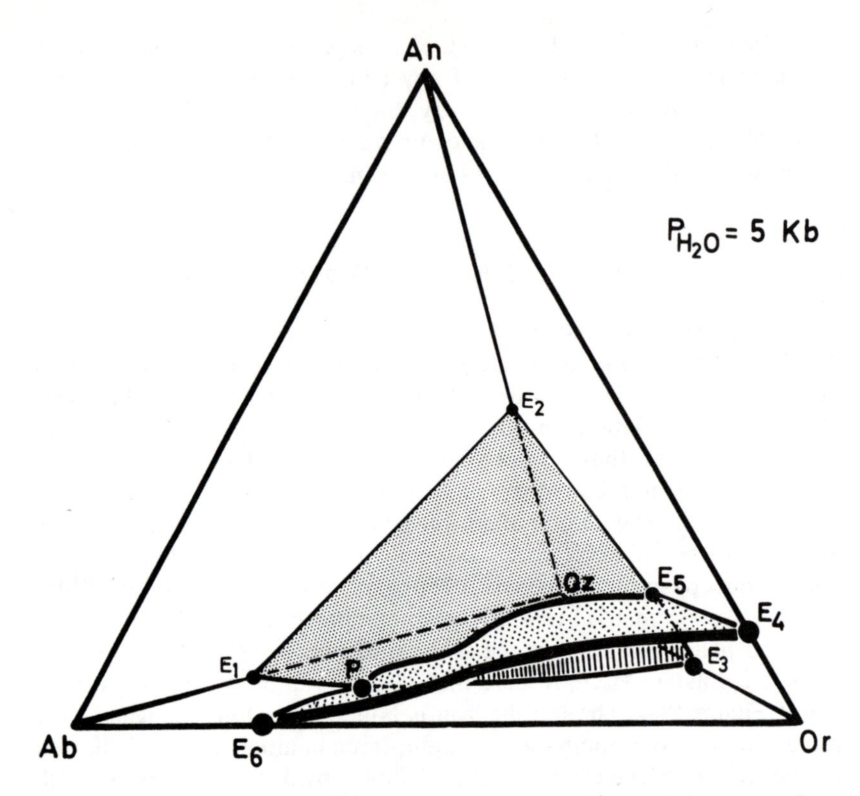

Fig. 18-4 *System Qz-Ab-Or-An.* Diagrammatic phase relations at a given H_2O pressure. Above approximately 3 kb point P is an eutectic point.

4. *Qz-Ab-Or-H_2O system.* This system is already known from the previous section. In Figure 18-4 three cotectic lines radiate from the eutectic P to the eutectics E_1, E_3, and E_6, respectively.

Table 18-1

Eutectic compositions of E_5				Temperatures
kb	Or	Qz	An	°C ±7°
2	51	40	9	730
4	49	39	12	708
5	48	38	14	700
7	44	38	18	686

Note that the small amount of An component that may substitute in alkali feldspar has not been taken into account in this presentation of the Qz-Ab-Or-An–H_2O system.

The space of the Qz-Ab-An-Or tetrahedron is divided by three cotectic surfaces. A large cotectic surface (E_1-E_2-E_5-P) separates the quartz space from the plagioclase space. A small cotectic surface (E_5-E_3-P) separates the quartz space from the alkali feldspar space. A further cotectic surface (E_4-E_6-P-E_5) separates the small alkali feldspar space from the large plagioclase space. Accordingly, there are three cotectic surfaces along which quartz + plagioclase + melt, quartz + alkali feldspar + melt, and alkali feldspar + plagioclase + melt coexist (together with the gaseous phase). These three cotectic surfaces intersect in a cotectic line, P-E_5, which indicates the compositions of melts coexisting with quartz + plagioclase + alkali feldspar + gaseous phase. The cotectic line begins at P at the eutectic of the system Qz-Ab-Or and passes with very gentle inclination through the An-poor region of the tetrahedron to the eutectic E_5 of the system Qz-An-Or. It should be noted that the cotectic line is located in a part of the tetrahedron characterized by a very low An content. For this reason, small contents of An component have a pronounced effect on the composition of cotectic melts.

Beginning of Melting

Let us visualize the situation of the beginning of melting in this system—the melt being in equilibrium with quartz, alkali feldspar, plagioclase, and a gaseous phase. A plane through the tetrahedron of Figure 18-4 connecting the points of quartz, of alkali feldspar composition, and of the coexisting plagioclase composition determines the base of a four-phase (irregular) tetrahedron, the apex of which represents the coexisting melt. The melt lies below the basal plane and is a point on the cotectic line P-E_5, as shown in Figure 18-5. This point represents the composition of the first melt formed when any mixture of the three solid phases defining the base of the mentioned tetrahedron is heated up. That point is therefore referred to as the *minimum-temperature melt* composition, or, more briefly, as the *minimum melt.* With rising temperature, the melt moves along the cotectic line P-E_5 in the direction of E_5 until one of the three solid phases has been completely melted; the coexisting feldspar phases thereby also change their compositions. The end of cotectic equilibria with *three* solid phases and melt depends on the compositions of the feldspars and on the amounts of the three solid phases; the end of cotectic equilibrium is only from a few degrees up to about

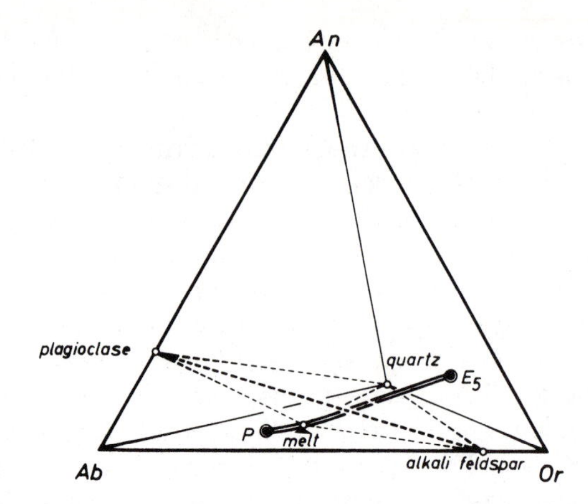

Fig. 18-5 Schematic representation of the coexistence of alkali feldspar, plagioclase, quartz, and cotectic melt. A water-rich gas phase is present as well.

25°C higher than the beginning of melting in common gneiss compositions.

Experimentally it is not possible, of course, to determine the exact composition of the minimum-temperature melt but of a melt which exists at a temperature about 10° to 15°C higher than at the true beginning of melting. Here, this will also be called "minimum-temperature melt," or "minimum melt," but notice that it has been *set in quotation marks*. This "minimum melt" then has a composition between that of the true beginning of melting and that of the end of coexistence of *three* solid phases and melt. With further rise in temperature, the melt composition leaves the cotectic line in Figure 18-4. Now only two solid phases are in equilibrium with the melt, and the compositon of the melt is represented by a point on one of the three cotectic surfaces in Figure 18-4.

Any late crystallizing melt and, vice versa, any melt formed at the beginning of anatexis of gneisses, which contain the two feldspar species and quartz, is situated on the cotectic line P-E_5, valid for the given water pressure. Bearing this in mind, a knowledge of the position of the isobaric cotectic line in the tetrahedron is necessary to understand the *beginning* of anatexis in gneisses that contain alkali feldspar, plagioclase, and quartz. The following points are noteworthy:

1. Gneisses never contain plagioclase very rich in anorthite component; commonly the plagioclase composition is in the range An_{10}–An_{40}. Therefore, the possible range of melt compositions formed on the cotectic line P-E_5 during the early stage of anatexis is restricted. The melt can never reach a position close to

E_5; its composition is restricted to the lower half of the cotectic line. Thus, during anatexis of a gneiss, at the condition of $P_{H_2O} = P_{total}$, plagioclase, alkali feldspar, quartz, and melt (and gas) can coexist only within a fraction of the total temperature range represented by the line P-E_5 (about 50° at 5 kb). In various experiments it has been shown (Winkler, 1967; Winkler and Breitbart, 1978) that the two feldspars, quartz, and melt coexist only within a temperature range of a few degrees up to about 25°C. Piwinski and Wyllie (1970) report a range of about 30°C. This experimental observation is now well understood.

2. An increase in pressure shifts the composition of E_5 toward somewhat higher An content (see Table 18-1). Increasing pressure also shifts point P mainly in the direction of Ab. Table 18-2 gives the compositions represented by point P at various pressures. The data are taken mainly from Tuttle and Bowen (1958) and from Luth *et al.* (1964). However, the stated temperatures for 2, 5, and 10 kb deviate slightly from those given by the authors. At 2 kb von Platen (1965) determined a value of 670°C (instead of 685°). At 5 kb a value of 640°C (instead of 650°) was determined in our laboratory, which agrees with the value given by Merrill *et al.* (1970). The temperature of 615°C at 10 kb is from the latter source.
3. The cotectic line P-E_5 stays in the region of low An content in the tetrahedron Qz-Ab-Or-An.
4. For petrogenetic considerations, compositions within the tetrahedron Qz-Ab-Or-An are commonly projected from the An apex onto the plane Qz-Ab-Or. The eutectic E_5 lying in the plane Qz-Or-An of the tetrahedron will then project onto the edge Qz-Or; this projection point E_5' is one of the two end points of the projected isobaric univariant cotetic line P-E_5'. It is noteworthy that

Table 18-2 Data for minima or eutectics in the system Qz-Ab-Or-H_2O at various pressures.

P_{H_2O}	Q	:	Ab	:	Or	T, °C
500	39	:	30	:	31	770
1,000	37	:	34	:	29	720
2,000	35	:	40	:	25	670
4,000	31	:	46	:	23	650
5,000	27	:	50	:	23	640
10,000	23	:	56	:	21	615

the ratio Qz:Or of the eutectic Qz:Or:An composition of E_5 is hardly influenced by pressure. The Qz:Or ratio of the projected point E'_5 on the line Qz:Or is 44:56 at 2, 4, and 5 kb and 46:54 at 7 kb.

As has been discussed under point 1, melt formed in the early stage of anatexis in plagioclase-alkali feldspar-quartz gneisses will lie somewhere on the left part of the cotectic line shown in Figure 18-4.

In order to know exactly the compositions of early melts along the cotectic line P-E_5, it is necessary to know not only the projection points, *i.e.*, the ratio of components Qz:Ab:Or, but the amount of the An component as well. In other words, the amount of all four components must be known. These have been determined by Winkler *et al.* (1975) for several points on the line P-E_5 at 5 and 7 kb. Table 18-3 gives the data which are also graphically represented in Figure 18-6. Note that at the site of the melt composition the amount of the An component is indicated.

The above data clearly show that in the space of the tetrahedron the isobaric cotectic line does not rise from P to E_5 with a constant slope. Rather, pronounced inflections are present. This is schematically shown in Figure 18-4 and it is well demonstrated (although in exaggerated manner) in the lower half of Figures 18-8 and 18-9.

A departure of the compositions of early-stage melts from the relevant cotectic line (supposing the pressure is known) seems possible only in the three following cases:

a. If some HCl is present in the fluid phase. According to experiments by von Platen (1965), the presence of HCl will shift the projection of the cotectic line P-E'_5 toward compositions with about 5% less Qz.

Table 18-3. Melt compositions (wt. %) coexisting with plagioclase + alkali feldspar + quartz + vapor at 5 and 7 kb. respectively.

	5 kb H_2O pressure						7 kb H_2O pressure				
Qz	27	27	32	34	35	38	25	26	31	34	38
Or	23	22	27	30	36	48	22	25	32	37	44
Ab	50	47	36	27	15	0	53	45	27	14	0
An	0	4	5	9	14	14	0	4	10	15	18
Temp.	640	655	660	670	685	700	630	640	655	670	686

Note: Compositions with zero An content from Luth *et al.* (1964) and compositions with zero Ab content from Winkler and Lindemann (1972) and Winkler and Ghose (1974).

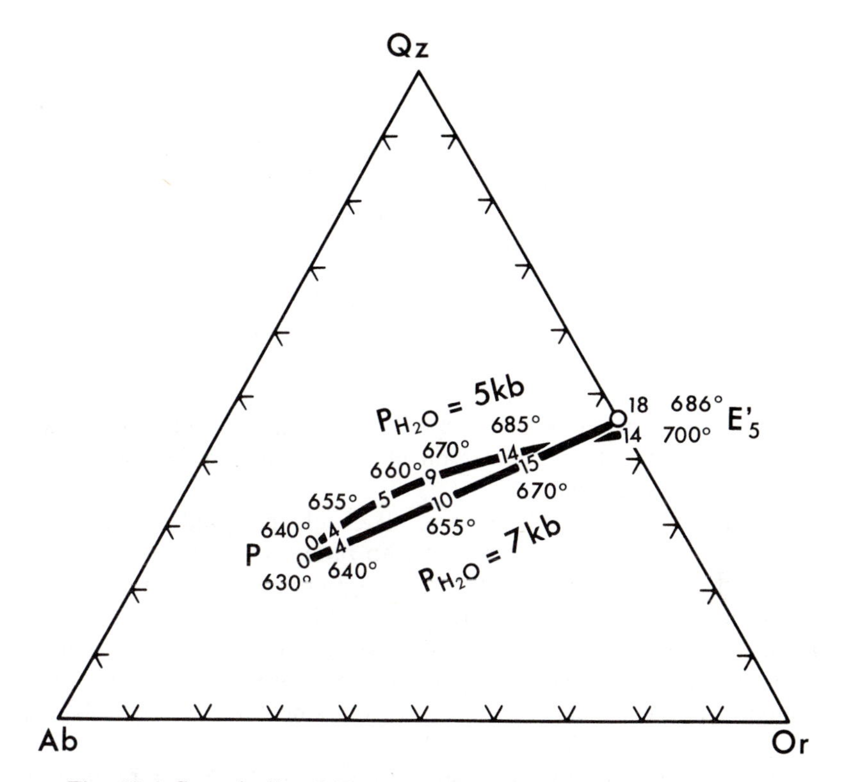

Fig. 18-6 Cotectic line P-E_5 at 5 and 7 kb projected downward radially through the An apex onto the side Qz-Ab-Or of the tetrahedron shown in Figure 18-4. The site of each determined melt composition is indicated by a number stating the amount of the An component. The An content is in weight per cent on the basis of Qz + Ab + Or + An = 100. The temperature in centigrade is also given for each melt composition on the two cotectic lines.

b. If the early stage of anatexis has already been surpassed, *i.e.*, when the melt no longer coexists with the *three* solid phases plagioclase, alkali feldspar, and quartz.
c. If the original gneiss did not contain alkali feldspar in addition to quartz and plagioclase. In this rather common case no melt can form that coexists with solid alkali feldspar, plagioclase, and quartz; therefore, its composition does not plot on the cotectic line P-E_5.

Von Platen (1965) has empirically found that rocks of various compositions and containing plagioclase, alkali feldspar, and quartz give rise to a "minimum melt" of practically the same composition if the various bulk rock compositions have the same Ab/An ratio. In other words, the

Table 18-4. Data valid for 2 kb.

Ab/An ratio	"Minimum melt" temperature, °C	Ratio of "minimum melt" composition Qz	:	Ab	:	Or
∞	670	34	:	40	:	26
7.8	675	40	:	38	:	22
5.2	685	41	:	30	:	29
3.8	695	43	:	21	:	36
2.9	695	44	:	19	:	37
1.8	705	45	:	15	:	40

Ab/An ratio of any rock that contains plagioclase, alkali feldspar, and quartz determines the composition of the "minimum" anatectic melt within rather narrow limits.

Using mixtures of plagioclase, alkali feldspar, and quartz having bulk compositions with different Ab/An ratios, von Platen determined "minimum melt" compositions,[3] that is, piercing points with different compositional planes of the cotectic line P-E_5 at 2 kb water pressure. These are listed in Table 18-4 together with their corresponding temperatures ("minimum melt" temperature) and Ab/An ratios of the bulk composition. The small amounts of An component and H_2O dissolved in the melt are not included in the table.

Table 18-4 clearly shows that a decrease of the Ab/An ratio in the system is accompanied by a decrease of Ab and an increase of Qz and Or contents in the "minimum melt." In other words, a higher An/Ab ratio in the bulk composition, essentially due to a higher An content of the plagioclase, shifts the composition of the "minimum melt" along line P-E_5 toward E_5. These empirical results are now well understood from the position of the isobaric cotectic line P-E_5 within the system Qz-Ab-Or-An-H_2O.

It follows from Table 18-4 that, even at the same depth (*i.e.*, at the same pressure), the melts first formed in gneisses ("minimum melts") do not originate at exactly the same temperature. In a gneiss complex, anatexis, due to the temperature rise in high-grade metamorphism, begins first in those layers having the most Ab-rich plagioclase. The higher the An content of plagioclase in a gneiss, the higher the temperature of the beginning of anatexis. This is clearly demonstrated by the experiments of Winkler and von Platen (1961), in which graywackes were subjected to high-grade metamorphic conditions. It was shown that temperature differences of 30°C are not unusual. For instance, if at

[3]Erroneously termed "eutectic" by von Platen.

P_{H_2O} = 2000 bars the temperature does not exceed 685°C, partial melting is not observed in paragneisses having an Ab/An ratio < 5.2, whereas in other gneisses (including many orthogneisses) an anatectic melt is formed (see also Table 18-4).

Furthermore, it must be taken into account that the gaseous phase, composed essentially of H_2O, may contain other components such as HCl and HF. As von Platen (1965) has shown, these components also influence the "minimum melt" compositions and the values of their temperatures in the granitic system. The presence of a small amount of HCl in the gaseous phase is very probable in metamorphic terrains. Many sediments contain salt solutions, and, in some cases, their concentration increases considerably with depth. The salt solution between mineral grains is preserved even in rocks compacted at depth. Experiments show that NaCl hydrolizes at metamorphic conditions. Some NaCl reacts with the minerals of the rocks and is combined in silicates primarily as Ab component, and an equivalent amount of HCl constitutes a component of the fluid phase present during metamorphism. Nothing is known about the concentration of HCl, but its presence must be accounted for in many cases. The effect of HCl during anatexis is such that the "minimum melt" compositions in sections through the system Qz-Ab-An-Or-H_2O-HCl are somewhat poorer in Qz and correspondingly somewhat richer in Or, as compared to the HCl-free system. Furthermore, the "minimum melt" temperatures are somewhat lower. The following examples are taken from the experimental investigation of von Platen (1965). The gas pressure is 2000 bars and the concentration of

Table 18-5

Ab/An ratio		"Minimum melt" temperature, °C	Ratio of "minimum melt" composition Qz	:	Ab	:	Or
∞	With HCl	660	29	:	38	:	33
	Without HCl	670	34	:	40	:	26
7.8	With HCl	655	34	:	35	:	31
	Without HCl	675	40	:	38	:	22
5.2	With HCl	680	38	:	30	:	32
	Without HCl	685	41	:	30	:	29
3.8	With HCl	690	39	:	23	:	38
	Without HCl	695	43	:	21	:	36
1.8	With HCl	700	40	:	15	:	45
	Without HCl	705	45	:	15	:	40

HCl is 0.05 moles/liter. The values are compared with those of the HCl-free system, only containing H_2O as a volatile component.

In order to understand the process of anatexis of gneisses, it must be realized that the composition of the gaseous phase is commonly only a minor factor, whereas water pressure and the ratio of components Ab/An are of major importance.

In the absence of an H_2O-rich gaseous phase between the mineral grains of a gneiss, anatexis cannot take place at the stated temperatures. Only volatile constituents, which under pressure can dissolve in the melt, are able to lower the melting temperatures drastically. If H_2O is absent, the temperatures required for partial melting of gneisses are higher (see p. 316). However, temperatures of 650° to 700°C are certainly reached and exceeded in high-grade metamorphism. In the presence of H_2O these temperatures are sufficiently high to induce anatexis at H_2O pressures as low as 4 to 2 kb. The maximum temperature attained in metamorphic terrains may be about 800°C. The amount of water present has no effect on the temperature at which the first melt is formed or on the composition of this melt. Only the amount of melt formed by anatexis depends on the amount of H_2O (+ HCl, etc.) available. A "minimum melt" or any other melt in the system here considered contains H_2O as well as silicate components and SiO_2. At 3 kb, the amount of H_2O in the melt is 8 weight percent if the melt is saturated with H_2O. If the ratio of components of a mineral mixture corresponds to the "minimum melt" composition, but only 2% H_2O instead of 8% or more is available, the total amount of minerals cannot be melted; the amount of H_2O is insufficient. A very considerable temperature rise is required to melt the remaining crystalline substances completely. Tuttle and Bowen (1958) give a fine experimental example of this, which is referred to in Figure 18-11.

As a general rule, the maximum amount of melt which potentially could be formed by anatexis can only be produced if sufficient H_2O is available. Observations in migmatite complexes suggest that sufficient H_2O is often present during anatexis. However, the writer (unpublished data) has investigated migmatite areas in the Black Forest (Germany) and in Southern Norway where only narrow portions of the gneisses underwent anatexis while the predominant part of the rocks remained unaffected. Determination of the solidus temperatures of the leucosomes, on the one hand, and of the unaffected gneisses in the immediate neighborhood, on the other, revealed in most cases identical solidus temperatures in the presence of water. This shows clearly that no water was available at the sites of the unaffected gneisses (the paleosomes) and thus anatexis was not possible at these sites.

Further Temperature Rise

So far, those principles have been discussed which are pertinent to the beginning-of-melting stage of anatexis in plagioclase + alkalifeldspar + quartz rocks. The early-stage melts have granitic compositons that are situated in the central field of Figure 18-6 or, more precisely, on the cotectic line P-E_5 which is valid for a given water pressure. However, a further temperature increase of about 10° to 30°C brings about the end of the early stage of melting, marked by the complete melting of one of three mineral phases; plagioclase, alkali feldspar, or quartz. As this phase disappears, the composition of the melt leaves the cotectic line P-E_5 and moves onto one of the three cotectic surfaces shown in Figure 18-4.

In many common cases of anatexis of paragneisses, the melt will now be situated on the plagioclase + quartz + melt + gas cotectic surface (as will be pointed out on p. 309ff). With only slight increases in temperature the melt will change its composition on that cotectic surface *but it will remain in parts of the surface representing relatively low-temperature melts.* Winkler *et al.* (1975 and 1977) have explored, at 5 and 7 kb water pressure, the low temperature regions in the system Qz-Ab-Or-An-H_2O. The authors outlined isotherms on the three cotectic surfaces. A perspective view is given in Figure 18-7 which is valid for 5 kb water pressure; it shows the isotherms 670°, 685°, and 700°C on the two cotectic surfaces plag + qtz + L + V and plag + alk feldsp + L + V. The exact data of the melt compositions that are situated on isotherms on all three cotectic surfaces at 5 kb are presented in Figure 18-8; Figure 18-9 is valid for 7 kb.

These two figures are believed to be significant in any discussion on genetic aspects of granitic and granodioritic rocks. The reader is therefore asked to make himself familiar with these representations. The following remarks should be useful.

Any composition which is situated within the space of the tetrahedron Qz-Ab-Or-An can only be properly represented in a plane by using a projection and by providing, in addition, for each point of the projection, a coordinate value indicating the amount of the fourth component which lies outside the triangular plane. Thus, the ratio of three components is shown in triangular coordinates, and the amount (in wt. %) of the fourth component is indicated by a number. This number has been drawn in the figures at the spot where the projected point would plot. From the ratio of three components (to be read off the projection) and the value of the fourth component, the complete melt composition can be calculated.

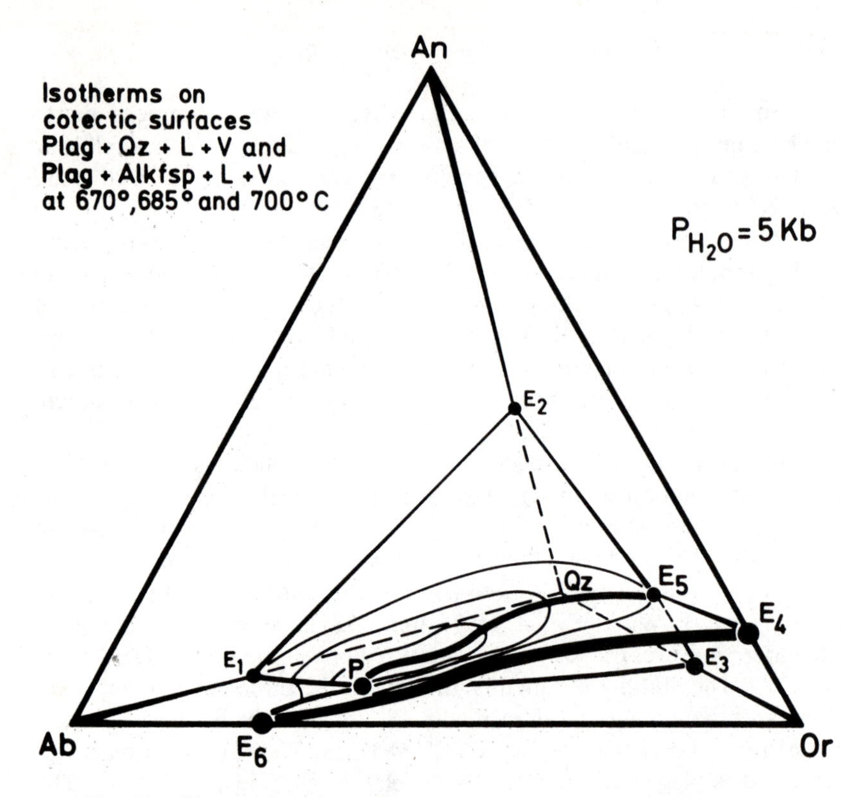

Fig. 18-7 Perspective view of the system Qz-Ab-Or-An-H_2O with low temperature regions, indicated by isotherms, shown on the cotectic surfaces plag + qtz + L + V and plag + alk feldsp + L + V.

Points on the cotectic surface plagioclase + alkali feldspar + melt + vapor are best represented by radial projection from the An apex downward onto the Qz-Ab-Or side of the tetrahedron. The number at a projection point indicates the An content of the melt in weight percent. Thin lines connecting points representing melt compositions that exist at a specific temperature are isotherms on the cotectic surface.

The cotectic surface between points P, E_5, and E_3 where a melt coexists with quartz, alkali feldspar, and vapor is roughly perpendicular to the previously mentioned cotectic surface (see Figure 18-4). Therefore, a different apex must be chosen for the projection. It is obvious from the arrangement of the surface quartz + alkali feldspar + melt + vapor in the space of the tetrahedron that points on this cotectic surface are best projected radially from the Qz apex onto the An-Ab-Or side. This is shown in the lower part of Figures 18-8 and 18-9. Again, numbers at the spot of analyzed melts indicate the amount of the fourth compo-

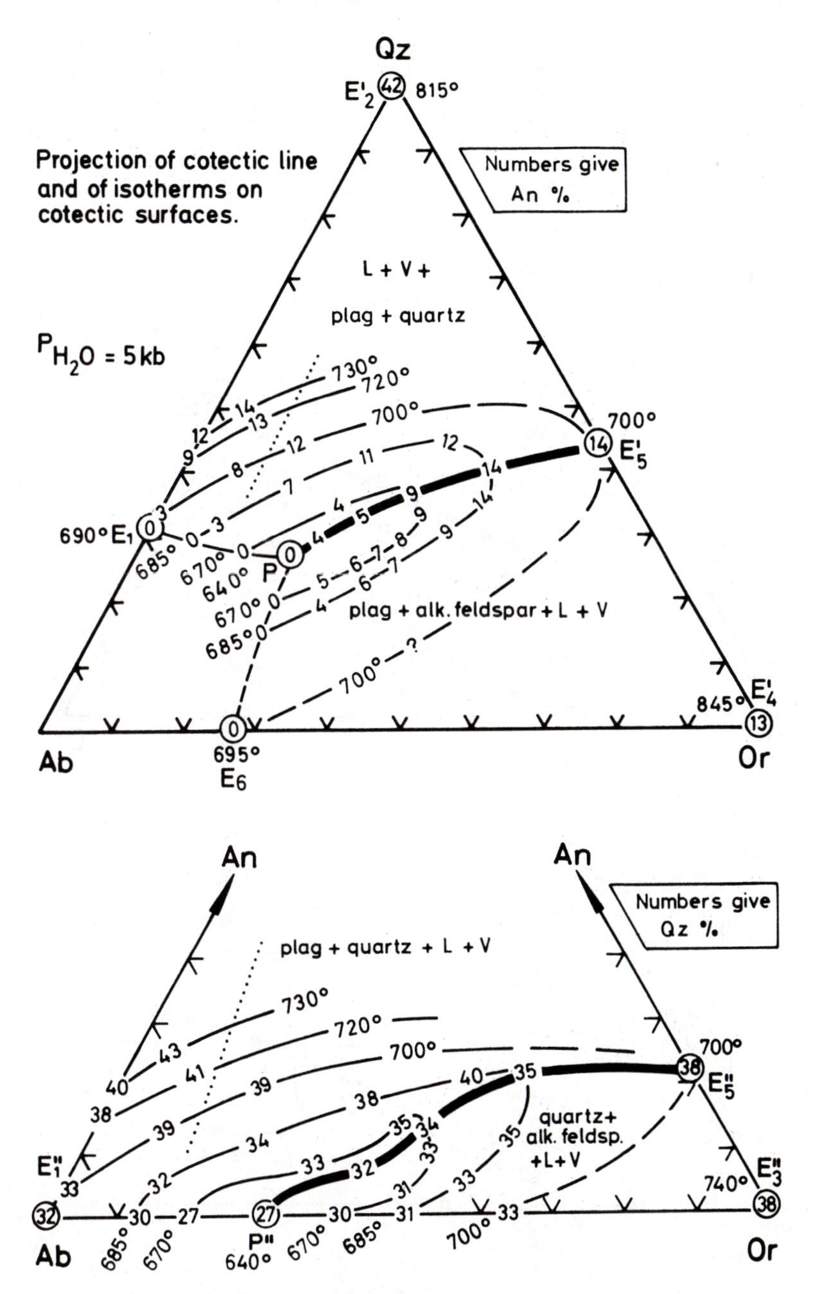

Fig. 18-8 Projection of the isobaric cotectic line P-E_5 and of isotherms on the three cotectic surfaces of the system Qz-Ab-Or-An-H_2O. Data valid for 5 kb water pressure. Numbers give percent An (top) and Qz (bottom) component. (For further explanation see text.)

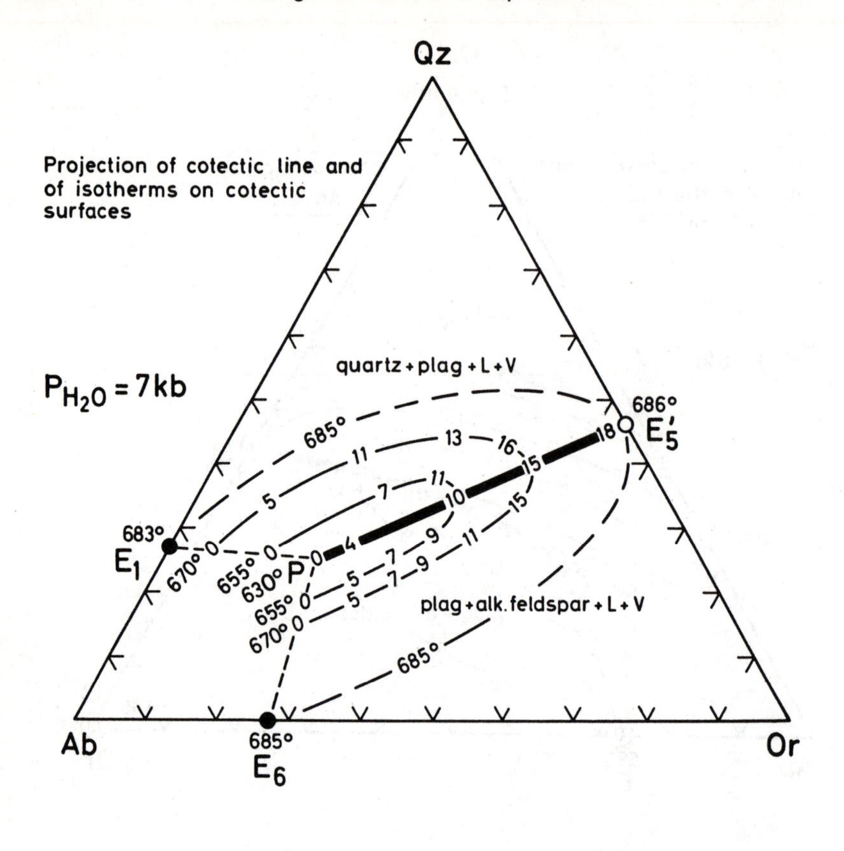

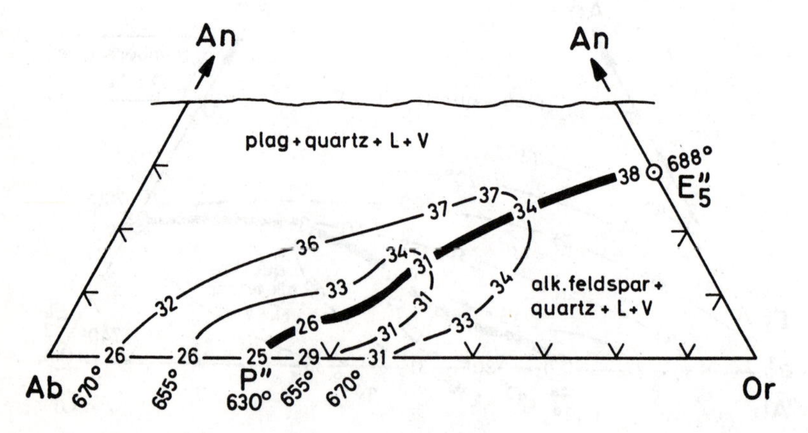

Fig. 18-9 Projection of the isobaric cotectic line P-E_5 and of isotherms on the three cotectic surfaces of the system Qz-Ab-Or-An-H_2O. Data valid for 7 kb water pressure.

nent in weight percent; *i.e.*, in this case the amount of Qz. The points are connected by isotherms shown as thin lines. The thick line is the cotectic line P-E_5. This cotectic surface is only slightly bent in a complex way.

The third cotectic surface where melts coexist with plagioclase, quartz, and vapor is strongly inclined towards the projection rays from the An apex as well as from the Qz apex. Therefore, it is best to show points that are situated on the cotectic surface plagioclase + quartz + melt + vapor in projection on side Qz-Ab-Or as well as on side An-Ab-Or. Thus, in the upper and lower parts of Figures 18-8 and 18-9, in the parts above the projection of the cotectic line (thick line), isotherms on the plagioclase + quartz + liquid + vapor cotectic surface are shown.

When the positions of the isotherms on the three cotectic surfaces are compared at 5 and 7 kb, *i.e.*, if Figure 18-8 on one hand and Figure 18-9 on the other are compared, it is obvious that the isotherm positions resemble each other very closely, provided a temperature reduction of 10° to 15°C is applied to the values obtained at 5 kb as compared with those at 7 kb. The compositions are only slightly different by not more than 1–2% of a component. The same result has been pointed out when comparing the positions of the cotectic line P-E_5 at 5 and 7 kb.

When we are not concerned about the exact value of temperatures but rather about all the possible melt compositions that may exist at relatively low temperatures, *i.e.*, at temperatures surpassing solidus temperatures by only a few tens of degrees, the isotherms on the three cotectic surfaces as determined at 5 kb H_2O pressure can be applied to evaluate melt compositions that have formed under very different pressures ranging from about 3 kb to very high pressures of 10 kb. Only at low pressures of 2 kb and less will deviations from the conditions determined at 5 kb be large enough to make a further correction necessary. This correction involves mainly a relative increase of the Qz component in low-temperature melts. It is to be expected that very close to the cotectic surfaces where melts coexist with only one solid phase, namely, with either plagioclase, quartz, or alkali feldspar, temperatures will have increased only slightly as compared to those for melt compositons nearby but which are located on the cotectic surface and coexist with two solid phases. However, when melt compositions that coexist with only one solid phase of our system are located not very close to but at a larger distance from a cotectic surface or from the cotectic line (at any given pressure), the temperature is expected to increase with the distance from the cotectic surface or line. This should be most pronounced in the case of melts situated in the quartz space and in the plagioclase space of the Qz-Ab-Or-An tetrahedron and less pronounced when melts are situated in the alkali feldspar space. Proof of the former cases has been given by Winkler and Breitbart (1978).

Relatively low temperature melts in the alkali feldspar space plot between (a) the low temperature part of the cotectic surface plagioclase + alkali feldspar + L + V, (b) the low temperature part of the cotectic surface quartz + alkali feldspar + L + V, and (c) the K-rich alkali feldspar + L + V field of the Qz-Or-Ab-H_2O system (the base of the tetrahedron as shown in Figure 18-4). Therefore, when a melt composition plots both in the upper and the lower part of Figure 18-8 in an area of the alkali feldspar field which is surrounded by a low temperature isotherm (say that of 685°C), this melt can also be taken as a relatively low temperature melt, although such a melt may have a temperature several tens of degrees higher than that melt which has an adjacent composition but which is located on one of the two cotectic surfaces.

For melts located in the plagioclase space or in the quartz space, the temperature increases very substantially with increased distance from a cotectic surface. Melts in the plagioclase space having about 2% more An content and thus lying still very close to the quartz + plagioclase + L + V cotectic surface have a temperature only 10° higher than melts on the cotectic surface in the low temperature area. But with larger departures from the cotectic surface a very marked temperature increase has been observed (Winkler and Breitbart, 1978).

It follows from this discussion that a rather large compositional variety of low temperature anatectic melts can be formed. These are situated in the An-poor part of the Qz-Ab-Or-An tetrahedron. When projected onto the Qz-Ab-Or composition triangle, it occupies a large central part, which, after considering a pressure range of about 2 to 10 kb, may well coincide with the frequency distribution of the composition of granitic-granodioritic rocks, as shown in Figure 18-14. However, attention is drawn to the discussion on pp. 333ff.

The Petrogenetic Significance

The significance of the above considerations becomes particularly evident when the previous petrogenetic interpretations of granitic compositions are reviewed.

Since the work of Tuttle and Bowen (1958), the system SiO_2-$NaAlSi_3O_8$-$KAlSi_3O_8$-H_2O has been considered by many workers as the theoretical framework of the petrogenesis of granitic rocks. Commonly, the weight percent of the three components designated as Qz, Ab, and Or were calculated from the chemical analyses of granitic rocks and plotted in the Qz-Ab-Or triangle. The minimum melting compositions and, at H_2O pressures higher than 2.3 kb, the eutectic points trace out a line in response to rising pressure. If rocks plotted close to this line, valid for the pressure range of 0.5 to 10 kb, it was assumed that they were magmatic granites, *i.e.*, that they had crystallized from a melt. In contrast, granites plotting at a greater distance from this line were thought to represent products of metasomatic transformation ("granitization") of preexisting rocks. However, these conclusions are not valid. The system Qz-Ab-Or-H_2O, sometimes referred to as the "granite system," provides no help in understanding the genesis of granitic rocks. The reasons for this statement, based on the previous discussion, are summarized below:

a. Granitic melts having the minimum melting or eutectic composition in the system Qz-Ab-Or-H_2O can be produced only in the early stage of anatexis of albite-bearing gneisses. Gneisses of this composition are rare.
b. A composition approximating the minimum melting composition exists only within a very narrow temperature range beyond the beginning of melting.
c. At any given pressure, late differentiates of calc-alkaline magmas and early anatectic melts of gneisses consisting of quartz, K feldspar, and plagioclase (not albite) are not located near the minimum melting or eutectic composition in the system Qz-Or-Ab-H_2O. Instead, they plot on the isobaric cotectic line of the system Qz-Or-Ab-An-H_2O. The exact position of the melt on that cotectic line is determined by the composition of both alkali feldspar and plagioclase coexisting with quartz, melt, and vapor at any particular T and P_{H_2O}.
d. The composition of melts formed at temperatures higher than those of the early stage of anatexis (or the late stage of crystallization) commonly departs considerably from the relevant cotectic line. The great variety of relatively low temperature melts are situated at a given pressure on one of the three cotectic surfaces of Figure 18-4 or in a narrow volume close to these surfaces. They have a granitic-granodioritic composition which can be produced by the anatexis of common gneisses and, consequently, they may form, after separation, large bodies of granitic igneous rock.
e. In order to ascertain whether a given granitic composition is on, close to, or further away from the cotectic line or a cotectic surface, it is necessary to know the amount of An component in addition to the ratio Qz:Ab:Or, and/or the amount of the Qz component in addition to the ratio An:Ab:Or. In other words, it is necessary to know the amounts of all four components Qz,

Ab, Or, and An that constitute all the quartz, plagioclase, and alkali feldspar of a granitic or granodioritic rock. These data can be obtained when the composition of the two feldspar species and the modal composition of the rock are known. The ratio Qz:Ab:Or:An can also be obtained by calculating the meso-norm from the rock analysis. Mielke and Winkler (1979) have used to advantage a revised method of meso-norm calculation, which differs somewhat from the methods of Barth (1959) and Parslov (1969).

From the position of a rock composition within the Qz-Ab-Or-An tetrahedron, the crystallization sequence of a melt having this composition can be deduced, as well as the reverse, namely the melting steps. Examples for this have been given by Winkler *et al.* (1975) and Winkler and Breitbart (1978).

The Trondhjemitic Part of the Qz-Ab-Or-An-H_2O System

In trondhjemites and tonalites, the amount of alkali feldspar relative to plagioclase is less than 1:10 (Streckeisen, 1973). Many trondhjemites and tonalites do not contain any alkali feldspar. Nevertheless, the chemical analysis of such rocks reveals a small amount of Or component, which is present as a constituent in plagioclase solid solution. Trondhjemites and tonalites cannot be adequately represented in the Or-free system Qz-Ab-An-H_2O; in fact, their composition plots in the Or-poor part of the system Qz-Ab-Or-An-H_2O.

The solubility limit of Or in plagioclase of compositions Ab_{85}/An_{15} and Ab_{75}/An_{25} has been determined approximately (Winkler *et al.*, 1977). The dotted lines in the projections of Figure 18-8, near the Ab-An side and near the Ab-Qz side approximately indicate the boundary between the one feldspar (plagioclase) field and the two feldspar (plagioclase and alkali feldspar) field. Thus, from a melt situated in the Or-poor part of the system, left of the dotted line, only quartz and plagioclase will crystallize. Therefore, during crystallization, the melt composition will move on the cotectic surface plagioclase + quartz + L + V in a direction of lower temperature but will never reach the cotectic line P-E_5; in other words, no alkali feldspar will crystallize. Similarly, melting a plagioclase-quartz rock not containing alkali feldspar cannot produce an early anatectic melt situated on the isobaric cotectic line P-E_5; it must lie on a low-temperature portion of the plagioclase + quartz + L + V cotectic surface. If, however, a trondhjemitic gneiss does contain a small amount of alkali feldspar, the earliest formed anatectic melt must lie on the isobaric cotectic line; the amount of this melt can only be small.

Experimental Anatexis of Rocks Composed of Alkali Feldspar, Plagioclase, and Quartz

When rocks consisting of alkali feldspar, plagioclase, and quartz are subjected to anatexis, the process of partial melting in the presence of excess water can be well understood in terms of phase relations in the system Qz-Ab-Or-An-H_2O. Anatexis will also be considered in situations where only a limited amount of water is available.

Anatexis with Water Available

Most experiments have been carried out with sufficient water to ensure the presence of a vapor phase consisting essentially of water. Thus P_{total} was very nearly equal to P_{H_2O} and the anatectic melts were saturated with water. Winkler and Winkler and von Platen were the first to investigate, at the condition of $P_{H_2O} = P_{total}$, the metamorphism and subsequent anatexis of clays (1957), NaCl-bearing clays (1958), $CaCO_3$-bearing clays (1960), and graywackes (1961). Later, the anatexis of gneisses was investigated by Steuhl (1962), von Platen and Höller (1966), and investigators of other laboratories.

Most of the earlier work was carried out at 2 kb. In nature, however, greater pressures commonly prevailed during anatexis. Therefore, experimental anatexis of common paragneisses has been carried out at 5 kb (Winkler and Breitbart, 1978). The unpublished data are presented here. As representative examples, two paragneisses are treated, one containing much alkali feldspar, in addition to plagioclase and quartz, and the other one containing only a small amount of alkali feldspar.

Example 1: A paragneiss from the southern part of the German Black Forest containing a large amount of alkali feldspar. Its modal composition is:

	Vol.%	Wt.%
Plagioclase, An_{28}	34	33
Alkali feldspar, Ab_{17}	20	20
Quartz	26	25
Biotite	20	22

Melting experiments carried out by raising the temperature from run to run in small increments of 5–10°C give the following results at 5 kb H_2O pressure:

655°C	Solidus
665°C	Alkali feldspar-out; residue plag + qtz + bio; about 60% melt
675°C	Quartz-out; residue plag + bio; about 70–75% melt
715°C	Plagioclase-out; residue bio; about 80% melt

The experiments show that a temperature rise of only 10°C above the solidus temperature (655°C at 5 kb) leads to melting of the total amount of alkali feldspar (20%) together with cotectic (almost equal) amounts of quartz and plagioclase. At temperatures above 665°C, only quartz and plagioclase (and biotite) are present as solids suspended in the melt. It is remarkable that a temperature rise of merely 10°C (from 655° to 665°C) causes about 60% of the rock to melt.

In the temperature interval from 665° to 675°C, the small amount of remaining quartz melts, together with an appropriate amount of plagioclase. At 675°C, only 20°C above the solidus temperature, the total amount of quartz and alkali feldspar and most of the plagioclase have formed a melt of granitic composition corresponding to 70–75% of the original rock. The remaining plagioclase (7–10%) and the practically unchanged amount of biotite (20%) form the solid residue.

A further temperature increase of 40°C, from 675° to 715°C, is required to melt the remaining small amount of plagioclase (7–10%). The melt still has a granitic composition and most of the biotite is present as a solid.

Generally, a temperature rise of 60°C above the solidus temperature causes part of the biotite to dissolve incongruently in the granitic melt, thereby adding to the amount of $KAlSi_3O_8$-component in the melt (see also p. 321ff). However, in this particular experiment, a decrease in the amount of biotite could not be detected. Possibly this resistance to melting is due to the unusually Mg-rich composition of the biotite.

The various steps in the anatexis of this paragneiss can be understood very well with the help of the granitic model system Qz-Ab-Or-An-H_2O. Figure 18-10 illustrates the process of anatectic melting of two different paragneisses. The sequence of projection points relevant for the paragneiss with 20% alkali feldspar is 4-6-10-12. Note that all the points lie inside the tetrahedron Qz-Ab-Or-An. The points are projected from the An apex onto the Qz-Ab-Or plane, and the number noted at each projection point indicates the height above that plane in terms of the percentage of An-component.

Melting starts at point 4 situated on the cotectic line. Alkali feldspar + plagioclase + quartz coexist with liquid and vapor along the cotectic line from point 4 at 655°C to point 6 at 665°C. As the temperature rises, the melt composition must leave the cotectic line, because all alkali feld-

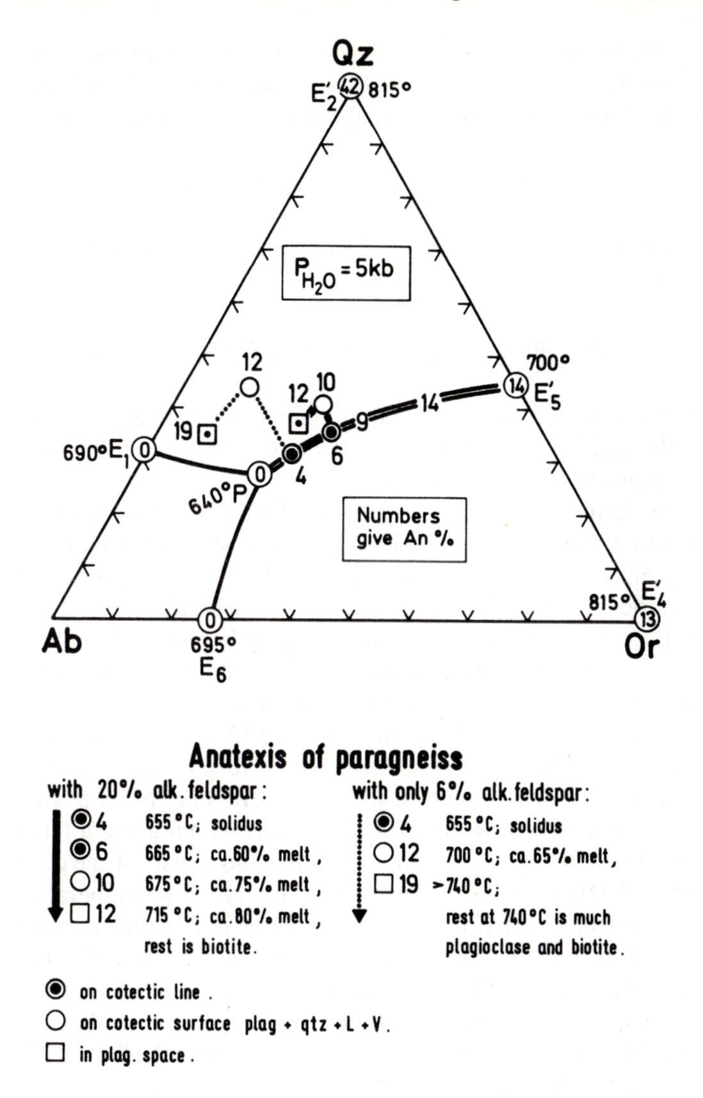

Fig. 18-10 Stages of experimental anatexis of two paragneisses as indicated in the legend.

spar has melted, together with cotectic amounts of plagioclase and quartz. From point 6 to point 10 only quartz and plagioclase coexist with liquid and vapor along the relevant cotectic surface. At point 10 (675°C) all quartz, together with a cotectic amount of plagioclase, has melted. At higher temperature, only some plagioclase remains; the melt is now situated in the plagioclase space of the granitic system. It takes a relatively large temperature rise from 675° to 715°C, as shown by the experiments, to dissolve the small amount (7–10%) of remaining plagioclase. Point 12 (square) gives the ratio of the components for the paragneiss

33% Qz : 35% Ab : 20% Or : 12% An;

hence, number 12 = 12% An designates the square which has been projected at

37% Qz : 40% Ab : 23% Or, where Qz + Ab + Or = 100.

Since the amount of biotite has remained unchanged, square 12 also gives the composition of the melt at 715°C; here, about 80% of the previous gneiss has become liquid.

The state of having 80% of the rock melted was reached at a relatively low temperature which is only 60°C above the solidus. It is even more remarkable that at a temperature only 20°C above the solidus, 70–75% of the rock was already melted. This behavior is generally expected in the case of all paragneisses with a relatively high percentage of alkali feldspar, approaching that of quartz and plagioclase. Indeed, the experimental results can be deduced from the composition of the rock in terms of Qz : Ab : Or : An in the granitic system. The square numbered 12 is situated inside the plagioclase space close to the plag + qtz + L + V cotectic surface and also close to the cotectic line. Such a composition must give rise to a large percentage of melt within a small temperature interval above the solidus, because melting along part of the cotectic line generates large amounts of melt within a very small temperature range. This is particularly obvious if a rock happens to have the composition of a point on the cotectic line. In this case, the whole rock will melt within a temperature range of a few degrees.

Example 2: A paragneiss from a locality close to that of the first paragneiss, but containing only a small amount of alkali feldspar. Its modal composition is:

	Vol.%	Wt.%
Plagioclase, An_{27}	59	58
Alkali feldspar, Ab_{17}	6	6
Quartz	26	26
Biotite	9	10

Melting experiments at 5 kb H_2O pressure gave the following data:

655°C	Solidus
≈ 657°C	Alkali feldspar-out; residue plag + qtz + bio, approx. 20% melt
700°C	Quartz-out; residue plag + bio, about 65% melt
740°C	Still appreciable plag + bio present, approx. 75% melt

As in the previous case, a decrease in the amount of biotite could not be detected, again suggesting a rather Mg-rich biotite. Due to the much smaller amount of alkali feldspar present in the paragneiss, the course of melting is distinctly different from that of the first example. Again, this is best shown in Figure 18-10 by the sequence of points 4-12-19.

When the solidus temperature has been reached at point 4 on the cotectic line, the small amount of alkali feldspar (6%) in the rock melts together with some plagioclase and quartz. Only a few degrees above the solidus temperature, the total amount of alkali feldspar melts and, therefore, the melt composition must leave the cotectic line P-E_5 and move onto the cotectic surface plag + qtz + L + V. With increasing temperature, it moves along this surface from about 657° to 700°C, at which temperature the total amount of quartz melts. Here, at point 12, the melt has incorporated all the alkali feldspar (6%) and quartz (26%) of the paragneiss, and 35–30% of the original plagioclase. About 65% of the rock has melted and 23–28% plagioclase together with 10% biotite of the original paragneiss constitute the solid residue at 700°C and 5 kb H_2O pressure.

During the process of anatexis, the melt composition changed from granitic to granodioritic to almost trondhjemitic over a temperature range of about 50°C. It is not surprising that at the stage of anatexis where the melt leaves the cotectic surface plag + qtz + L + V, at 700°C and 5 kb, about 25% plagioclase is still present. This is simply due to the fact that the gneiss contained a very large amount of plagioclase (58%), more than twice as much as the amount of quartz.

Above 700°C, the melt composition is situated inside the plagioclase space of the granitic system and a very considerable temperature increase will be required to dissolve appreciable amounts of the remaining plagioclase. This is evident from the fact that square 19 is situated very high above the plag + qtz + L + V cotectic surface within the plagioclase space. It is unlikely that metamorphic temperatures would have been sufficiently high to dissolve all of the remaining 25% plagioclase. Therefore, the position of point 19 (square) in the granitic system, giving the ratio 29% Qz : 46% Ab : 6% Or : 19% An of the paragneiss, would never represent the composition of a liquid, *i.e.*, of magma with dispersed biotite as the only solid phase. Rather, an appreciable amount of solid plagioclase in addition to biotite remains suspended in the melt.

After crystallization, the total rock will, of course, have a composition given by the above ratio. This represents a trondhjemite (tonalite) of quite common composition.

The preceeding considerations are of great petrogenetic significance because they bear on the question of whether granitic to tonalitic magmas are totally liquid or not.

The two examples of anatexis of common paragneisses demonstrate two important points:

a. Anatexis of the same gneiss can produce melts of different compositions, depending on the temperature reached during high-grade metamorphism.
b. The trend of the melt composition generally does not follow a simple path because it is governed by phase relations in the system Qz-Ab-Or-An-H_2O.

Other aspects of anatexis are demonstrated in an experimental study of four paragneisses derived from graywackes in the same area. The results are summarized in Table 18-6, taken from Winkler and von Platen (1961), with some corrections applied. The experiments were carried out at 2 kb water pressure, and the amounts given in the table should be regarded as estimates rather than exact quantities. The four rocks have a very similar mineralogical composition and contain only a small amount of alkali feldspar. However, the An content of the plagioclase increases from rocks (a) to (d) and the ratio of quartz to plagioclase varies. Due to these differences in chemical composition, the rocks exhibit a very different behavior during experimental anatexis.

The following points are evident from Table 18-6:

c. Melting begins at progressively higher temperatures as the An content of the plagioclase increases.
d. The amount of melt formed at a given temperature is very different from rock to rock.
e. The type and amount of solids remaining even at rather high temperature are very different.

These observations can be understood in terms of phase relations in the Qz-Ab-Or-An-H_2O system. Given the ratios of the components Qz:Ab:Or:An from the chemical analyses of the rocks, it can be deduced from the granitic system that, at the same temperature, rock (a) will have only a small amount of quartz as residual solid, rocks (b) and (c) will have considerably more residual quartz, and rock (d) will have plagio-

Table 18-6. The anatexis of four paragneisses produces different amounts of melt at the same temperature and pressure: P_{H_2O} = 2000 bars.

Original rock: graywacke	Ab/An ratio	Beginning of melting ± 5°C	Amount of melt at different temperatures, °C 700	720	740	770
(a) IV/25	83 : 17	685	48	59	68	73
(b) IV/29	85 : 15	685	20	49	68	68
(c) IV/16	69 : 31	700	—	31	48	63
(d) 1 d	62 : 38	715	—	30	50	70

Mineralogical composition (in weight percent) of the highest-grade gneisses:

Original rock	Quartz	Plagioclase; An content given in ()	K feldspar	Cordierite	Biotite	Sillimanite	Opaque minerals
(a) IV/25	31	31 (13)	7	8	11	8	4
(b) IV/29	47	28 (18)	5	8	5	5	3
(c) IV/16	52	31 (30)	4.5	7	5	—	2
(d) 1 d	28	44 (40)	9	4	10	—	4

Approximate amounts (in weight percent) of the unmolten residual minerals in comparison to the amount of melt at 770°C:

Original	Quartz	Plagioclase	Cordierite	Biotite	Sillimanite	Opaque minerals	Amount of melt
(a) IV/25	3	—	11	4	5	4	73
(b) IV/29	23	—	4	—	2	3	68
(c) IV/16	30	—	5	—	—	2	63
(d) 1 d	—	15	6	5	—	4	70

clase but no quartz as residual solid, in addition to minerals like cordierite, biotite, etc. For readers wishing to convince themselves, the ratios of components are given below. It should be remarked that the system valid for 5 kb (Figure 18-8) can be used (disregarding the absolute temperatures) to interpret the experimental results obtained at 2 kb.

	Qz :	Ab :	Or :	An	Qz :	Ab :	Or
(a) IV/25	49	37	6.5	7.5	53	40	7
(b) IV/29	62	29	4	5	66	30	4
(c) IV/16	61	25	3	11	68	28	4
(d) 1/d	38	32	10	20	47	40	13

Minerals like cordierite, garnet, ore, sillimanite, and hornblende, either previously present in the rock or formed during anatexis, will remain as a crystalline residue at any stage of anatexis, because only small amounts, if any, can dissolve in the anatectic melt. The special role of biotite will be discussed later. However, if much biotite is present in a rock being subjected to anatexis, biotite is a major constituent of the crystalline residue and it may be associated with hornblende. In addition to these mafic minerals, one or two of the minerals quartz, alkali feldspar, and plagioclase may be present as well in the crystalline residue; see Table 18-6. Plagioclase *or* quartz are the most common felsic minerals in the "melanosome" of migmatites, but some instances are known where both quartz and plagioclase are present. The former statement is very well substantiated by extensive petrographic observation. Thus, quoting Mehnert (1968), "it is remarkable that melanosomes exist that contain as the sole light mineral only quartz, and again others contain only feldspar (plagioclase or potash feldspar)." This is now well understood in the light of the system Qz-Ab-Or-An-H_2O: The anatectic melt was in equilibrium with either quartz or plagioclase and only rarely was its composition still situated on the cotectic surface where quartz and plagioclase could be in equilibrium with the melt and thus form part of the crystalline residue.

By the process of anatexis, a gneiss or quartz- and feldspar-bearing mica schist is "split" into a melt portion which consists essentially of feldspar components and quartz component, thus having a granitic, granodioritic, or tonalitic composition, and into a crystalline residue enriched in Mg, Fe, Al, and possibly Ca. The portion of anatectic melt is very large in paragneisses derived from graywackes.

The anatexis of gneisses derived from clays and shales gives rise to a smaller amount of melt, but the melt still comprises about half of the original paragneiss if the temperature of metamorphism exceeds by 50° to 70°C the temperature of the beginning of anatexis. In the presence of free water, the temperature of the beginning of melting decreases with increasing pressure (see Table 18-2). The lowest possible temperatures of the beginning of melting in gneisses at various water pressures are shown by curve 1 in Figure 18-11.

Anatexis with Limited Amount of Free Water Available

In the presence of free water, the anatexis of gneisses composed of alkali feldspar, quartz, and plagioclase begins at the temperatures given in Table 18-2 and shown in Figure 18-11, if the plagioclase is pure albite. However, if the plagioclase contains An component, the temperature of the beginning of melting is increased (see Table 18-4). The lowest possi-

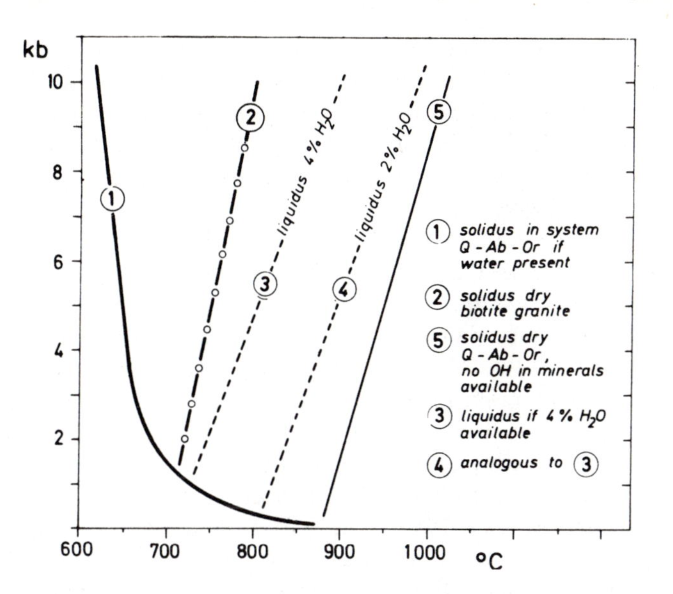

Fig. 18-11 Various solidus and liquidus relations in granitic rocks (see text).

ble temperatures of the beginning of anatexis in gneisses are identical with the beginning of melting in the system Q-Ab-Or-H_2O, shown as curve 1 in Figure 18-11.

As has been pointed out (Table 18-6), the *amount of melt* produced at a given temperature, say 50°C higher than the beginning of anatexis, strongly depends on the proportion of alkali feldspar, plagioclase, and quartz and on the compositions of the feldspars in a gneiss. If the bulk composition corresponds to a melt on the cotectic line P-E_5 at a given pressure, all minerals would melt well before a temperature rise of 50°C had taken place.

Let us assume that 50% of the composition of a gneiss corresponds to a melt composition on the cotectic line P-E_5 valid for the acting water pressure. This gneiss will yield 50% melt within the initial 10° to 30°C of anatexis. However, this statement is correct only if enough free water is available to saturate that amount of melt. If anatexis takes place at 15 km depth, corresponding to about 4 kb pressure, one half of 9% = 4.5 weight percent water is required to saturate the melt. If this amount of water is not available, a smaller amount of melt will be formed if the temperature remains constant. As Tuttle and Bowen (1958) pointed out, "the amount of water and other volatiles available to flux the silicates determines the amount of liquid formed at any depth at which the melting temperature and pressure have been reached."

Therefore, if during anatexis only a limited amount of free water is available, only a portion of the cotectic mixture in the gneiss can be melted at a given temperature and pressure. In other words, in addition to the leucosome and melanosome (residue portion), a migmatite may have portions unaffected by anatexis (the so-called paleosome). A higher An content of the unaffected portions, requiring a higher temperature for melting, could be the reason for the lack of anatexis. However, in the absence of such mineralogical differences, it is most likely that not enough water was available to liquefy all portions of a gneiss that potentially could melt, given sufficient water. This point deserves close scrutiny in petrographic, chemical, and experimental work, and should also take into account the following: The probability of inadequate amounts of water for maximum melting at any given *P-T* conditions increases with pressure because more water is needed to saturate a melt at higher pressures. For saturation a granitic melt requires about 8 weight percent water at 3 kb, 12% at 5 kb, and 20% at 9 kb.

Tuttle and Bowen (1958, p. 122) illustrated that the lack of a sufficient amount of water is an effective obstacle to complete melting of granites. This consideration also applies to the early-stage melts produced by the anatexis of gneisses. According to Tuttle and Bowen (1958, p. 123), the anhydrous chemical composition of most granites is such "that given the required quantity of volatiles they would melt completely, or nearly completely, at the temperature of the beginning of melting."[4] The beginning of melting of a "minimum" granite takes place at 4 kb and 655 C, and at slightly higher temperature the granite melts completely if about 10 weight percent water is available. However, if only 2% water is present, the temperature has to be increased to 870°C in order to bring about complete melting; if 4% water is available, the temperature required for complete melting is about 780°C. This is illustrated in Figure 18-11, where the liquidus curve for granite with 2% water available has been taken from the work of Tuttle and Bowen and the equivalent curve for 4% water has been estimated.

Recently, Brown and Fyfe (1970) have suggested that free water may not be available during anatexis in granitic terrains. In this case the OH-bearing minerals of a gneiss will supply water on decomposition and thus initiate the formation of a melt. Commonly, the OH-bearing minerals are biotite and muscovite. Brown and Fyfe performed experiments to determine the beginning of melting in a dry granite and in a dry tonalite with biotite, muscovite, or hornblende as OH-bearing minerals.

---

[4]We now know that this statement of Tuttle and Bowen is not entirely correct. Indeed, it is known that some granites melt completely at a temperature only slightly above the solidus, but many granitic rocks behave differently, *e.g.*, as shown by Winkler and Breitbart (1978).

Most relevant to natural occurrences is the melting behavior of the biotite granite; its solidus is shown as curve 2 in Figure 18-11. Curve 2 has a positive slope like curve 5 of a dry mixture of quartz + alkali feldspar + plagioclase (without biotite).

The melts formed in the temperature range 20°C above the beginning of melting have granitic-granodioritic compositions, similar to those obtained when excess water is available. However, the temperatures necessary to form melts in the absence of free water are approximately 100°C higher at medium pressures and 150°C higher at high pressures. The melts thus formed are undersaturated in water, which is easily seen by comparing the solidus curve 2 with that of curve 1, valid for water saturation. In the case of water undersaturation, P_{H_2O} is always lower than P_{total}.

Only a small amount of biotite decomposes incongruently at the beginning of melting, and a considerable temperature rise is required to decompose a significant amount of biotite and thus increase the amount of melt. Therefore, very high temperatures are necessary to melt appreciable amounts of paragneiss commonly containing about 15 to 25% biotite. If at most only 2% water can be supplied by the decomposition of biotite, the temperature of melts must exceed 900°C at pressures greater than 5 kb. These temperatures seem to be unrealistically high for most migmatite areas. Therefore, we believe that partial melting in the absence of any free water is not of petrogenetic significance. However, the situation of water being available in limited amounts only, particularly in some layers within a paragneiss, will be common. The facts dealt with on page 300 clearly support this: there are large volumes of unaffected paragneiss (paleosomes) in migmatite areas.

Experimental Anatexis of Rocks Composed of Plagioclase and Quartz but Lacking Alkali Feldspar

In the preceding account, anatexis of rocks containing alkali feldspar has been considered. However, there are many paragneisses and quartz-plagioclase-mica schists which do not contain K feldspar. What changes occur in these rocks during anatexis? It might be thought at first that such metamorphic rocks would not give rise to a granitic or granodioritic melt during anatexis because K feldspar, which constitutes an essential component of such melts, is absent. Experiments, however, have shown that these considerations are not correct. In fact, metamorphic rocks marked by the absence of K feldspar may be the source of anatectic melts containing an appreciable amount of K feldspar component.

The antexis of rocks consisting of plagioclase + quartz + biotite + muscovite will be discussed first and then muscovite-free rocks composed of plagioclase + quartz + biotite, which are even more common, will be considered.

The equilibrium,

$$\text{muscovite} + \text{quartz} = \text{K feldspar} + Al_2SiO_5 + H_2O$$

is of particular significance in the melting behavior of muscovite-bearing rocks. The reaction curve of this equilibrium is shown in Figure 7-3. Its position indicates that the stability field of K feldspar in the presence of Al_2SiO_5 and H_2O is rather limited, *i.e.*, the stability field of muscovite + quartz is very large and, at H_2O pressures greater than 3 to 4 kb, it overlaps the pressure-temperature field of anatexis. The stability of muscovite + quartz extends as far as curve 2 in Figure 7-3. It has been observed, however, that in the presence of plagioclase, muscovite + quartz become unstable at a lower temperature (see also Figure 7-3). In this case, anatexis causes the disappearance of muscovite. The lower the An content of the plagioclase, the lower the temperature of anatexis.

As an example, the anatexis of a muscovite-plagioclase-quartz gneiss will be examined. At P_{H_2O} = 5 kb and 680°C an anatectic melt will begin to form. At this stage it is observed that muscovite disappears and the melt contains K feldspar component (Winkler, 1966). This component, together with the Ab and Qz and a small amount of An component, forms a granitic melt. Except for the effect of higher H_2O pressures causing somewhat different compositions of the anatectic melts, the results are the same as those obtained at lower H_2O pressure in the case of rocks composed of alkali feldspar in addition to plagioclase and quartz and some Al_2SiO_5. The process may be described as follows: instead of

$$\text{muscovite} + \text{quartz} \rightarrow \text{K feldspar} + \text{sillimanite} + H_2O$$

the reaction is

$$\text{muscovite} + \text{quartz} + \text{plagioclase} \rightarrow \text{K feldspar-plagioclase-quartz components in the anatectic melt} + \text{An-richer plagioclase} + \text{sillimanite} + H_2O$$

If biotite is present as well as muscovite in association with quartz and plagioclase, the muscovite breakdown reaction is succeeded at slightly higher temperature, when a melt has already formed, by the following reaction (von Platen and Höller, 1966):

2 biotite + 6 sillimanite + 9 quartz → 2 K feldspar component
+ 3 cordierite + 2 H_2O

At pressures sufficiently high to allow the formation of almandine in addition to cordierite or instead of cordierite, the following reactions take place:

2 biotite + 8 sillimanite + 13 quartz → 2 K feldspar component
+ 3 cordierite + 2 almandine + 2 H_2O

or

biotite + sillimanite or kyanite + 2 quartz → K feldspar component
+ almandine + H_2O

(The last reaction has not yet been verified by experiment but it is probably valid.)

The Al_2SiO_5 required in these reactions is produced by the previously discussed breakdown of muscovite in the presence of quartz. Some Al_2SiO_5 may have been present in the metamorphic rock in addition to muscovite + biotite prior to anatexis. (Other reactions involving biotite but not Al_2SiO_5 will be discussed later.)

The essential feature is that biotite as well as muscovite constitutes a source of K feldspar component which is produced in the presence of quartz and plagioclase at the beginning of anatexis. Plagioclase, quartz, muscovite, and/or biotite supply the components of the anatectic melts of granitic or granodioritic composition.

According to the investigations of von Platen and Höller (1966), the anatectic melts produced from quartz-plagioclase-muscovite-biotite gneisses not containing any K feldspar have the same "minimum melt" composition as melts produced from rocks in the system Qz-Ab-An-Or-H_2O having the same Ab/An ratio as the gneisses. The reactions during anatexis are different, however, if a metamorphic rock consists of biotite only, besides plagioclase and quartz, *i.e.*, if muscovite is absent. This mineral assemblage is common in paragneisses. In such rocks, the total amount of K_2O is contained in biotite. Knabe (1970b) has investigated the anatexis, at P_{H_2O} = 2000 bars, of biotite + quartz + plagioclase mixtures and of metamorphosed graywackes of corresponding composition. It is certain that, unlike muscovite, only a small amount of the biotite disappears at the beginning of anatexis. The amount of biotite diminishes as the temperature is increased, and even at temperatures 70° to 100°C higher than the temperature of the beginning of anatexis, an appreciable portion of the biotite is still preserved. Over a considerable temperature range, melt + biotite + plagioclase + quartz + gaseous phase are in equilibrium. As a rule, during anatexis a portion of the biotite forms part of the crystalline residue, together with other minerals like garnet, sillimanite, cordierite, etc. The rest of the biotite, by incongruent melting and reaction with other minerals, supplies K feldspar component for the anatectic melt.

In addition to the previously mentioned reactions producing K feldspar component from biotite and Al_2SiO_5 in the presence of quartz and plagioclase, the following experimentally observed reactions are of petrogenetic importance (Knabe, 1970a):

$$\text{Al-rich biotite} + \text{quartz} \rightarrow \text{K feldspar component} + \text{gedrite} + H_2O$$

The formation of orthorhombic gedrite has been observed at 2 kb water pressure when using as starting material Al-rich biotite (from biotite-quartz-plagioclase metagraywacke) which had $(FeO + Fe_2O_3 + MnO)/(MgO + FeO + Fe_2O_3 + MnO) = 0.5$. If this ratio is lower, *i.e.*, if the biotite is Mg richer, formation of hornblende is observed according to the following reaction:

$$\begin{gathered}\text{biotite} + \text{plagioclase} + \text{quartz} \rightarrow \\ \text{K feldspar component and albite component} + \text{hornblende}\end{gathered}$$

If the biotite involved in the reaction has a high content of Ti and Fe, ilmenite is formed during experimental anatexis. Hornblende is a reaction product as well, even though the biotite is Fe rich rather than Mg rich. At higher oxygen fugacity, magnetite is formed which incorporates an appreciable amount of Ti.

In natural occurrences, sphene + hornblende are typical reaction products, in addition to K feldspar, when biotite-plagioclase-quartz gneisses are subjected to anatexis leading to the formation of migmatites. This is demonstrated by Büsch (1966, 1970) whose petrographic studies lead him to postulate the following reaction:

$$\begin{gathered}\text{biotite} + \text{plagioclase} + \text{quartz} = \\ \text{hornblende} + \text{K feldspar} + \text{sphene}\end{gathered}$$

The breakdown of biotite in the presence of quartz and plagioclase, as indicated by the schematic reaction equations, provides a source of K feldspar component if the rock is heated through an appreciable temperature range extending beyond the beginning of anatexis. At the beginning of anatexis, only a fraction of the biotite is consumed; therefore, the amount of K feldspar component, relative to the amount of plagioclase and quartz components, is generally small. For this reason, at the beginning of anatexis, a melt of granodioritic composition is generally formed. A small increase in temperature causes very little change in composition, in contrast to the melt formed by the anatexis of paragneiss containing K feldspar (Figure 18-10). This is due to the fact that no crystalline phase melts completely at a temperature only slightly higher than that of the beginning of melting. As the temperature is increased, more biotite reacts and provides additional K feldspar component, which, together with appropriate amounts of plagioclase and quartz component, increases the amount of anatectic melt without appreciably changing the composition of the melt. Accordingly, a melt of granodioritic composition produced by anatexis of a quartz-plagioclase-biotite gneiss maintains this composition even if the temperature is increased by several tens of degrees. The rocks designated 1, 2, and 4, which are further described in the legend of Figure 18-12, gave rise to such anatectic melts of granodioritic composition. The ratio Qz:Ab:Or of the melt is shown in Figure 18-12. The plotted composition of the anatectic melt produced from rock 3 indicates that not only granodioritic melts but occasionally granitic melts may be produced as well. This happens in the case of rocks containing very Fe-rich biotite, which melts incongruently to a greater degree in comparison to other biotites at a given temperature and pressure.

If a separation of the melt from the crystalline residue occurs, as is the case in migmatites, the anatectic melts of granitic or granodioritic composition give rise to mineral assemblages of granites and granodiorites on crystallization. Also, tonalitic or trondhjemitic melts may be produced by the anatexis of K_2O-poor plagioclase-quartz-biotite gneisses, as our experiments have shown. It appears reasonable that the melt segregates not only in lenses and veins but also collects in the form of large magmatic bodies.

This process leads to a concentration of the crystalline residue consisting of biotite + hornblende + garnet ± cordierite + ore + commonly some quartz or plagioclase. Such anatectic "restites," together with metamorphic rocks which do not melt at the temperatures of anatexis, would be expected to underlie the large granitic and granodioritic plutons. Metamorphic rocks which remain solid during anatexis include amphibolites, plagioclase-free quartz-biotite schists, and, in more subordinate amounts, calc-silicate rocks, marbles, and pure quartzites.

The formation of large amounts of granitic, granodioritic, trondhjemitic, and tonalitic melts by anatexis is a process having a sound physicochemical basis. In terrains of high-grade metamorphism, the formation of such melts is inevitable if quartz-feldspar-mica gneisses and H_2O are present. Melting begins, independent of the amount of free H_2O

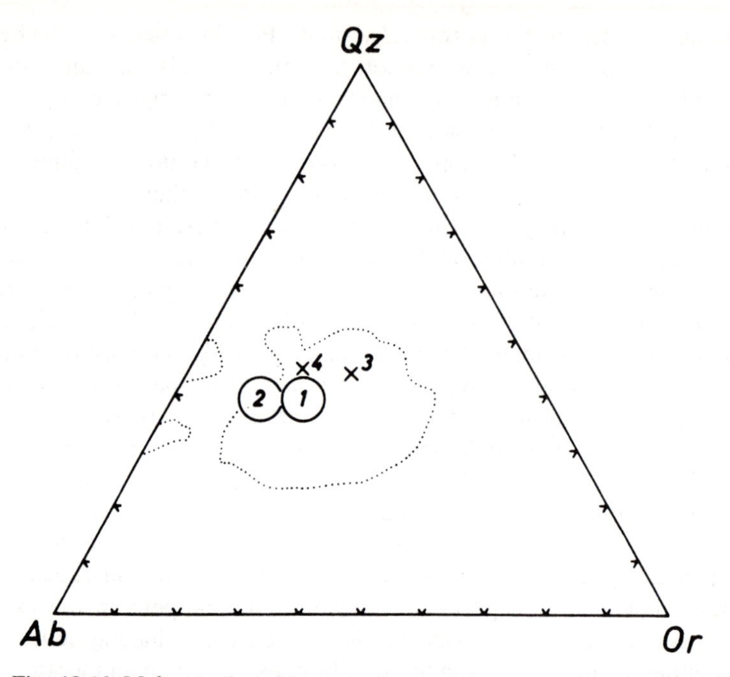

Fig. 18-12 Melts produced by experimental anatexis at P_{H_2O} = 2000 bars. Starting materials were rocks composed of quartz, biotite, and plagioclase. Typically they contained neither muscovite nor K feldspar. Circles 1 and 2 indicate the composition of anatectic melts at 720°C and crosses 3 and 4 at 760°C. The dotted line marks the composition field of granitic, granodioritic, and trondhjemitic rocks taken from Figure 18-11. Composition of the rocks:

Rock	Quartz	Biotite	Plagioclase (An content)	Beginning of anatexis	Amount of melt
(1)	20	37	40 (An 25)	715 ± 10°C	About 45% at 720°C
(2)	21	30	45 (An 21)	690 ± 10°C	About 50% at 720°C
(3)	38	34	28 (An 13)	670 ± 10°C	About 55% at 690°C About 75% at 760°C
(4)	38	34	28 (An 13)	690 ± 10°C	About 55% at 720°C About 60% at 760°C

The very low temperature of the beginning of melting of rock (3) is due to its large Ab/An ratio and the fact that its biotite is very Fe rich. The mole fraction FeO/FeO + MgO of biotite from rock (3) is about 0.8; from rocks (1) and (2) it is about 0.5; and from rock (4) it is only 0.3. The compositions of rocks (4) and (3) differ only in containing biotite of different composition.

available, at temperatures commonly attained during high-grade metamorphism.

Formation of Migmatites[5]

Occasionally the opinion is advanced that the formation of the leucocratic, granitic-granodioritic portions of the heterogeneous *in-situ* migmatites is due to a metasomatic process, *i.e.*, to the addition of "emanations" from depth.[6] The emanations, although not precisely described, are supposed to have supplied alkalis, in some terrains especially Na, whereas in others mainly K. On the other hand, anatexis is generally recognized at present as an essential process in the formation of migmatites, but the nature of this process was not known until the experiments discussed in the previous sections had been carried out. No clear conception existed regarding the nature of the melts produced, so that the formation of migmatites without an addition of alkalis was deemed impossible. Therefore, it was to the great merit of Mehnert (1953) that he ascertained the following: "In simple cases, where a quantitative estimation was possible, a summation of (leucocratic, granitoid) metatect + (dark-colored) residue indicated a constancy of material." Supplementing this observation, Mehnert and Willgallis (1957) came to the following conclusion: "The granitization (formation of migmatites) of the intermediate paragneisses of the Black Forest took place, on the whole, without the addition of alkalis." In this area which has been very carefully studied petrographically, there is no indication of a general alkali supply from depth. Merely a rearrangement, a separation into leucocratic, granitic portions and into biotite-rich, so-called restite portions, which also contain cordierite, sillimanite, and in many cases quartz, has taken place *in situ*. The composition of migmatites, integrated over a certain area about the size of a quarry, is the same as that of the unmigmatized rocks.[7] This same conclusion has also been reached by Suk

[5]See also the introduction to Chapter 18. A book by K. R. Mehnert (*Migmatites and the Origin of Granitic Rocks,* Amsterdam, 1968) is highly recommended for further study. It presents a different treatment, emphasizing petrographic aspects and field relationships of migmatites.

[6]See brief historical survey in Barth (1952).

[7]This statement applies, of course, only to migmatites formed *in situ*. The other type of migmatites, called injection migmatites, is formed by the injection of granitic melt, *e.g.*, into fractured portions of amphibolite, older granitic, or other rock. Such injection migmatites, which, in comparison to *in-situ* migmatites, are of subordinate occurrence, pose no genetic problem, and are not considered in the discussion presented here.

(1964) in Bohemia and by Büsch (1966) who, continuing Mehnert's work, has investigated migmatites that have leucosomes of trondhjemitic compositions.

This result is to be expected if the anatexis of gneisses, which is inevitable in terrains of high-grade metamorphism, was responsible for the formation of migmatites. Anatexis produces *in situ, i.e.,* within the gneiss complex itself, mobile melts of granitic, granodioritic, and trondhjemitic composition which initially segregate as lenses and veins and thus separate, on a scale of a few centimeters to tens of centimenters, from the crystalline residue. Naturally, very extensive material transport takes place over only a small distance, as a consequence of the segregation of mobile melt and less mobile residual portions of the rocks, but, viewed on a large scale, the composition of the rock volume, *i.e.,* the summation of both portions, remains the same.

Migmatites exhibit a great variety of structures. Migmatites with phlebitic (vein) structure and with stromatic (layered) structure contain various amounts of granitic material (leucosome) in the form of segregations in the gneiss, mainly produced *in situ.* Agmatites resemble breccias; angular fragments of darker gneiss are surrounded by more or less homogeneous granitic material. Schlieren structure is characterized by biotite-rich segregations. If migmatites are nearly homogeneous they are said to have a nebulitic structure. All gradations exist between the various structural types. Excellent drawings and photographs of migmatite structures are given in Mehnert's book (1968).

The origin of such manifold forms can be understood if it is kept in mind that a rock complex subjected to metamorphism and subsequently to anatexis consists of various sediments such as graywackes, shales, sandstones, tuffs, etc. In order to illustrate the main principle, it suffices to restrict the discussion to one rock group of metagraywackes. We refer to Table 18-6 where the results of high-grade metamorphism and anatexis of four different paragneisses derived from graywackes are documented.

The four gneisses, although having qualitatively a very similar mineralogical composition, exhibit a very different behavior during anatexis. The temperatures of the beginning of anatexis are appreciably different because of the different Ab/An ratio of the gneisses. Gneisses (a) and (b) begin to melt at 685°C, gneiss (c) at 700°C, and gneiss (d) not until 715°C, because its plagioclase has the highest An content (the original graywacke contained 5% calcite). As a consequence of these different temperatures of initial melting, partial melting at, *e.g.,* 700°C can only take place in layers of the gneiss complex having a composition of gneisses (a) and (b), whereas other layers remain completely solid. Even in the parts of the gneiss complex where anatexis has started, the amounts of melt at 700°C are very different, *e.g.*, in gneiss (a) the amount of melt is 48%, whereas in gneiss (b) it is about 20%. If thin layers of composition (c) or (d), which are completely solid, are separated by the thick layers of composition (a) containing almost 50% melt, it is readily visualized that, in response to tectonic deformation, the thin solid layers break and angular fragments are embedded in the anatectic melt from layers of composition (a). Thus, agmatites are formed. The liquid portion of (a) can also penetrate into fractures and schistosity planes of (c) and (d). On the other hand, in gneisses of composition (b), the anatectic melt constitutes only 20% of the rock, and veined or layered migmatites are formed. The amount of melt increases as the temperature rises, but the extent of this increase varies from rock to rock, because of differences in quantitative mineralogical composition.

At 720°C, an anatectic melt is formed in all four gneiss layers, constituting 60% of gneiss (a), but only 30% of gneiss (d). It is to be expected that the mechanical behavior of layers with greatly differing amounts of melt will vary considerably during deformation and give rise to diverse structures, such as folds of various sizes and shapes, etc.

At 770°C the melt produced in all four gneiss layers constitutes two-thirds to three-quarters of the rock. At this stage, an extensive mixing and homogenization may take place (nebulites). At 770°C (and P_{H_2O} = 2000 bars), the residual portions of gneisses derived from these graywackes may contain either considerable or small amounts of quartz or plagioclase in addition to the other minerals listed in the table, in which predominantly Mg, Fe, and much Al are fixed.

The formation of the different types of migmatites has been discussed using graywackes as an example of original rocks. If clays, arkoses, sandstones, and tuffs of a sedimentary sequence are taken into consideration, the process of anatexis remains the same in principle. At a given temperature, some layers will remain solid, whereas others will contain various amounts of melt. Layers consisting of quartzite or amphibolite remain solid during anatexis. If they fracture, the fragments are engulfed by more or less homogeneous anatectic melts.

As mentioned previously, it is a worldwide observation that in areas of high-grade metamorphism, migmatites are always present as well. Experiments have made the reason for this association obvious. Regional metamorphism is induced by a supply of heat. At the highest temperatures of metamorphism, regional anatexis and the formation of migmatites are inevitable if some H_2O is present.

Depending on the amount of melt formed, two categories of migmatites are sometimes distinguished: metatexites and diatexites. These are considered by Mehnert (1968) to be the products of the following anatectic stages:

Metatexis refers to incipient partial melting, when molten and unmolten portions can still be distinguished petrographically (*i.e.*, parent rock with metatects ± restites).

Diatexis refers to complete or nearly complete melting, when molten and unmolten portions can no longer be distinguished (*i.e.*, schlieric, nebulitic, or nearly homophanous rocks of plutonic habit).

The amount of melt formed in a given gneiss at a given pressure has been shown to depend on the temperature, on the composition of plagioclase and alkali feldspar, on the ratio of quartz:alkali feldspar:plagioclase, and on the availability of water.

A good illustration of an anatectic massif is given by Mehnert (1968); his figure is reproduced here as Figure 18-13. High-grade para-

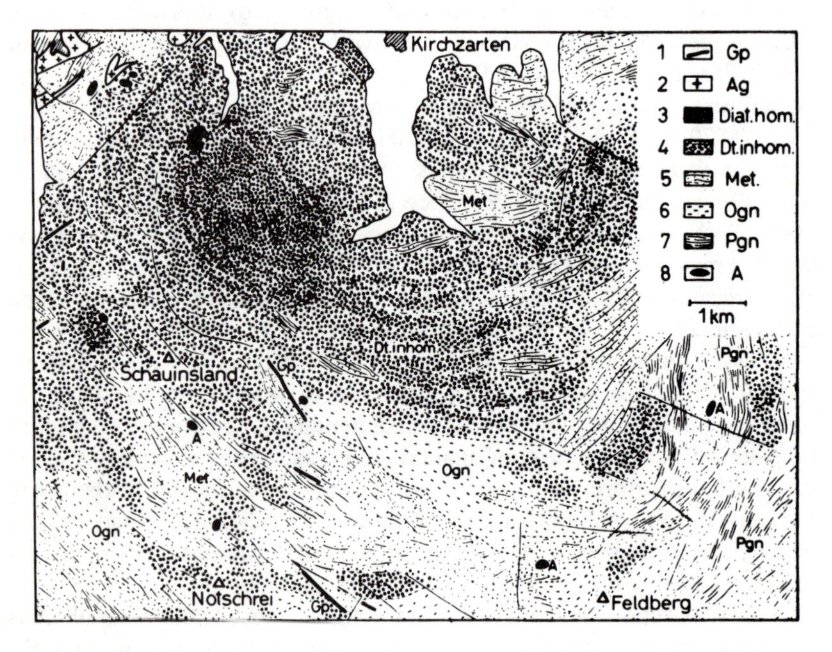

Fig. 18-13 Anatectic massif in the southern Black Forest (from Mehnert, 1968).

1: granite porphyry; 2: aplitic granite,
3: homogeneous diatexite from the center,
4: inhomogeneous diatexite from outer zones,
5: metatexite = phlebitic and similar migmatite,
6: orthogneiss as relics,
7: paragneiss as relics,
8: amphibolite.

gneisses have been converted to metatexites, *i.e.*, migmatites with phlebitic or stromatic structure. Toward the higher temperature part of the massif, this type of migmatite progressively gives way to inhomogeneous "diatexites" and then to homogeneous "diatexites" with prevailing nebulitic structures. Mainly in the southeast part of the massif, some areas of nonmigmatitic paragneiss persist, and two areas of nonmigmatitic orthogneiss are present in the south and southwest of the massif. Probably, these areas have remained unaffected by anatexis due to a lack of water. This reason is especially plausible in the case of the orthogneiss derived from a granite because its metamorphism would not produce any water.

Formation of Granitic Magmas by Anatexis

The commonest sediments are shales, but in geosynclinal basins graywackes constitute a considerable portion of the sediments as well. It is thought that in Precambrian times, graywackes were particularly common. The large migmatite terrains and granite complexes are generally very old, constituting huge parts of the basement complexes. It is impossible to imagine that these immense granite bodies were formed from melts produced by fractional crystallization of gabbroic magma. A different possibility of forming granitic magma had not been examined in its quantitative aspects until recently. For this reason, about 40 years ago, granitization hypotheses were advanced which attempted to explain granites as products of transformation of sediments by a process of metasomatism in the absence of a melt. For a long time, the origin of granite has been vehemently discussed by "transformists" and "magmatists." In this connection, the publications by Gilluly (1948), Read (1957), Raguin (1965), and Mehnert (1959) are very interesting. Mehnert undertook the task of reviewing the state of the granite problem in 1959, citing about 300 publications. His 1968 book is a more recent treatment of the subject.

Experimental investigations of the anatexis of gneisses derived from clays, shales, and graywackes lead to the conclusion that large amounts of granitic, granodioritic, and (to a lesser extent) tonalitic or trondhjemitic magmas must be produced by the anatexis of paragneisses and quartz-feldspar-bearing micaschists. In high-grade metamorphic terrains or orogenic belts, anatexis takes place on a large scale. The anatectic melts may constitute 50% of the rock and, in the case of graywackes, 70% or even 95%. Large volumes of these melts may accumulate. The higher the temperature of anatexis, the higher the temperature of the magma produced. The maximum temperature attained by anatectic melts is probably about 800°C. The anatectic magmas of granitic-granodioritic composition may crystallize at the same level on which they originated, or they may rise to higher levels in the crust, *i.e.*, they may intrude. The level to which a melt, separated from its crystalline residue, is able to rise depends on the difference between the temperature of the melt and the solidus temperature of the melt. A granitic magma saturated with H_2O but having a temperature not appreciably higher than its solidus temperature cannot rise at all. It must remain at the place of its formation, because a decrease of pressure on rising would immediately raise the solidus temperature and, therefore, cause the crystallization of the magma (see Winkler, 1962; Harris *et al.*, 1970). This effect is particularly strong when the water pressure is smaller than about 3kb. On the other hand, a granitic magma having attained a temperature considerably higher than its solidus temperature for a given water pressure can very well rise to higher levels in the crust, whether saturated or undersaturated with water (compare Figure 18-11).

Read (1957) recognizes several groups of granitic rocks on the basis of their relation to the country rocks. The preceding discussion provides a reasonable explanation of this subdivision. (a) Autochthonous granites have remained at the place of their formation and are closely associated with migmatites and metamorphic rocks. (b) Parautochthonous granites have moved somewhat from the place of their formation and their contacts with the country rocks are diffuse. (c) Intrusive granites have left the place of their formation in regionally metamorphosed terrains and, at present, are found to cut the metamorphic zones and to have sharp contacts with the country rocks. (d) Granite plutons have risen to a shallow depth in the crust (shallow-seated plutons).

It is probable that some granitic magmas are differentiation products of gabbroic magmas, but their amount is very subordinate. These magmas have temperatures of about 950°C and therefore can even reach the surface of the earth as rhyolitic lavas or obsidian. However, by far the larger amount of granitic magmas are products of anatexis. In the opinion of the author, granites, like the granitic portions of migmatites, have crystallized from magmas. Only locally, in the vicinity of intrusive contacts, have rocks become "granitelike" by feldspar metasomatism. This is commonly observed on a local scale.

The experimental work of Tuttle and Bowen (1958) was a most important contribution to the study of granite speaking in favor of the "magmatists." Experimental studies of anatexis and of the granite system, which have been discussed, further consolidate the position of the "magmatists." The following two points summarize the most significant arguments: (1) The composition of granitic rocks (*i.e.*, granites, adamellites, granodiorites, trondhjemites, and tonalites) with respect to the main chemical components Qz, Ab, and Or is restricted to a relatively small range. The compositions do not show a random scatter but are clearly grouped about a small field of most frequent compositions. Figure 18-14 shows the frequency distribution of the normative Qz:Ab:Or ratio of 1190 granitic rocks. The frequency maximum lies in the central (black) field. Within the black field and the surrounding finely stippled field lie 53% of all granitic rocks. Such a systematic frequency distribution could hardly be the result of a metasomatic process. (2) On comparing the frequency distribution of granitic rocks (Figure 18-14) with Figure 18-15, it is apparent that the compositions of melts formed by experimental anatexis coincide with the field of granitic-granodioritic compositions in Figure 18-14. This is particularly true if it is realized that most of the melts indicated by dots have been produced at 2 kb water pressure and that the composition of melts formed at higher pressures will shift somewhat in the direction indicated by the arrows in Figure 18-15.

In Figure 18-15 the significant parts of the cotectic lines P-E_5 at 2 and 10 kb are also shown in projection. These lines bound the area where

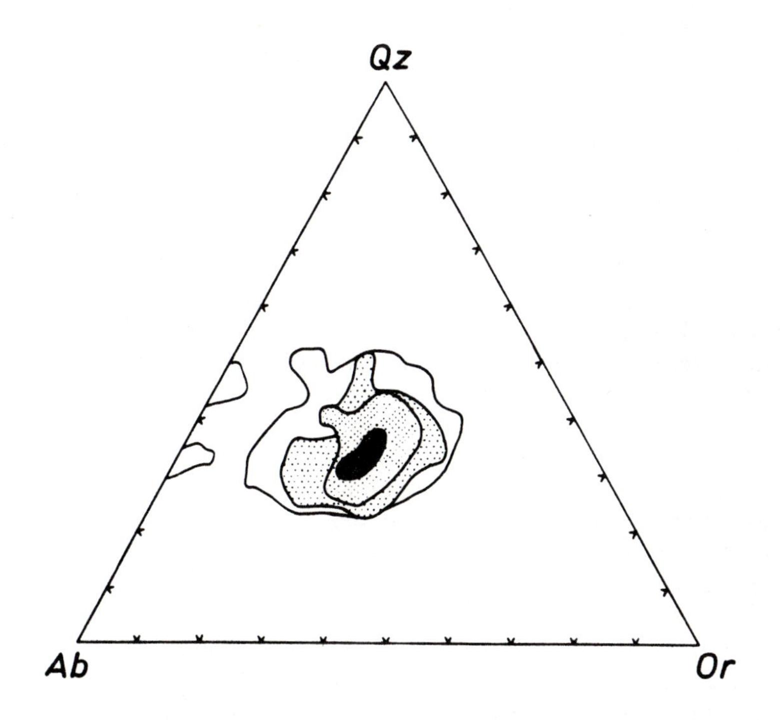

Fig. 18-14 Frequency distribution of the normative Qz:Ab:Or ratios of 1190 granitic rocks (from Winkler and von Platen, 1961). The fields bounded by the outermost line include 86% of all granites; the three patterned fields, 75%; the finely-stippled field and the black field, 53%; the black field alone, 14%. The frequency maximum lies within the black field.

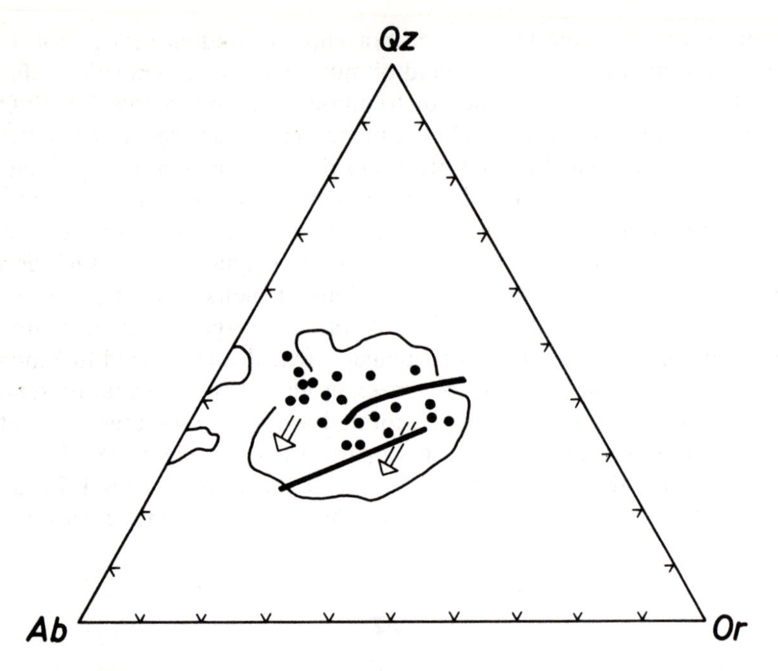

Fig. 18-15 Normative Qz:Ab:Or ratios of anatectic melts produced experimentally, mostly at 2 kb water pressure. The arrows indicate the direction in which the projection points of melts shift when water pressure is greater than 2 kb. Data taken from previous figures and from Winkler and von Platen (1961).

melts must form in the early stage of anatexis of rocks composed of plagioclase, alkali feldspar, and quartz. This area is petrogenetically significant. However, the composition of melts formed at relatively low temperatures will lie not only on and close to the cotectic line, valid for a given water pressure, but also further away as demonstrated in Figure 18-15 and in Figure 18-10.

The points representing the compositions of anatectic melts in Figure 18-15 do not, of course, indicate any frequency; however, on the basis of the average composition of shales and graywackes, it may be said that the commonest anatectic melts will lie in the central part of the shown field.

It is now evident that granites crystallized largely from melts produced by the anatexis of gneisses. It is difficult to imagine metasomatism being effective *over great distances* and causing migmatite formation and "granitization" in rock volumes of many cubic kilometers by means of the introduction of an "ichor" of high alkali and silica concentration. Furthermore, in view of experiments by Luth and Tuttle (1969), the formation of a fluid phase of appropriate composition is not possible (except at very high pressures of about 10 kb). Their experiments show that the generation of a gas phase which can deposit feldspars, in addition to quartz, is possible only in the presence of granitic magma but not when melt is absent. Therefore, metasomatic transformation of rocks into granite-looking rocks can be expected only in the proximity of a granitic magma and only on a restricted scale; large scale metasomatic transformation is not possible. On the other hand, the process of anatexis must lead to the formation of migmatites and large amounts of granitic magma.

A word of caution which, however, does not invalidate the general genetic conclusions must be mentioned: It is known that the generation of granitic, granodioritic, and tonalitic melts cannot be properly understood without taking into account the An component in addition to the Qz, Ab, and Or components. The amount of An component is not indicated in Figure 18-15; in fact, the An component was not determined in our early work. However, most experimentally produced melts form at a temperature not very much above the solidus, *i.e.*, the melts are situated on or close to a cotectic surface of the system Qz-Ab-Or-An-H_2O at 2 kb. On the other hand, the points representing rock compositions that have been used to draw the frequency distribution of Figure 18-14 may be situated on a cotectic surface or at a considerable distance from a cotectic surface. In the latter case, an appreciable amount of solid plagioclase or quartz would be suspended in the melt, and the ratio Qz:Ab:Or:An of the total rock would not give the composition of the liquid part of the magma. Therefore, a frequency distribution calculated on the basis of total rock analyses can be misleading.

This discussion makes it evident that we must inquire as to whether granitic magmas generally were totally liquid at relatively low temperatures or whether variable amounts of crystals were suspended in the melt.

The Physical State of Granitic Magmas

Since the introduction of Bowen's view of granitic magmas as a late product of crystallization differentiation of gabbroic magma, all discussions of magmatic genesis of granitic rocks have taken for granted that the composition of a granitic rock represents the composition of the liquid magma (*e.g.*, Tuttle and Bowen, 1958). It is commonly thought that only the crystallization of a granitic magma will produce minerals which may remain suspended in the melt.

Are granitic magmas completely molten? When asking this question we are not concerned with minerals like garnet, cordierite, sillimanite, or andalusite as constituents of some granitic rocks; it is generally accepted that these minerals represent unmelted portions of the original rock subjected to anatexis. To answer the question of whether a granitic, granodioritic, or tonalitic magma was completely liquid, Winkler and Breitbart (1978) used an approach based on the following reasoning: Leucosomes of *in-situ* migmatites (not of injection migmatites) constitute the only easily accessible place where crystallized granitic, granodioritic, or tonalitic magma is still found at its site of generation. In these migmatites, the leucosomes have been formed by partial melting (anatexis) of gneisses and certain schists under high-grade metamorphic conditions. By means of melting experiments at small temperature intervals, it can be determined whether such leucosomes were completely liquid under anatectic high-grade metamorphic conditions or not. In the latter case, temperatures much higher than those prevailing during metamorphism would be required to achieve complete melting of the leucosome. Our experiments were carried out at a pressure of 5 kb and in the presence of excess water.

Leucosomes of *in-situ* migmatites from various European and African localities were collected and a selection of nine leucosomes was made to ensure a great variety of leucosome compositions. The total analyses are given by Winkler and Breitbart (1978). The ratios of components Qz:Ab:Or:An are listed in Table 18-7 and graphically shown in Figure 18-16.

In our selection, leucosomes of granitic, granodioritic, and trondhjemitic/tonalitic compositions are represented. Figure 18-16 shows most clearly the diversity of the leucosome compositions. Although phase relations in the system Qz-Ab-Or-An-H_2O indicate that low temperature melts of very different compositions may be generated, the observed melting behavior of samples 723, 30bI, NoA1, and 18c was not expected.

Sample 723, consisting of alkali feldspar and an abundance of quartz, represents an extreme composition of very rare occurrence.

Table 18-7. Ratio of components of leucosomes.

	633	630	723	44L	CeA1	3.13	18c	NoA1	30bI
Qz	33	40	66	28	35	29	27	38	46
Ab	28	26	4.5	37	46	48	60	37.5	32
Or	35.5	32	29.4	21	15	8	3	3.5	3
An	3.5	2	0.1	14	4	15	10	21	19
	granitic				granodioritic		trondhjemitic/ tonalitic		

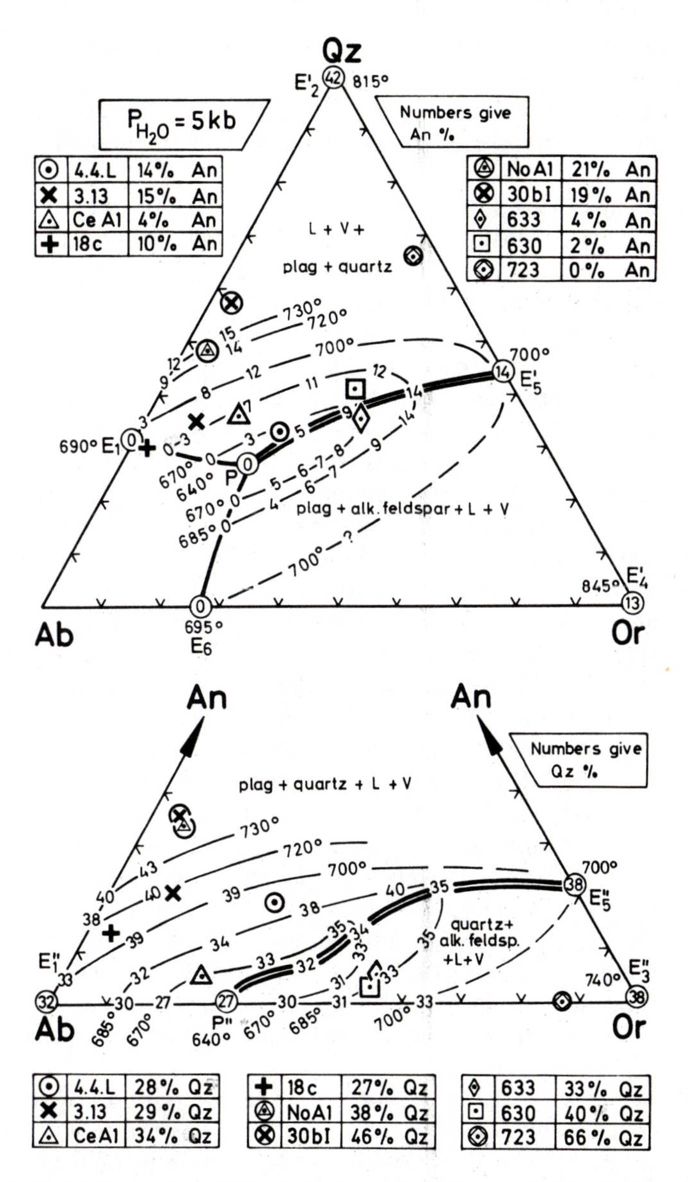

Fig. 18-16 Projection of the isobaric cotectic line and of isotherms on the three isobaric cotectic surfaces of the system Qz-Ab-Or-An-H_2O at 5 kb water pressure. Shown are also the compositions of nine leucosomes (see Table 18-7) plotted within the Qz-Ab-Or-An tetrahedron. Numbers give percent An (top) and Qz (bottom) component of melt compositions situated on cotectic surfaces and the cotectic line.

However, the other three leucosomes of tonalitic composition are believed to be quite common.

As can easily be deduced from Figure 18-16, the types of melting behavior of the leucosomes can be grouped into four categories:

a. Samples CeA1, 630, and 633. The composition of the leucosome is situated close to or on a cotectic surface within a low temperature region. For example, the composition of sample 633 falls on the cotectic surface quartz + alkali feldspar + L + V. Consequently, these leucosomes were completely liquid at low temperatures. The experiments show that at 5 kb the melt composition left the cotectic surface at the following temperatures: sample CeA1 at 680°C with 4% quartz as residue; sample 630 at 675°C with 11% quartz as residue; sample 633 at 680°C with no residue. A slight temperature rise will lead to the complete melting of the felsic constituents of samples CeA1 and 630.
b. Sample 30bI. The composition of the leucosome is situated close to a cotectic surface within a high temperature region. Consequently, a small amount of solids is present in the melt even at relatively high temperatures. The experiments show that, even at 730°C, the melt has not left the cotectic surface quartz + plagioclase + L + V; at this high temperature about 5% plagioclase and 3% quartz remain suspended in the melt.
c. Samples 44L and 18c. The composition of the leucosome is situated far away from a cotectic surface, within a low temperature region. Consequently, a small amount of solids is present in the melt even at relatively high temperatures. Both samples lie within the plagioclase + L + V volume. Experiments show that the melt composition left the cotectic surface at the following temperatures: sample 44L at 675°C with 20% plagioclase as residue; sample 18c at 695°C with 25% plagioclase as residue. At 730°C, 8% and 5% plagioclase remained solid in samples 44L and 18c, respectively.
d. Samples 3.13, NoA1, and 723. The composition of the leucosome is situated far away from a cotectic surface within a high temperature region. Consequently, a large amount of solids is present in the melt even at high temperatures. In order to dissolve such a large amount of solids, a considerable temperature increase would be necessary, so much so that this commonly will not be encountered in high-grade regional metamorphism. Samples 3.13 and NoA1 lie within the plagioclase + L + V volume, and sample 723 lies within the quartz + L + V volume. Experiments show that the melt composition left the cotectic

surface at the following temperatures: sample 3.13 at 705°C with 21% plagioclase as residue; sample 723 at 715°C with 47% quartz as residue; sample NoA1 at >730°C with 20% plagioclase and 3% quartz as residue. It is obvious that very high temperatures must be attained to dissolve the remaining amounts of solids.

The leucosomes in category (a) show that magmas formed by anatexis may be completely liquid at low temperatures, if mafic minerals are disregarded. However, this is rather the exception than the rule. In most cases, represented by categories (b) and (c), there will be a small amount of plagioclase and/or quartz suspended in the melt even at the relatively high temperatures attained in high-grade regional metamorphism. The leucosomes of category (d) show most clearly that an anatectically generated magma may never have been completely liquid but may have included crystals of plagioclase and/or quartz amounting to 15%, 25%, and rarely even a greater percentage, at 730°C and 5 kb water pressure; such large amounts will not completely melt in response to a geologically reasonable temperature increase.

Summing up: From the results of our investigation it is clear that, in most cases, small to appreciable amounts of felsic solids were present in granitic, granodioritic, and tonalitic/trondhjemitic magmas. These solids could not separate from the anatectically formed melt of the leucosomes. Indeed, a complete separation of solids from the melt cannot be expected. Thus, part of the plagioclase or/and quartz constitutes "resisters," such as garnet, cordierite, and sillimanite which are occasionally found in granitic rocks generated by anatexis (see Mehnert, 1968, p. 244, Figs. 89 and 90). However, some of these minerals concentrate at the borders of leucosomes in the form of melanosome seams, as does a great deal of the biotite. Nevertheless, a portion of the undissolved biotite can not separate from the melt; it remains, together with some plagioclase or/and quartz, suspended in the melt. Hornblende exhibits a similar behavior.

Previous to our experiments, it was thought that the felsic minerals in granitic rocks had been completely molten and that they had all crystallized from the melt. Now it is known that this is only an exceptional case and not the rule. It may be appropriate to recall the following definition: A magma is a naturally occurring, mobile liquid within the earth that may contain suspended crystals or rock fragments as well as dissolved or exsolved gasses and that can form a rock (Yoder, 1976, p. III). The "suspended crystals" cannot be ignored in discussing the nature of granitic, granodioritic, and tonalitic magmas because they may have been nonseparated "resisters" of a partially molten paragneiss.

The bulk composition of a granitic, granodioritic, or tonalitic rock commonly does not represent the composition of the liquid part of magma from which the rock crystallized. It is clear that this fact has to be considered in any petrogenetic argument which is based on the chemical and/or modal composition of granitic rocks. However, all previous petrogenetic discussions of granitic rocks did not distinguish between rock composition and the composition of the liquid part of the magma.

References

Ashworth, J. R. 1976. *Mineral Mag.* **40:** 661–682.
Barth, T. F. W. 1952. *Theoretical Petrology*. John Wiley & Sons, New York.
Barth, T. F. W. 1959. *J. Geol.* **67:** 135–152.
Brown, E. H. and Fyfe, W. S. 1970. *Contr. Mineral. Petrol.* **28:** 310–318.
Buddington, A. F. 1963. *Geol. Soc. Am. Bull.* **74:** 353.
Büsch, W. 1966. *Neues Jahrb. Mineral. Abhand.* **104:** 190–258.
——— 1970. *Neues Jahrb. Mineral. Abhand.* **112:** 219–238.
Gilluly, J., ed. 1948. *Geol. Soc. Am. Memoir 28.*
Harris, P. G., Kennedy, W. Q., and Scarfe, C. M. 1970. In G. Newal *et al.,* eds. *Geol. J. Special Issue* **2:** 187–200.
James, H. L. and Hamilton, D. L. 1969. *Contr. Mineral. Petrol.* **21:** 111–141.
Knabe, W. 1970a. *Geol. Jahrb.* **88:** 355–372.
——— 1970b. *Geol. Jahrb.* **89:** 1–32.
Luth, W. C., Jahns, R. H., and Tuttle, O. F. 1964. *J. Geophys. Res.* **69:** 759–773.
——— and Tuttle, O. F. 1969: Geol. Soc. America, Memoir *115,* 513–548.
Mehnert, K. R. 1953. *Geol. Rundschau* **42:** 4–11.
——— 1959. *Fortschr. Mineral.* **37:** 117–206.
——— 1962. *Neues Jahrb. Mineral. Abhand.* **98:** 208–249.
——— 1968. *Migmatites and the Origin of Granitic Rocks.* Elsevier, Amsterdam.
——— and Willgallis, A. 1957. *Neues Jahrb. Mineral. Abhand.* **91:** 104–130.
Merrill, R. B., Robertson, J. K., and Wyllie, P. J. 1970. *J. Geol.* **78:** 558–569.
Mielke, P. and Winkler, H. G. F. 1979. *N. Jb. Miner. Mh.,* (in press).
Morse, S. A. 1970. *J. Petrol.* **11:** 221–251.
Parslov, G. R. 1969. *Mineral Mag.* **37:** 262–269.
Piwinski, A. J. and Wyllie, P. J. 1970. *J. Geol.* **78:** 52–76.
von Platen, H. 1965. *Contr. Mineral. Petrol.* **11:** 334–381.
——— and Höller, N. 1966. *Neues Jahrb. Mineral. Abhand.* **106:** 106–130.
Raguin, E. 1965. *Geologie du Granite.* Paris, 1957. English translation, John Wiley & Sons, New York.
Read, H. H. 1957. *The Granite Controversy.* Interscience Publishers, John Wiley & Sons, New York.
Seck, H. A. 1971. *Neues Jahrb. Mineral. Abhand.* **115:** 140–163.
Sederholm, J. J. 1908. *Bull. Comm. Geol. Finlande* **23:** 110.
Shaw, H. R. 1963. *Am. Mineral.* **48:** 883–896.

Sörensen, H. 1960. *Intern. Geol. Congr. Copenhagen.* Pt. 26, pp. 54–78.
Steuhl, H. H. 1962. *Chem. Erde* **21:** 413–449.
Stewart, D. B. 1957. *Carnegie Inst. Year Book* **56:** 214–216.
——— 1967. *Schweiz. Mineral. Petrog. Mitt.* **47:** 35–59.
Streckeisen, A. 1973. *Neues Jahrb. Mineral. Monatsh.* **1973:** 149–164.
Suk, M. 1964. *Krystallinicum* **2:** 71–105.
Tuttle, O. F. and Bowen, N. L. 1958. *Geol. Soc. Am. Memoir No. 74.*
Winkler, H. G. F. 1957. *Geochim. et Cosmochim. Acta* **13:** 42–69.
——— 1961. *Geol. Rundschau* **51:** 347–457.
——— 1962. *Beitr. Mineral. Petrog.* **8:** 222–231.
——— 1966. *Tschermaks, Miner. Petrogr. Mitt.* **11:** 266–237.
——— 1967. *Petrogenesis of Metamorphic Rocks.* 2nd edit. Springer-Verlag, New York.
———, Böse, M., and Marcopoulos, T. 1975. *Neues Jahrb. Mineral. Monatsh.* **1975:** 245–268.
——— and Breitbart, R. 1978. *Neues Jahrb. Mineral. Monatsh.* **1978:** 463–480.
———, Das, B. K., and Breitbart, R. 1977. *Neues Jahrb. Mineral. Monatsh.* **1977:** 241–247.
——— and Ghose, N. C. 1974. *Neues Jahrb. Mineral. Monatsh.* **1974:** 481–484.
——— and Lindemann, W. 1972. *Neues Jahrb. Mineral Monatsh.* **1972:** 49–61.
——— and von Platen, H. 1958. *Geochim. Cosmochim. Acta* **15:** 91–112.
——— 1960. *Geochim. Cosmochim. Acta* **18:** 294–316.
——— 1961. *Geochim. Cosmochim. Acta* **24:** 48–69, 250–259.
Yoder, H. S. 1968. *Carnegie Inst. Year Book* **66:** 477–478.
——— 1976. *Generation of Basaltic Magma.* National Academy of Sciences, Washington D.C.
———, Stewart, D. B., and Smith, J. R. 1957. *Carnegie Inst. Year Book* **56:** 206–216.

Appendix

Nomenclature of Common Metamorphic Rocks

Magmatic rocks are usually named after some locality. Only in rare cases does the rock name give any indication about the fabric and mineralogical composition of the rock. The names of magmatic rocks have to be memorized like words of a foreign language. Fortunately, this difficulty is not encountered in the nomenclature of most metamorphic rocks. It is only necessary to learn a few names of rock groups, which are characterized by a certain fabric and/or mineralogical composition. Furthermore, the presence of the main or critical minerals is indicated by placing their names in front of the group name. For instance, there is the group of marbles, all of which contain well-crystallized carbonates as their main constituent. A particular marble may be designated as dolomite marble, diopside-grossularite marble, tremolite marble, etc. Thus, the nomenclature of most metamorphic rocks is clear and easily understood.

A more elaborate nomenclature based on quantitative mineralogical composition was proposed by Austrian petrographers after a discussion with colleagues from other countries.[1] This nomenclature is recommendable and is to a large extent adopted here.

Names of Important Rock Groups

Phyllite. Fine-grained and very finely schistose rock, the platy minerals of which consist mainly of phengite. Phengite sericite gives an overall silky sheen to the schistosity planes. The grain size is coarser than in slates but finer than in mica schists.

---

[1]"Ein Vorschlag zur quantitativen und qualitativen Klassifikation der kristallinen Schiefer" (a symposium). *Neues Jahrb. Minerals Monatsh.*: 163–172 (1962).

In phyllites the amount of phyllosilicates (phengite + some chlorite ± biotite) exceeds 50%. The other most abundant constituent is quartz. If the amount of quartz exceeds the amount of phyllosilicates, the rock is called a *quartz phyllite*. In both phyllites and quartz phyllites, albite may amount to as much as 20%.

An exact designation of the rock is achieved by placing the name of subordinate constituents in front of the rock name, beginning with that mineral present in the smallest amount. Minerals constituting less than 5% of the rock are generally not taken into consideration. Examples are chloritoid-chlorite-albite phyllite and phlogopite-calcite phyllite. If amounts smaller than 5% are considered significant this can be designated by using an adjective form such as "graphite-bearing."

Schists. Medium- to coarse-grained rock, the fabric of which is characterized by an excellent parallelism of planar and/or linear fabric elements (schistosity). The individual mineral grains can be recognized megascopically (in contrast to phyllites). If mica, chlorite, tremolite, talc, etc., constitute more than 50% of a rock, the corresponding rock is called a mica schist, chlorite schist, tremolite schist, talc schist, etc. Phengite-epidote-chlorite-albite schists are known as greenschists.

If a schist contains more quartz relative to the sum of the phyllosilicates, the rock is called quartz-mica schist. A further subdivision of schists is effected according to the same rules as in the case of phyllites.

The cited symposium gives 20% as the maximum amount of feldspar in a schist. If rocks contain more feldspar, they are designated as gneisses rather than schists. It is true that schists commonly contain less than 20% and gneisses more than 20% feldspar, but this distinction is generally not valid. The most characteristic difference between schists (or quartz schists) and gneisses is not the mineralogical composition but the fabric. This distinction between schistose and gneissic fabric was clearly stated by Wenk (1963): "When hit with a hammer, rocks having a schistose fabric (schists) split perfectly parallel to 's' into plates, 1–10 mm in thickness, or parallel to the lineation into thin pencil-like columns." Schists split into thinner plates than gneisses.

Gneiss. Medium- to coarse-grained rock having a gneissic fabric, *i.e.*, it "splits parallel to 's' generally along mica or hornblende layers, into plates and angular blocks, a few centimeters to tens of centimeters in thickness, or parallel to B into cylindrical bodies (pencil gneisses). The prevalent light-colored constituents (feldspar + quartz) have interlocking boundaries and provide, as compared to schists, a better coherence and a coarser fissility to the rock; nevertheless, the fissility in many cases creates an almost perfect plane" (Wenk, 1963). Some prefer a definition of gneiss based not only on fabric but also on mineralogical features. Thus Fritsch *et al.* (1967) advocated the use of the term gneiss for a rock with recognizable parallel structure consisting predominently of quartz and feldspar—feldspar amounting commonly to more than 20% and mica to at least 10%.

Two groups of gneisses are recognized. Orthogneisses are formed from magmatic rocks, such as granites, syenites, diorites, etc. On the other hand, paragneisses are derived from sediments, such as graywackes, shales, etc. The particular mineralogical composition is indicated according to the same rule as in the case of phyllites, *e.g.*, kyanite-staurolite-garnet-biotite gneiss.

Amphibolite. A rock consisting predominantly of hornblende and plagioclase, which is produced by metamorphism of basaltic magmatic rocks, tuffs, or marls. The hornblende prisms lie within the plane of schistosity if this is developed. The fissility generally is not as well developed as in schists. Amphibolites contain only small amounts of quartz or none at all.

Marble. A rock consisting predominantly of fine- to coarse-grained recrystallized calcite and/or dolomite. Other minerals present are indicated in the usual manner, *e.g.*, muscovite-biotite marble.

Quartzite. A rock composed of more than 80% quartz. The interlocking boundaries of the quartz grains impart a great strength to the rock. Metamorphic quartzites must be distinguished from unmetamorphosed, diagenetically formed quartzites.

Fels. Fels is a term referring to massive metamorphic rocks lacking schistosity, *e.g.*, quartz-albite fels, plagioclase fels, calc-silicate fels. Generally, in English books, the term "rock" is used for such metamorphic rocks, *e.g.*, lime-silicate rock (Harker, 1932, 1939). It is suggested that "fels" be used instead.

Hornfels. Nonschistose and fine-grained rock, which splinters on impact. The edges of thin rock chips occasionally are translucent like horn. The rock has a granoblastic fabric, *i.e.*, it is a mosaic of equidimensional small mineral grains, in which larger porphyroblastic minerals (or relics) are frequently embedded. Hornfelses are typically produced by contact metamorphism of clays, fine-grained graywackes, etc., and occasionally by regional metamorphism.

Granulite, Granolite, and Granoblastite. See p. 256ff.

Eclogite. See p. 276.

Prefixes

Meta-. This prefix designates metamorphosed igneous or sedimentary rocks in which the original fabric still can be recognized; *e.g.*, metabasalts, metagraywackes. Others use the prefix "meta-" in a more general sense to designate metamorphic rocks according to the type of original rock from which they are derived. *Example:* Meta-graywacke or metadiorite = rock derived from graywacke or diorite.

Ortho-. This prefix indicates that the metamorphic rock originated from a magmatic rock, *e.g.*, orthogneiss, orthoamphibolite.

Para-. This prefix indicates that the metamorphic rock originated from a sedimentary rock, *e.g.*, paragneiss, para-amphibolite.

Classification

A quantitative classification of common metamorphic rocks is shown in Figures A-1 and A-2 taken with slight modification from the cited symposium (1962). The objections of Wenk regarding the distinction between gneiss and schist or phyllite should not be ignored; therefore, the boundary between the two groups, shown as a broken line in the two figures at 20% feldspar, should not be taken as critical in assigning a name to a rock. The distinction between gneiss and schist or phyllite is not based on mineralogical composition but on the character of fissility. This distinction is particularly significant if the mineralogical composition is the same.

The classification shown in Figures A-1 and A-2 applies to rocks predominantly composed of either quartz, feldspars, and phyllosilicates, or quartz, phyllosilicates, and carbonates. In many metamorphic rocks, these minerals are the main constituents. Figure A-2 is valid for rocks of

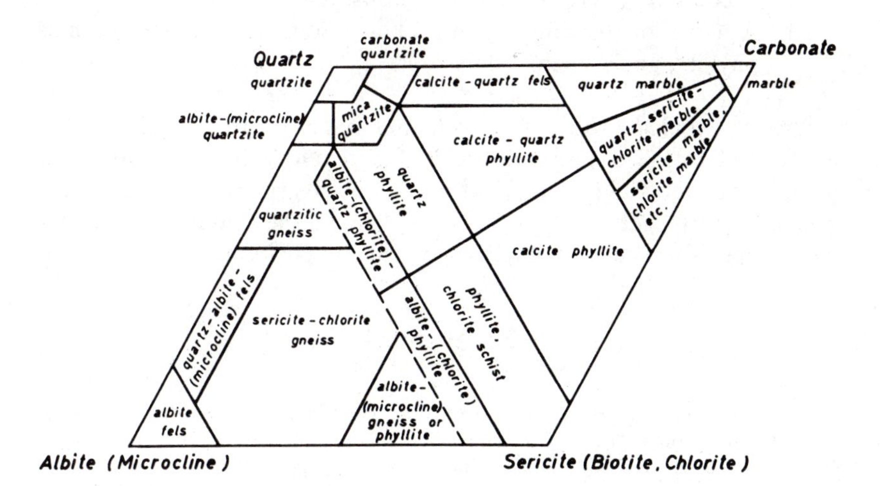

Fig. A-1 Composition of metamorphic rocks of lower temperature ranges in terms of certain main constituents as indicated in the diagram.

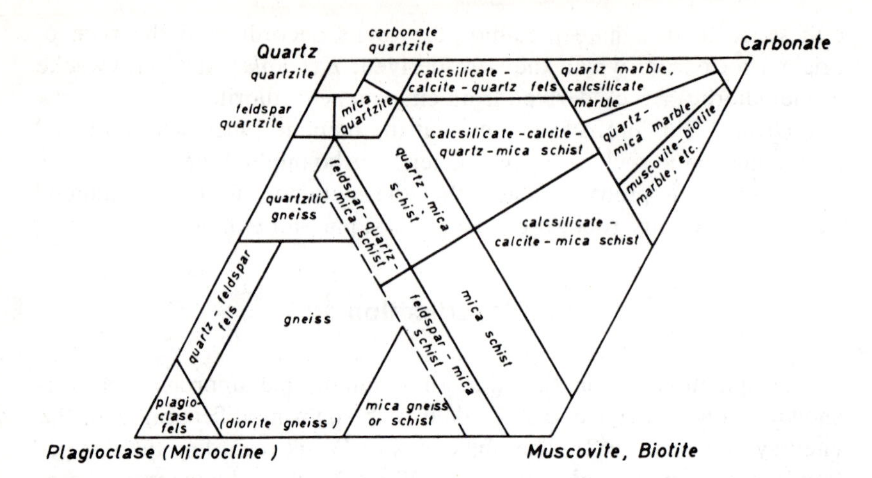

Fig. A-2 Composition of metamorphic rocks of higher temperature ranges in terms of certain main constituents as indicated in the diagram.

lower temperature and Figure A-2 for rocks formed at higher temperature. In higher-grade metamorphic rocks, schists take the place of phyllites, and calc-silicates such as diopside and grossularite, which are not found in rocks of low temperature, are present, *e.g.,* in marbles (silicate marble).

The names of most metamorphic rocks consist of compound terms:

a. A combination of the names of constituent minerals;
b. A name for the category of rock according to its fabric, such as phyllite, gneiss, schist, fels.

Commonly, rocks with the fabric characteristics of gneiss, schist, etc., are formed in the appropriate field of mineralogical compositions as given in the preceding figures, but this is not invariably so. In any case, the name gneiss, schist, etc. is to be used only if the characteristic fabric is developed, irrespective of mineralogical composition.

References

Fritsch, W. Meixner, H., and Wieseneder, H. 1967. *Neues Jahrb. Mineral. Monatsh.* **1967:** 364–376.

Harker, A. 1932, 1939. *Metamorphism*. Methuen, London.

Wenk, C. 1963. *Neues Jahrb. Mineral. Monatsh.* **1963:** 97–107.

Index